T0292771

Multivariate Biomarker Discovery

Data Science Methods for Efficient Analysis of High-Dimensional Biomedical Data

Multivariate biomarker discovery is increasingly important in the realm of biomedical research and is poised to become a crucial facet of personalized medicine. This will prompt the demand for a myriad of novel biomarkers representing distinct "omic" biosignatures, allowing selection and tailoring of treatments to the various individual characteristics of a particular patient. This concise and self-contained book covers all aspects of predictive modeling for biomarker discovery based on high-dimensional data, as well as modern data science methods for the identification of parsimonious and robust multivariate biomarkers for medical diagnosis, prognosis, and personalized medicine. It provides a detailed description of state-of-the-art methods for parallel multivariate feature selection and supervised learning algorithms for regression and classification, as well as methods for proper validation of multivariate biomarkers and predictive models implementing them. This is an invaluable resource for scientists and students interested in bioinformatics, data science, and related areas.

DARIUS M. DZIUDA, PhD, is Professor of Data Science and Bioinformatics at Central Connecticut State University (CCSU) with both academic and biotechnology industry experience. His research focuses on multivariate biomarker discovery for medical diagnosis, prognosis, and personalized medicine. Dr. Dziuda is also designing and teaching courses for two specializations of CCSU's graduate data science program: *Bioinformatics* and *Advanced Data Science Methods.*

Multivariate Biomarker Discovery

Data Science Methods for Efficient Analysis of High-Dimensional Biomedical Data

DARIUS M. DZIUDA
Central Connecticut State University

Shaftesbury Road, Cambridge CB2 8EA, United Kingdom

One Liberty Plaza, 20th Floor, New York, NY 10006, USA

477 Williamstown Road, Port Melbourne, VIC 3207, Australia

314–321, 3rd Floor, Plot 3, Splendor Forum, Jasola District Centre, New Delhi – 110025, India

103 Penang Road, #05-06/07, Visioncrest Commercial, Singapore 238467

Cambridge University Press is part of Cambridge University Press & Assessment, a department of the University of Cambridge.

We share the University's mission to contribute to society through the pursuit of education, learning and research at the highest international levels of excellence.

www.cambridge.org
Information on this title: www.cambridge.org/9781316518700

DOI: 10.1017/9781009006767

First published 2024

A catalogue record for this publication is available from the British Library

Library of Congress Cataloging-in-Publication Data
Names: Dziuda, Darius M., author.
Title: Multivariate biomarker discovery : data science methods for efficient analysis of high-dimensional biomedical data / Darius M. Dziuda.
Description: Cambridge, United Kingdom ; New York, NY : Cambridge University Press, 2024. | Includes bibliographical references and index.
Identifiers: LCCN 2023041772 (print) | LCCN 2023041773 (ebook) | ISBN 9781316518700 (hardback) | ISBN 9781009006767 (ebook)
Subjects: LCSH: Biochemical markers–Research–Statistical methods. | Medicine–Research–Statistical methods. | Multivariate analysis.
Classification: LCC R853.B54 D95 2024 (print) | LCC R853.B54 (ebook) | DDC 610.72/7–dc23/eng/20231213
LC record available at https://lccn.loc.gov/2023041772
LC ebook record available at https://lccn.loc.gov/2023041773

ISBN 978-1-316-51870-0 Hardback

To Dorota and Mateusz

Contents

Preface

For almost twenty years now, I have been designing and teaching graduate bioinformatics and data science courses; yet I could not find a single book that would be appropriate as the primary text for any of them. My lecture notes prepared for my very first bioinformatics course, *Data Mining for Genomics and Proteomics*, have been extended into the book with the same title and published by Wiley in 2010. Since then, I have continued to develop and teach multiple other bioinformatics and data science courses. This book is a result of the expansion and consolidation of my lecture notes for courses on multivariate analytics, biomarker discovery, and advanced data science methods for estimation and classification.

Consequently, it can be used as a text for such bioinformatics or data science courses at the graduate, postgraduate, and advanced undergraduate levels. It can also be used as a monograph by scientists interested in these and related areas of research. This includes researchers and practitioners involved in predictive modeling for biomarker discovery, as well as anybody interested in state-of-the-art data science methods for multivariate analysis of high-dimensional "omic" data. Furthermore, the methods and approaches described in the book are applicable to predictive modeling based on any high-dimensional data within and without the biomedical field.

All aspects of predictive modeling for biomarker discovery based on high-dimensional data, as well as modern data science methods for the identification of parsimonious and robust multivariate biomarkers for estimation and classification, are covered in this book. One of its main goals is to explain and show how to identify multivariate biomarkers that represent real patterns in the target population rather than spurious patterns that are present in the training data but not in the target population from which the data were sampled.

It is typical for graduate students entering my bioinformatics classes to assume that all published papers, and especially those in reputable journals, present valuable scientific results. After introducing them to common misconceptions in biomarker discovery – and, more generally, in predictive modeling based on high-dimensional data – my students are astounded at how easily they can find papers that use inappropriate methodologies (and thus are not worth reading), and that those are the same kinds of papers they would be relying upon if they did not know what to look for. It is my hope that from this book, the reader will not only learn about such misconceptions but also be introduced to proper methodologies at each and every step of predictive modeling for multivariate biomarker discovery based on high-dimensional data.

While writing this book, one of my priorities was to facilitate the reader's understanding of the presented concepts and methods. Although such an approach seems rather obvious, one can recently observe a disturbing tendency to promote the notion that anyone can undertake bioinformatics or data analysis projects as long as they are told which functions of which software to use and what sequence of operations to perform – with no fundamental understanding of the methods or concepts necessary!

The book is divided into five parts, which together comprise seventeen chapters. Part I (Chapters 1–5) provides a general context for, and description of, all elements of predictive modeling for biomarker discovery. Chapter 1 focuses on terminology and basic concepts of the area, and places multivariate biomarker discovery in the context of biomarker studies and personalized medicine. Chapter 2, while continuing with the concepts important for multivariate analysis of high-dimensional data, adds considerations related to the curse of dimensionality and descriptions of common misconceptions. Chapter 3 provides an overview of all elements of the predictive modeling process, from the selection of training and test data sets, parallel multivariate feature selection experiments and deciding on an optimal multivariate biomarker, to building, tuning, validating, and testing predictive models implementing the optimal biomarker. Discussed are also such topics as bias–variance tradeoff, segmentation models, and committees of predictive models. Chapter 4 provides detailed coverage of methods for evaluation of predictive models: the methods applicable to regression models implementing estimation biomarkers, as well as methods evaluating binary and multiclass classification models. Discussion of resampling techniques is accompanied by accentuating the danger of information leakage and by emphasizing the paramount importance of avoiding internal validation. Discussion of metrics for the evaluation of classification biomarkers includes the issue of proper and improper interpretation of sensitivity and specificity, illustrated by an example of a screening biomarker targeting a population with low prevalence of the tested disease. For such biomarkers, positive predictive value may be unacceptably low even when the biomarker has a very high specificity and sensitivity. Discussed in Chapter 4 are also misclassification costs and incorporating them into cost-sensitive classification. Chapter 5 is dedicated to the most important part of predictive modeling for biomarker discovery based on high-dimensional data – multivariate feature selection. When dealing with sparse biomedical data whose dimensionality is much higher than the number of training observations, the crucial issue is to overcome *the curse of dimensionality* by using methods capable of elevating signal (predictive information) from the overwhelming noise. One way of doing this is to perform many (hundreds or thousands) parallel feature selection experiments based on different random subsamples of the original training data and then aggregate their results (for example, by analyzing the distribution of variables among the results of those parallel experiments). Two designs of such parallel feature selection experiments are discussed in detail: one based on recursive feature elimination, and the other on implementing stepwise hybrid selection with T^2. The chapter also includes descriptions of three evolutionary feature selection algorithms: simulated annealing, genetic algorithms, and particle swarm optimization.

Part II (Chapters 6–9) is dedicated to regression methods for estimation biomarkers. Chapter 6 starts with the description of multiple regression. Even if it is unlikely for multiple regression to be used as the primary method for multivariate biomarker discovery based on high-dimensional data, presenting this classical method provides the necessary background for regression analysis and highlights the weaknesses of multiple regression, which will be addressed by the subsequently presented methods. This chapter also presents partial least squares regression, which by performing supervised dimensionality reduction addresses some weaknesses of multiple regression; however, by not performing any feature selection, PLSR does not reduce noise that is typically abundant in high-dimensional data. Chapter 7 is dedicated to regularized regression methods, which, by penalizing models that are too complex, are capable of providing a reasonable tradeoff between bias and variance. Ridge regression implements L_2 regularization, which results in more generalizable models but does not perform any feature selection. L_1 penalty used by the lasso allows, however, for simultaneous regularization and feature selection. The elastic net algorithm combines the two approaches by applying both L_1 and L_2 penalties, which allows for solutions combining the advantages of both ridge regression and the lasso. Chapter 7 concludes by discussing a general class of L_q-regularized least squares optimization problems. Chapters 8 and 9 present random forests for regression and support vector regression – relatively newer supervised learning algorithms for predictive regression modeling, which, at least in some situations, may outperform the least squares-based methods.

Part III (Chapters 10–13) focuses on classification learning algorithms. Chapter 10 covers random forests for classification. Chapter 11 presents classification with support vector machines (SVMs) and also includes a discussion of hyperparameters, variable importance measures, and cost-sensitive SVMs. Chapter 12 presents discriminant analysis – a classical (and powerful) supervised learning approach for classification. Discussed are Fisher's discriminant analysis, as well as Gaussian linear, quadratic, and regularized discriminant analysis. The chapter concludes with a discussion of partial least squares discriminant analysis, which is still popular in some application areas, even if its application to high-dimensional data is likely to result in solutions that are suboptimal in terms of predictive abilities and interpretability (alternative approaches are recommended). Chapter 13 discusses neural networks and deep learning; included is the presentation of deep convolutional networks that seem to have a great potential in the classification of medical images.

Part IV (Chapters 14–15) describes a method for the identification of parsimonious and robust multivariate biomarkers that may also have the best chance for plausible biological interpretation. The method is based on multistage signal enhancement and identification of essential patterns. The essential patterns are identified by analyzing distributions of groups of variables (with similar patterns) among a large number of optimal-size biomarkers generated by parallel feature selection experiments. A similar approach leads to the subsequent identification of the essential variables of those essential patterns. As a result, the final multivariate biomarker identified via this method is most likely to represent a real population pattern associated with biological processes underlying changes in the investigated response variable. Furthermore, having the variables of the final biomarker associated with their respective essential patterns facilitates biological interpretation of the biomarker.

Finally, Part V (Chapters 16–17) presents two real-life multivariate biomarker discovery studies. The goal of the first study – which implements the method presented in Part IV – is to identify essential gene expression patterns and a multivariate biomarker common for multiple types of cancer. This study is based on the TCGA RNA-Seq data for 3,528 patients and 20,530 gene expression variables; the data represent five tumor types of five different tissues. The second study identifies multivariate biomarkers for liver cancer; it is performed in an R environment, and R scripts for all of its steps are provided.

Acknowledgments

I would like to thank Dr. Daniel S. Miller and Mateusz M. Dziuda for reading the manuscript and for their invaluable comments and suggestions, which definitely improved this book. Thanks also go to executive publishers in Cambridge University Press, Dr. Susan Francis and Dr. Katrina Halliday, as well as to my day-to-day contact in the Press, Aleksandra Serocka – thank you all for making this project run so smoothly.

Finally, a very special thanks to my wife Dorota for her patience and support during this venture.

Part I

Framework for Multivariate Biomarker Discovery

1 Introduction

Even though multivariate biomarker discovery has already made its debut in the realm of biomedical research, it is poised to become a crucial facet of personalized medicine, which will prompt the demand for a myriad of novel biomarkers representing distinct "omic" biosignatures. So, what is a *multivariate biomarker*? Stating that it is a set of more than one variable (such as genomic variations, gene or protein expression levels, or metabolite concentrations), whose combined pattern of values can be used for predicting the value of a target (or response) variable (such as a disease state, response to treatment, or therapy outcome) would not be sufficient. A very important characteristic of a truly multivariate biomarker is that it has to be identified with the use of *multivariate methods*, that is, methods that evaluate each variable in the context of other variables (and thus, consider correlations and interactions among the variables). That means that a set (of variables) identified by selecting its members via univariate methods – for example, by evaluating the correlation of each variable with the target – should not (and herein, will not) be considered a multivariate biomarker. We are interested in the association between a *set of variables* and the target variable, whereas members of a multivariate biomarker may or may not be individually associated with the target variable.

One may insist that there could exist efficient biomarkers that are composed of several variables, which were selected in a univariate way. That is, of course, possible. However, if a set of univariately-identified variables is used as a biomarker, then it is virtually guaranteed that a better (and truly multivariate) biomarker could have been identified were a multivariate approach used.[1]

Another important characteristic of a properly-identified multivariate biomarker is its parsimony – not only to keep Ockham happy. An optimal multivariate biomarker should consist of as few variables as possible. As will later be discussed, such parsimony is extremely important for biomarkers based on high-dimensional data.[2] For such data, it is

[1] The only exception would be a situation where the variables are independent of each other (i.e., where there are no correlations among them), which is definitely not the case for high-dimensional biomedical data (such as gene or protein expression data).

[2] One may argue that a group of variables (such as genes) that are highly correlated should be included in a biomarker if any one of them is selected into the biomarker. Although such an argument may be supported by the view that including groups of highly correlated variables in a multivariate biomarker (instead of only one of them) would allow for easier biological interpretation of the biomarker, we will see, in Part IV, that there are better approaches, such that will allow both for the identification of parsimonious biomarkers as well as facilitating their biological interpretation.

quite trivial to identify a multivariate biomarker that would overfit the training data, but much more difficult to find one that would be well generalizable to unseen data. Accordingly, one of the main goals of this book is to explain and show how to identify multivariate biomarkers that represent the real patterns in the target population, rather than the easily identifiable spurious patterns that are present in the training data but not in the target population from which the data were sampled. The principle of parsimony is one of the crucial aspects of this approach and it requires that we apply a combined criterion of minimizing the size of the biomarker (its cardinality) while simultaneously maximizing its predictive power; hence, such optimization will represent a compromise between these two seemingly contradictory criteria. It may appear that increasing the number of variables included in a multivariate biomarker would increase its predictive power; this, however, is a misconception – such a perceived increase in predictive power, when improperly estimated on the training data would, actually, decrease the generalizability of the biomarker.

There are other misconceptions in performing predictive modeling, which, if followed in a biomarker discovery project, would render its results – at the very best – suboptimal. In this book, we will discuss such misconceptions, as well as describe proper methodologies for each and every step of the predictive modeling process for multivariate biomarker discovery based on high-dimensional biomedical data.

1.1 Biomarkers and Multivariate Biomarkers

There seems to be some confusion and inconsistency in using terminology and interpreting concepts related to biomarkers and biomarker discovery. This was a reason for the FDA-NIH to decide in 2015 to create "The Biomarkers, EndpointS, and other Tools (BEST) Resource", which is intended to be a dynamic glossary that will be periodically updated and revised. A consequence of this initiative, for example, is that the general definition of a biomarker has evolved from:

> *A characteristic that is objectively measured and evaluated as an indicator of normal biological processes, pathogenic processes, or pharmacologic responses to a therapeutic intervention.* (Biomarkers Definitions Working Group 2001),

to that provided by the BEST resource in November 2021:

> *A defined characteristic that is measured as an indicator of normal biological processes, pathogenic processes, or biological responses to an exposure or intervention, including therapeutic interventions. Biomarkers may include molecular, histologic, radiographic, or physiologic characteristics. A biomarker is not a measure of how an individual feels, functions, or survives.* (FDA-NIH Biomarker Working Group 2021).[3]

[3] www.ncbi.nlm.nih.gov/books/NBK338448/#IX-B

Though such initiatives should be applauded, the confusion about terminology related to the emerging *multivariate* biomarker research remains, nevertheless, abundant. It seems quite likely that the main reason for this confusion is the still enduring (and often followed, consciously or not) traditional, historical one-gene-at-a-time (or, more generally, one-variable-at-a-time) approach to discovering and describing biomarkers. However, **a multivariate biomarker is *a biomarker***. It is not a "set of biomarkers", nor a "panel of biomarkers", nor any similar univariately-biased concept.[4] A properly identified multivariate biomarker is a set of variables, which together – as a set – constitute *a single biomarker*. If any of the variables included in the set were a biomarker for the researched goal, there would be no need for the multivariate biomarker. This also means that a proper biological interpretation of a multivariate biomarker should interpret it as a set; interpretation of the individual genes, proteins, or metabolites that compose the biomarker does not translate into the biological interpretation of the set. Hence, a proper multivariate interpretation of biological processes underlying the predictive abilities of a multivariate biomarker is a challenging task and still quite uncharted territory.

Included in the categories of biomarkers listed in the BEST resource are:

- Diagnostic biomarkers
- Prognostic biomarkers
- Predictive biomarkers
- Response biomarkers
- Monitoring biomarkers
- Safety biomarkers
- Risk biomarkers

Diagnostic biomarkers are used to recognize – or confirm the presence of – a specific disease, or to support differential diagnosis among subtypes or stages of a disease. *Prognostic biomarkers* estimate the probability of a specific future clinical event, such as relapse, disease progression, or recovery. *Predictive biomarkers* evaluate a patient's sensitivity, as well as the probability of a positive response to specific exposure, such as ionizing radiation or a particular chemotherapy. *Response biomarkers* evaluate a patient's response to a medical product or therapy. *Monitoring biomarkers* are those that are repeatedly measured to assess a patient's status, such as disease progression or response to a treatment, which may be expected to change over a period of time. *Safety*

[4] Such names are appropriate only in situations when independently identified and validated univariate biomarkers are used together to achieve a more comprehensive view of the patient's condition. However, even in such situations, they are still independent single-variable biomarkers that could be used individually or in combinations with other biomarkers or other diagnostic or prognostic evaluations, and they do not constitute a multivariate biomarker. Combinations of such independent biomarkers are sometimes called *composite biomarkers*. On the other hand, multivariate biomarkers, especially the omic ones, have been recently also called *biosignatures* ('omic' biosignatures, molecular signatures), which is a valid alternative description of them.

biomarkers are used to evaluate the probability or presence of adverse reactions to a drug or other treatment. *Risk* (or risk profiling) *biomarkers* indicate predisposition, or risk, to develop a specific disease or medical condition.

Although taxonomies of biomarkers may be useful, one may observe that there are overlaps between these biomarker categories, and thus a particular biomarker may be classified to more than one of them. For example, a diagnostic biomarker that assigns a patient to one of the disease subtypes may, by this very assignment, play an additional role of a prognostic biomarker. Furthermore, in the context of data science and predictive modeling for biomarker discovery, biomarkers are the means to predict the value of the target variable (such as a disease state or the level of toxicity). Consequently, in this context, they are all *predictive* vehicles, regardless of the BEST category (or categories) to which they would be assigned. On the other hand, some terms that are commonly used to describe some types of biomarkers are not explicitly included in the BEST list; for example, *screening* biomarkers, which target at-risk populations of asymptomatic individuals, in the hope of the early detection of specific conditions, for which early treatment may be crucial for positive outcomes.[5]

1.2 Biomarkers and Personalized Medicine

The main goal and promise of *personalized medicine* (also called individualized or – more recently – precision medicine) is to tailor treatment to the various individual characteristics of a patient, such as the patient's genomic, transcriptomic, proteomic, epigenetic, or metabolomic profiles. More generally, personalized medicine will try to determine an optimal approach to a particular patient's care (prediction, prevention, diagnosis, or therapy) by integrating comprehensive information about the patient's condition, which will include specific profiles identified by specific biomarkers. Hence, moving toward this goal will require discovery, validation, and clinical implementation of many new biomarkers; it is quite likely that most of them, especially the omic ones, will be multivariate biomarkers.

Although the paradigm of Western medicine can be seen as treating all patients with a specific diagnosis in the same – or very similar – way, one may nevertheless argue that conventional medicine has been – at least to some extent – personalized for a long time, as some individual characteristics of a patient were always taken into account and, thus, that personalized medicine is not a novel concept. However, developments in molecular biology and recent technological advances allow for the detection or quantification of not only many new physiological or pathological characteristics of patients, but also – and foremost – new types and scopes of such information (like those based on

[5] They are treated by the BEST biomarker taxonomy as a subtype of diagnostic biomarkers that are used for screening. Although this may be reasonable, the term *screening biomarkers* has been and probably will continue to be used to describe this specific type of biomarker, for which a low prevalence of the screened condition requires focusing on biomarkers with a very high specificity (see the example in Chapter 4).

the analysis of whole genome or transcriptome data, or on the simultaneous analysis of large numbers of proteins, metabolites, or non-coding regulatory RNAs). This facilitates the "personalization" of a patient's care in a much more specific and individualized way. And this process has just begun. How are (and will) such high-throughput technologies be utilized? By facilitating the discovery of new biomarkers. Since it is likely that we will need, and be able, to identify profiles associated with more and more complex biological processes, it is quite safe to assume that many, if not most, such new biomarkers will be multivariate ones.[6]

A step toward truly personalized medicine is *stratified medicine*, which assigns patients to specific subpopulations (or strata) based on molecular profiles as well as clinical and environmental variables characteristic for those subpopulations. Patients with similar "signatures" that, for example, may include similar genotype, lifestyle, and environmental factors, are assigned to a particular stratum and thus, considered for similar treatment options.

1.3 Biomarker Studies

Although the design of biomarker studies may differ from study to study,[7] there are some general steps that may be associated with any of them:

- Sample acquisition and preparation
- Processing samples in a laboratory (technologies may include genomic, proteomic or transcriptomic microarrays, next generation sequencing (NGS), liquid chromatography, mass spectrometry, nuclear magnetic resonance (NMR) spectroscopy, etc.)
- Low-level data preprocessing (depends on technology)
- Quality control (which may include batch and confounding factors adjustment)
- Normalization
- Presenting data in a form ready for analysis (usually, as a data matrix of variables times samples)
- Exploratory data analysis (to gain some general insight into the data)

[6] Personalized medicine has many aspects; for example, wearable devices transmitting real-time data from patients (and possibly from healthy individuals) as well as various "online medicine" tools and services (which may be associated with the idea of personalized medicine, but are better described as "personalized healthcare"), raise not only privacy concerns, but also concerns about the ethical, social, legal, and political implications of personalizing medicine (Nuffield Council on Bioethics 2010; Prainsack 2017). Here, we are focusing on personalized medicine only in the context of biomarker discovery.

[7] Biomarker discovery studies are primarily retrospective studies (using data collected from patients, for whom the outcome was known before the study was designed). Their main goal is to identify an optimal biomarker and design a predictive model implementing the biomarker. However, after the biomarker has been discovered, a prospective study (enrolling patients for whom the outcome is unknown, and who are then observed over a period of time) using the biomarker may follow, to validate and evaluate the biomarker predictions in clinical settings.

- Multivariate biomarker discovery (applying data science methods to identify an optimal biomarker for a particular target variable)
- Initial clinical evaluation
- Development of the test implementing the biomarker and the predictive model
- Independent clinical validation (predictive accuracy, reproducibility, relevance to clinical application)

The focus of this book is on the *multivariate biomarker discovery* step that involves application of appropriate data science methods in order to identify (and test) an optimal parsimonious multivariate biomarker for a particular target variable. Nevertheless, every step of this process is very important, and mistakes or deficiencies in any of them may severely impact the overall quality of a study.

For example, the collection of biological samples (of solid tissue, blood or other biofluids) should be preceded by a careful selection of patients (and controls) to promote homogeneity of the training data with regard to all variables except the target variable. Of course, absolute homogeneity is impossible, but as many factors as possible (such as age, gender, and ethnicity) should be taken into account in order to increase the probability that the main differences between the observations in the study are in fact due to the researched target variable.[8] The number of patients (or observations) included in the study is also a very important factor. Generally, the larger this number, the better and more robust the results expected. Although it may depend on what target is researched, we should not treat seriously those biomarker studies that are based on a dozen or so observations, as their results may be – at best – anecdotal. With the availability and the continually decreasing costs of high-throughput omic technologies, the *preferable* number of observations (or observations in a class, in the case of classification) should be at least in the hundreds. Furthermore, if the goal of a biomarker is to differentiate among specific conditions (or classes), then patients included in the study must be diagnosed into one of those classes with a very high level of confidence. Decisions on the type of collected samples should also depend on the prospective use of the biomarker. For example, if the biomarker is to be used for screening large populations, it should preferably be based on samples collected in the least invasive way possible (such as saliva, urine, or peripheral blood). Sample preparation, laboratory processing, and low-level data preprocessing (if necessary) depend heavily on the type of biological samples and the technology used. Normalization (as well as batch adjustment) may be necessary to make the data – which may have been prepared and processed in different labs, at different times, and not necessarily under the exactly the same conditions – comparable.

Analysis of the data may start once the good quality data is presented in the typical format of a data matrix with rows representing variables (such as the gene or protein expression level) and columns representing observations (patients, cell lines,

[8] In cases when this would be difficult to achieve – and when heterogeneity of the target population is not an intended aspect of the study design – we may decide to pursue segmentation predictive models based on separate biomarkers identified for different segments (or strata) of the target population (see Chapter 3 for more on this subject).

compounds, etc.). Each observation, say a patient, is also associated with the value of the target variable, whose prediction is the goal of the study. For classification biomarkers, this is the value of the categorical target variable, that is, a class label (such as the name of one of the differentiated disease states or subtypes). For estimation biomarkers, this is the value of the continuous target variable (such as the toxicity level). Data analysis typically starts with the exploratory data analysis (EDA), which – in addition to providing basic statistical information about the data – may identify some issues with the data quality, if they were not identified and corrected at the earlier steps. As a part of EDA, unsupervised data science approaches (such as clustering, principal component analysis, or self organizing maps) may be used to *visualize* the data and to look at how the observations may be grouped. If, for example, clusters of observations are aligned with the different laboratories the data were processed in, then this would indicate a serious problem with the data quality.

Multivariate biomarker discovery will be described in detail throughout this book; here, we will take a look at its main components:

- Identifying training and test data sets.

  It would be preferable if the training and test sets are independent statistical samples from the target population. However, if we only have one data set available for the analysis, then the data are randomly split into training and test sets. The latter (which, in this situation, is also called the holdout set) is set aside and not used or "seen" during the analysis. Hence, in either case, the test set will only be used to test the final predictive model resulting from the analysis.[9]

- Performing many multivariate feature selection experiments (with each of them using a different random subsample of the training data).

  This is the most important step of biomarker discovery based on high-dimensional data.

- Aggregating results of the feature selection experiments in order to:
  - identify an optimal size of the multivariate biomarker, and
  - select an optimal subset of variables to be included in the biomarker.
- Building and evaluating a predictive model (or models) that implements the identified multivariate biomarker.
- Testing the model on the test data set (that was never "seen" or used during the entire analysis).
- Facilitating biological interpretation of the multivariate biomarker.

A multivariate biomarker discovery project results in a multivariate biomarker (an optimal set of variables) with properly estimated predictive power and generalization

[9] Sometimes it may be convenient or necessary to split the available data into training, validation, and test sets. For example, the validation set may be used for tuning parameters of a predictive model implementing an already identified biomarker, or to choose from among a number of candidate multivariate biomarkers before the final biomarker is tested on the test data.

abilities, as well as a predictive model implementing the biomarker. Preferably, the results should also include information that could facilitate biological interpretation of the biomarker.[10] If the identified biomarker and the predictive model have sufficiently high predictive power (which may be evaluated, for example, by sensitivity, specificity, and accuracy for classification biomarkers, or by the mean squared error for estimation biomarkers), then subsequent stages of the biomarker study may be initiated.[11]

To have any clinical value, a biomarker must have sufficient predictive performance, must be reproducible, and must be acceptable and relevant for clinical use; its biological interpretation may also play an important role in its acceptance. Hence, each new biomarker and a test implementing it have to be subject to thorough independent clinical validation. It appears, however, that the major obstacle in the adoption of new biomarkers is not the fact that some of them are failing the clinical validation stage, but the fact that many of them are not even entering this stage. One of the possible reasons for this situation is that biomarker discovery projects have a good chance of being published; however, clinical validation of their results may be seen as having limited scientific value and thus, lower chances for funding and publication (Kumar and Van Gool 2013).

Furthermore, if a multivariate biomarker discovery project was based on outdated paradigms, such as employing the one-variable-at-a-time (univariate) approach, or basing the results only on their statistical significance, then such intrinsic deficiencies would render its results unlikely to pass thorough validation. The algorithms and data science methods that are appropriate for multivariate biomarker discovery – and free from such deficiencies – will be described and discussed in detail throughout this book.

1.4 Basic Terms and Concepts

Although the terms and concepts used in the book are explained in places where they are introduced or discussed, short descriptions of a few of them are, nevertheless, provided here for ease of reference.

Multivariate Biomarker

A multivariate biomarker is a set of variables (representing, for example, the gene or protein expression levels), whose combined pattern of values can – with high accuracy – predict the value of the dependent variable (such as disease state or the probability of relapse) for new or future observations. To be considered a multivariate biomarker, the set has to be identified via multivariate methods. Such methods consider sets of variables, rather than individual ones. Hence, individual relations between each independent variable and the dependent variable are irrelevant. It is also very important to understand that ***a multivariate biomarker* is a *biomarker***. It is not a "set of biomarkers"

[10] Which may potentially answer the questions of why and how the biomarker works.

[11] Perhaps assuming also that the predictive power of the new biomarker is significantly higher than that of existing biomarkers (if any) for the same target variable.

and should not be referred to by using such univariately-biased terms. The variables included in the multivariate biomarker – all of them together, as a set – constitute a single biomarker. None of them individually is a biomarker; if any of them were a biomarker for the researched goal, there would be no need for the multivariate biomarker. Therefore, in this book, whenever the terms *biomarker* or *biomarker discovery* are used without qualifications, they should be treated as synonymous with *multivariate biomarker* and *multivariate biomarker discovery*, respectively.

Optimal Multivariate Biomarker

An optimal multivariate biomarker should be parsimonious and well generalizable. This means that it should consist of as few variables as possible, while simultaneously trying to maximize its predictive power. Hence, its optimality should be based on a compromise therebetween.

Independent Variables

Independent variables are characteristics of the observations (say, patients) included in the training data (for example, the gene expression levels for 20,000 or so genes). Only some of them will be included in the multivariate biomarker and will be used to predict the value of the dependent variable.[12] It has to be emphasized that the term ***independent variables*** should be understood only in the context of independent variables versus the ***dependent variable***. It does not provide *any* information about relations and correlations among the independent variables. Whenever the term "variables" is used, in this book, without qualification, it will refer to independent variables.

Dependent Variable

Predicting values of the dependent variable is the goal of biomarker discovery (or, more generally, of any predictive modeling). If the dependent variable is categorical, we are solving a classification problem; if it is continuous, we are solving a regression problem. The synonyms for the dependent variable are ***target*** variable and ***response*** variable.

Biomarkers for Classification

If the dependent variable is categorical, the goal of biomarker discovery is to identify a biomarker that will be able to classify new patients (or, more generally, new observations) into one of the differentiated categories (such as two or more disease states). In this case, predictive modeling is using classification learning algorithms; the

[12] This is why we do not call all of them "predictors" (as they are often referred to in the literature, especially in statistical texts). With thousands of variables in typical high-dimensional data, most represent noise, and only a few of them will be selected into a multivariate biomarker and eventually used for prediction.

identified multivariate biomarker can be called a ***classification biomarker***, and the predictive model that implements it – a ***classification model***.

Biomarkers for Estimation

If the dependent variable is continuous, the goal of biomarker discovery is to identify a biomarker capable of predicting (estimating) the value of this dependent (target, response) variable (for example, the probability of relapse). In this case, predictive modeling is using regression learning algorithms; the identified multivariate biomarker can be called an ***estimation biomarker***, and the predictive model implementing it – a ***regression model***.

Personalized Medicine

Personalized medicine attempts to determine an optimal and individualized approach to a patient's care (prediction, prevention, diagnosis, or therapy) by integrating comprehensive information about the patient's condition, especially by considering the patient's omic profiles (such as genomic, transcriptomic, proteomic, or metabolomic signatures). *Stratified medicine* is a step toward personalized medicine – patients' omic signatures (and treatment options) are not yet considered at a personalized level, but are instead matched to the profiles characteristic for specific subpopulations (or strata).

Predictive Modeling

Although, in the context of this book, *predictive modeling* can be seen as synonymous with *multivariate biomarker discovery*, one may argue that the goal of biomarker discovery is to find an optimal biomarker, and that predictive modeling also includes building a predictive model (or models) based on such an optimal biomarker. Therefore, even if in practice both of these goals are essential parts of a properly designed biomarker discovery process, we will also, to avoid any confusion, use such phrases as *predictive modeling for biomarker discovery*.

Multivariate Methods

Multivariate methods consider all independent variables simultaneously; thus, interactions and correlations among the variables are taken into account. Even if a multivariate algorithm sometimes focuses on a specific variable (for example, a feature selection method may consider which of the variables should be added or removed from the currently considered subset of variables), such a variable is always evaluated in the context of other variables.

Univariate Methods

Univariate methods consider only one independent variable at a time, and evaluate the relationship between each of them and the dependent variable individually, ignoring any and all relationships among the independent variables. As such, univariate methods are not used for multivariate biomarker discovery.

Supervised Learning Algorithms

Supervised learning algorithms are methods used for predictive modeling (thus, including biomarker discovery). They are supervised by the values of the dependent variable that are associated with the observations in the training data, and they are used to identify a multivariate biomarker and predictive model (implementing the biomarker), which can then be used to predict the value of the dependent (response) variable. The goal of supervised learning is to maximize the accuracy of predicting the value of the response variable (which may be achieved via maximizing class separation – for classification, or via maximizing the proportion of the explained variation in the response variable – for regression) for *new* observations.

Unsupervised Learning Algorithms

Unsupervised learning algorithms are used to identify groups of similar observations or groups of similar variables. They can provide visualization of high-dimensional data (and the grouping results) in a low-dimensional space, and are thus valuable tools for exploratory data analysis. However, unsupervised methods are blind to the dependent variable and are therefore inappropriate for biomarker discovery. It should be stressed – as this is still a quite common misconception – that they should not be used for decreasing the dimensionality of the training data that would be then used for biomarker discovery.

Training Data

The training data set is used to perform all steps of predictive modeling for biomarker discovery. It includes N observations (e.g. patients) and p variables (e.g. gene or protein expression levels). Each of the observations may be represented by a $p \times 1$ vector $\mathbf{x}_i = [x_{1i}, \ldots, x_{pi}]^T$, and the value of the response variable, y_i, $i = 1, \ldots, N$, associated with the observation. Hence, the entire training data may be represented by a $p \times N$ matrix $\mathbf{X}$ and a $N \times 1$ vector $\mathbf{y}$.

Test Data

The test data set is used exclusively for testing the performance of the optimal multivariate biomarker identified by the predictive modeling analysis. It has to be unavailable during this analysis. Ideally, the test data would be independent of the training data (for example, collected in a different geographical region and processed in a different lab). If such independent data are unavailable, then the data available to the project need to be randomly split into the training and test data, and the test data set aside, and used only after the analysis is completed.

Target Population

Both training and test data sets are statistical samples from the target population, which is the population for which we are building a predictive model. For classification modeling, when the dependent variable is categorical – and, for example, represents several disease states – each category could be considered a separate population; however, while keeping this in mind, we will often refer to all of them by the singular term.

High-dimensional Data

High-dimensional biomedical data sets, for which the number of variables is greater, or often much greater, than the number of biological samples, are routinely generated by current high-throughput omic technologies. Applying – to such data – traditional statistical or predictive modeling methods that have been successfully used in low-dimensional settings will virtually guarantee overfitting. It is very likely that most of the variables in such high-dimensional data sets represent noise, and to be able to extract a true signal therefrom, more sophisticated heuristic multivariate search algorithms or regularization methods need to be used.

Feature Selection

The goal of feature selection for multivariate biomarker discovery is to identify a small subset of independent variables whose combined pattern of values allows for the accurate prediction of the value of the response variable for new observations. In this context, feature selection is synonymous with ***multivariate feature selection***. Furthermore, since we are interested in biomarkers consisting of some of the original variables (rather than any "engineered" features), these terms should also be understood as being synonymous with *variable selection*, and *multivariate variable selection*.

Parallel Feature Selection Experiments

Performing only a single feature selection experiment, when analyzing high-dimensional data, will virtually guarantee overfitting (that is, finding a spurious pattern that exists in the training data, but does not exist in the target population). To overcome this curse of dimensionality, we should perform many parallel feature selection experiments, with each of them using a random subsample of the original training observations. By aggregating the results of such feature selection experiments, we can identify a multivariate biomarker that is much more likely to be generalizable (to the target population) than those resulting from a single feature selection run.

Hyperparameters

In the context of predictive modeling, the term *parameter* may refer to a predictive model's parameter as well as to its *hyperparameter*; however, a distinction between the two is usually made. *Parameters* are such internal characteristics of a predictive model that are estimated directly from the training data. *Hyperparameters* are, however, external to the model, and their values are either set manually or require tuning. Hence, we can say that hyperparameters are *tunable* parameters.

Bias-Variance Tradeoff

The bias-variance tradeoff refers to finding an optimal balance between fitting the model to the training data and its ability to be generalizable to the target population. If a predictive model perfectly fits the training data, it is likely that the identified pattern is a spurious one, which exists in the training data by chance, but does not exist in the target population. In such situations, the model performance in predicting new observations would be poor, and the model would have low bias and high variance. A high-variance model is very sensitive to changes in the training data, while the opposite is true for a high-bias model. The goal of a proper tradeoff is to find a model, which is not too complex to overfit the training data, but complex enough to provide accurate predictions for new observations from the target population.

2 Multivariate Analytics Based on High-Dimensional Data: Concepts and Misconceptions

2.1 Introduction

Although *Multivariate Analysis* courses included in statistics curricula and those designed for students of data science or bioinformatics may cover the same multivariate methods, their foci are usually quite different. Statistics courses mainly focus on statistical inference and theoretical aspects of the methods; typical examples – used in such courses – would be low-dimensional illustrations of the methods, and thus they may have little practical value for biomarker discovery based on high-dimensional data. For biomarker discovery, pure statistical-inference conclusions based on such measures as, for example, *p*-value, are rarely satisfactory – the fact that a solution is not likely to represent a chance event does not necessarily mean that the solution (for instance, a multivariate biomarker that can be used to build a classification model) has sufficient predictive power to be suitable for any practical use. In contrast, data science approaches to biomarker discovery consist of data-driven discovery-based algorithmic methods aimed at identifying novel multivariate patterns that are capable of accurately predicting the value of the target variable (such as the class label associated with one of the differentiated disease states for classification, or the real value of the continuous target variable for regression).

Hence, whether a multivariate biomarker or a predictive model implementing the biomarker is statistically significant is quite irrelevant. A model that is statistically significant may be completely useless for the purpose of prediction. On the other hand, models that are not statistically significant are rather unlikely to show any performance better than by random chance.

In this chapter, we will discuss some of the important aspects of multivariate approaches, as well as provide information and warnings about some common misconceptions. The following topics will be addressed: high-dimensional data and the curse of dimensionality, multivariate and univariate approaches, and supervised and unsupervised approaches.

2.2 High-Dimensional Data and *the Curse of Dimensionality*

Data sets with a large number of variables, and especially high-dimensional data for which $p > N$ or $p \gg N$ (where p is the number of variables and N the number of biological samples, or more generally, observations) are now an everyday reality in biomedical research, as well as in many other applications of data science. Therefore, it is important to realize that many of the well-established methods and procedures that have been successfully used to analyze traditional low-dimensional data (such as large business-type data sets with the number of observations N larger or much larger than the number of variables) will almost certainly fail in high-dimensional situations, especially when $p \gg N$. In such situations, traditional predictive modeling approaches are virtually guaranteed to overfit the training data; they may produce models with perfect *training-data*-based metrics (such as accuracy or R^2), but those models would be useless for predicting new observations.

It is a typical situation for high-dimensional data that many (or even most) variables represent noise – at least from the point of view of the goal of our biomarker discovery project. Since the training data set is sparse (relatively few data points in a high-dimensional space), a chance solution that fits the noise (and, at the same time, perfectly or near perfectly fits the training data) is the most likely outcome when traditional (or inappropriate) methods are used. To properly extract real predictive information from an overwhelming amount of noise, more sophisticated methods must be used, among them: repeated heuristic multivariate feature selection, regularization, and the ensemble approach. Often, a combination of some (or all) such methods may be necessary to find an optimal solution.

Let us first take a look at one – very important – element of proper predictive modeling: multivariate feature selection (Chapter 5 is dedicated to this topic). The goal of multivariate feature selection is to find a small subset of variables that maximizes predictive power and minimizes noise – we will call it an optimal set of variables (optimal according to some criteria, which should include predictive power, as well as a parsimonious solution). With a data set having thousands of (or even just a hundred) variables, evaluating all possible subsets of variables would be infeasible. Nonetheless, even if this were possible, the best subset – that is, the global optimum for the training data – would quite certainly overfit the training data and thereby be useless for predicting new observations. This creates difficulties that are known collectively as ***the curse of dimensionality*** (Bellman 1961). Appropriate heuristic multivariate search algorithms have to be used for the efficient identification of an optimal solution.

Even though the multivariate analysis of high-dimensional data has already entered the mainstream of bioinformatics research, there still seems to be some confusion about how to properly deal with the curse of dimensionality. When we try to analyze high-dimensional data using the methods that have worked in other settings, we are very likely to find that when the resulting biomarkers are tested on independent data, their predictive abilities are not necessarily better than random choice. To illustrate just a few problems with traditional data analysis approaches that may work well for low-dimensional data, but fail for high-dimensional settings, let us consider two hypothetical scenarios. Assume – also hypothetically – that data analysts whose experience has been limited to $p < N$ (or, more likely, $p \ll N$) low-dimensional data (such as, again, traditional business-type data) are asked to perform two predictive modeling projects, which are described in the following two scenarios:

Scenario One:

Training data consisting of 50,000 observations (say, customers) and 20 variables need to be used to create a classification model differentiating two classes of customers represented in the training data. Our experienced analysts (*EAs*) decide to use all 20 variables to build a classifier. After deciding on the learning algorithm to be used, they build a classifier and then estimate its performance using 10-fold cross-validation (CV). As they had done many times in the past, for each fold a classifier is built on a randomly selected 90% portion of the 50,000 training observations and the remaining 10% are used to test the classifier. After CV results are averaged over the folds, sensitivity, specificity, and accuracy of the final classifier are estimated, and let's say that, for the sake of simplicity, all three measures are equal to 99%. Seems like a success. Now the classifier is tested on new data, and its performance in classifying new customers is similar to that estimated. Success!

Scenario Two:

Training data consisting of 200 observations (say, patients) and 10,000 variables (say, genes represented by their expression levels) need to be used to create a classification model differentiating two classes of patients represented in the training data. The same *EAs* are charged with this project. Though they have no experience with high-dimensional data, they know that the data with 10,000 variables must include many uninformative or redundant variables as well as a lot of noise; thus, feature selection is necessary. The *EAs* decide on a feature selection algorithm (say, recursive feature elimination),[1] perform it once on the entire training data, and identify a subset of 10 genes. Then, they build a classifier – using only this subset of 10 genes – and estimate its performance using 10-fold CV. As they had done many times in the past, for each fold a classifier is built on a randomly selected 90% portion of the 200 training observations and the remaining 10% are used to test the classifier. After CV results are averaged over the folds, sensitivity, specificity, and accuracy of the final classifier are estimated, and let's say that, for the sake of simplicity, all three measures are equal to 99%. Seems like a success. Now the classifier is tested on new data, and its performance in classifying new patients is hardly better than that of a coin toss. What went wrong?

A simple answer is: everything. If we repeated Scenario Two using data that are entirely randomly generated noise, with randomly assigned class labels (so, the data include no discriminatory information that would be useful for differentiation between the two classes), we would most likely find a classifier with estimates of its performance as high as they were for the original project. Why? Because this process was first fitting noise, and then was reusing the same training data to estimate the noise-fitting classifier's performance.

[1] Recursive feature elimination is described in Chapter 5.

The first problem was that the feature selection search was performed only once, and thus – for high-dimensional data –was likely to result in a spurious pattern. The second problem was the use of the *internal* cross-validation.[2] *What is internal CV?* – one could ask. *Wasn't cross-validation performed in the same manner in both scenarios?* Well, in *Scenario One*, an *external* CV method was used, which means that for each classifier built during the CV process, the 10% of observations used as a test sample had in no way influenced the design of the classifier, so those observations were *external* from the classifier's point of view. In *Scenario Two*, however, all training observations had already been used during the feature selection step; and since each classifier evaluated in the course of the CV process was built on the subset of 10 genes whose selection was based on the information from all of the 200 training observations, none of these observations (whether selected into 90% or 10% of the folds) can be considered unseen by the classifier. Such *internal* cross-validation is as reliable (or rather as unreliable) as reclassifying the training data.

Bear in mind that the deficiencies related to feature selection and cross-validation are only some of the many aspects of the fallacy of taking the same approaches that are commonly (and successfully) used for low-dimensional data and applying them to high-dimensional data. Most of them will not work and thus may result in failure. "*Why don't they work?*" The simplest answer is that they were not designed to overcome the curse of dimensionality. Another general, but more meaningful, explanation is that when we deal with low-dimensional data that include 50,000, or 100,000, or sometimes millions of observations, and only 20 or 50 or so variables, such data can be treated more like a *population* rather than a small or relatively-small sample randomly selected from the target population. And when we deal with populations, no sophisticated data science or statistical methods are really needed. What this means is that in low-dimensional situations essentially *any* method would work. Hence, with only slight exaggeration, we may say that *any* method may be successfully used to analyze low-dimensional data; if our target is of the size of a barn and we are standing three feet from it, any method will find the target . . . we can blindly throw a shoe and it will hit the target.

However, when dealing with high-dimensional data, such as typical gene or protein expression data, we may have a few hundred patients – we feel lucky when we have more than 1,000 patients in the training data – and we may have 10,000 or 20,000 or even more variables. If we imagine (if we can) a 20K-dimensional space and only a couple hundred patient data points therein, the data are very very sparse. Our target multivariate patterns are now buried under, or hidden within, possibly *billions and billions of random patterns*. Those random patterns are the patterns that exist in the training data by chance, and thus do not exist in our target patient population. How can we identify the real patterns that could represent important new multivariate biomarkers? Definitely, throwing a shoe would not work. The general idea for the solution is conceptually quite simple: *remove noise and elevate signal*, then remove more noise and elevate signal some more, and then continue this signal-enhancement process until

[2] Methods for evaluating the performance of predictive models (including cross-validation) are covered in Chapter 4.

we can identify some robust and generalizable patterns that have the best chance for representing real patterns in the target population. Though this is a simple idea, its proper and effective implementation requires rather sophisticated data science methods; these methods will be described throughout this book.

2.3 Multivariate and Univariate Approaches

A multivariate method considers all variables simultaneously; even when we sometimes look at a specific variable, it is always evaluated in the context of other variables in the training data set. Hence, multivariate methods consider correlations and interactions among variables. In contrast, a univariate method considers only one variable at a time, each of them independently of the other variables in the training data, which means that relationships among the variables are simply ignored. For example, a *t*-test or the ANOVA *F*-test (which test whether the means of two or more populations, or classes, are significantly different), are univariate methods. As such, they do not belong to the area of multivariate biomarker discovery.[3] More generally, no univariate method should be used to select variables for a multivariate biomarker. This means, for example, that methods that first evaluate each independent variable by its individual correlation with the target variable, and then apply some multivariate approach only to the variables whose correlation to the target is above some threshold, should not be considered for multivariate biomarker discovery (as they are univariately-biased methods).[4] The reason for the exclusion of not only univariate methods but also univariately-biased multivariate ones (hence, not truly multivariate) is that variables that are individually insignificant (whether their significance is measured by univariate tests, such as the mentioned *t*-test or *F*-test, or by their correlation to the target) may be – and, in practice, some of them are – very important when combined with other variables.

To illustrate this, consider two very simple classification examples with only three variables and two differentiated classes (see Figure 2.1). None of these three variables can individually differentiate between the two classes. However, in either of these examples, the set of all three of these univariately-insignificant variables can perfectly separate the classes. Of course, in such a simple low-dimensional situation, we can not only visualize the classes in the entire space of independent variables, but we can also consider all possible combinations of the variables. However, this would not be possible in a real biomarker discovery study with hundreds or thousands of variables; only multivariate approaches can identify such solutions.

[3] Together with unsupervised learning algorithms, univariate methods may be used for exploratory data analysis (to gain some insight into the training data), but they should not be driving the biomarker discovery process.

[4] To emphasize: performing any univariate analysis to limit the number of considered variables to those that are individually significant, and *then* performing multivariate analysis using only those univariately significant amounts to a serious error in study design. This may lead either to suboptimal study results or to negative results even if the training data would have supported the identification of an efficient biomarker were a proper multivariate method used.

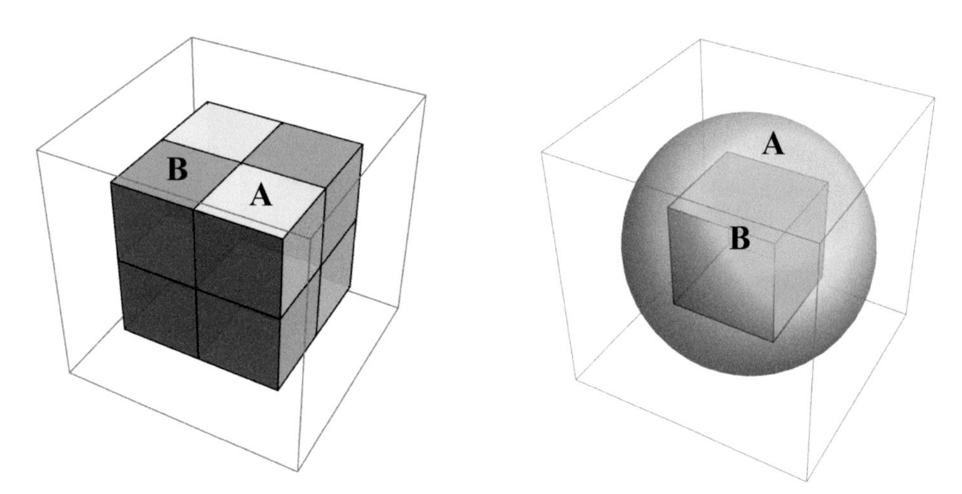

Figure 2.1 Two classification examples with three independent variables and two differentiated classes (say, disease states or subtypes, A and B). In the left panel, class A is represented by four yellow cubical areas, and class B by four green cubical areas. In the right panel, class A is represented by a spherical "shell", within which there is a cube corresponding to class B. None of the three independent variables can individually distinguish between the classes. However, the set of all three of these univariately-insignificant variables can perfectly separate the classes, in either example.

All of this leads, again, to the main tenet of multivariate biomarker discovery: to identify an optimal multivariate biomarker, which would be both parsimonious and highly predictive, we have to evaluate *sets* of variables. Individual significance of any variable is irrelevant; it would be important only if we wanted to find a single-variable biomarker, which would be a completely different (and rather trivial) task.

2.4 Supervised and Unsupervised Approaches

The main *statistical* application of multivariate methods is statistical inference that considers many independent variables. *Data science* applications of multivariate methods – learning algorithms – have two main goals: either predictive modeling (supervised methods), or the identification of groups of similar entities (unsupervised methods).

It is important to understand that the goals of supervised and unsupervised approaches are very different. Therefore, neither of these approaches may be considered *better* or *worse* – they perform different tasks and are not interchangeable. Supervised learning algorithms are used for predictive modeling – learning from the training data, they build classification or regression models for predicting the value of the response variable for new observations. In contrast, unsupervised learning algorithms search for some internal structures in the data represented only by the independent variables; they are blind to the dependent (response) variable. Consequently, unsupervised methods have no interest in the response variable, have no use for it, and their results are in no way aligned or associated with the response.

What supervises the supervised methods? It is the ***response variable***, or more precisely, its values that are associated with training observations. Hence, supervised methods require *training data* (that include the value of the response variable for each observation) in order to learn about patterns associating observations (represented by the vectors of values of the independent variables) with the values of the response variable. For predictive modeling to be successful, the patterns learned have to be generalizable, that is, the resulting predictive models have to accurately predict the value of the response variable for new (or future) observations from the target population (i.e. for the observations that were not included in the training data). The goal of supervised learning – that is, the accurate prediction of the response variable – can be achieved by identifying the directions that either maximize the *variation in the response variable* (for regression modeling) or maximize class separation (for classification modeling).

The goal of unsupervised methods (such as clustering or principal component analysis) is to elucidate intrinsic groupings of similar entities. Such groupings may give us some insight into the relationships among the observations or among the variables. Let's focus on grouping observations. Each of them is represented by a vector of values of the independent variables. Recall, however, that unsupervised methods are blind to the response variable; so, even if its values are known, they are not used; thus, there is *no supervision*. Let's focus further on Principal Component Analysis (PCA), an unsupervised dimensionality reduction method. PCA finds a set of orthogonal directions (with each of them being a linear function of all independent variables) that enables a low-dimensional visualization of the data. PCA finds those directions (principal components, or PCs) by maximizing the *variation of the data*. By visualizing the data in a low-dimensional space of the first few PCs, PCA is a valuable tool for exploratory data analysis.

However, a common misconception is to use PCA (or any other unsupervised method) to decrease the dimensionality of the training data used in a biomarker discovery study. Biomarker discovery – or, more generally, predictive modeling – is a supervised task. The key information driving the discovery is the value of the response variable associated with each training observation (say, patient). However, since PCA is blind to the response variable, the directions of its principal components are *random* from the point of view of the goal of biomarker discovery. A quite bizarre justification that is sometimes provided is that PCA preserves most of the *information* in the data. However, "information" without qualification is meaningless. PCA preserves most of the *variation* in the data represented only by the independent variables. That is not what we are interested in for biomarker discovery; for example, for classification we are interested in the *discriminatory information* that would allow us to maximize class separation (by, for example, maximizing the *ratio* of the between-to-within class variation). This means that by replacing the original data with the data represented by some number of top principal components, we may discard most – or even all – discriminatory information that would be crucial for achieving the goal of biomarker discovery.

A simple two-dimensional example (Figure 2.2) illustrates this misconception by focusing on the different goals of supervised and unsupervised dimensionality reduction

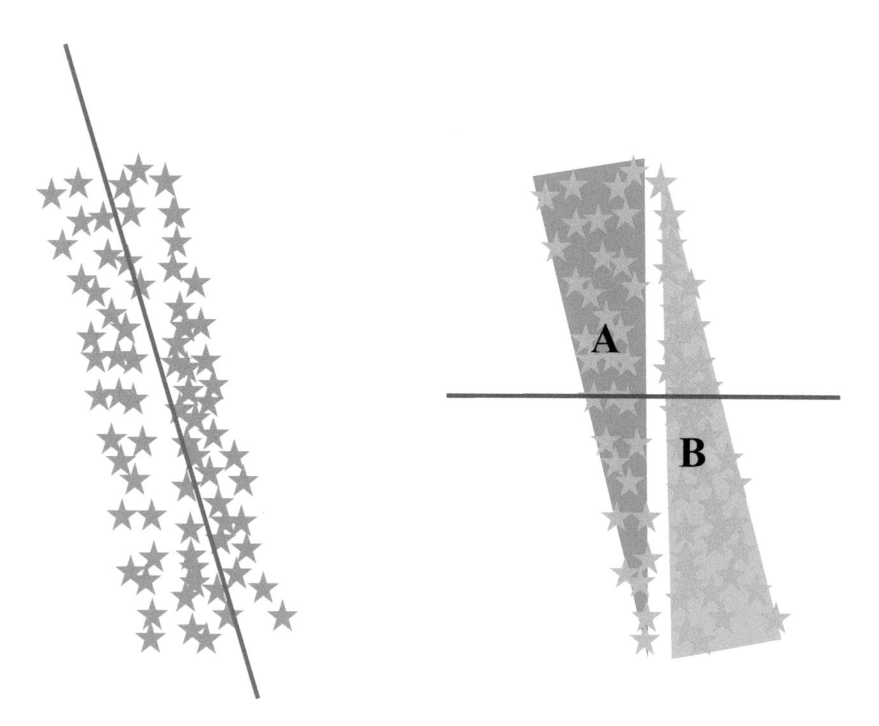

Figure 2.2 The two ways in which unsupervised methods and supervised methods perceive (and analyze) the same training data. This is, again, a very simple classification example with only two independent variables ($p = 2$) and two classes – say, disease states A and B (which are categorical values of the target, or dependent, variable). Each data point (a gray star) represents a patient from the training data set, and each patient is associated with the diagnosis (either A or B). **The left panel** illustrates implementation of an unsupervised dimensionality reduction method, in this case Principal Component Analysis (PCA). Like any unsupervised method, PCA is blind to the target variable (the diagnosis) and considers only the two independent variables. This method seeks and finds the direction that maximizes the variation in the data; this direction is represented by the blue line. If the PCA results were used to decrease the dimensionality of the problem (from the two original dimensions to one), the data points would be projected onto the blue line. It is easy to see that this would eliminate all discriminatory information contained in the data. If data prepared this way were then used for biomarker discovery, there would be no way to distinguish between the two classes. **The right panel** illustrates the way in which a supervised dimensionality reduction (using, for example, Fisher's Discriminant Analysis, or Partial Least Squares) would be performed. For a supervised method, the most important information about the data points is the value of the target variable associated with each of them (the patient's diagnosis). Such a method will seek and find the direction that maximizes separation between classes; this direction is represented by the red line. If we now decrease the dimensionality of the problem by projecting all of the data points onto the direction represented by the red line, the two classes would be perfectly separated in this one-dimensional space.

methods, and by demonstrating the different ways, in which such methods perceive the same data. In this example, there are two independent variables and two classes (disease states A and B). The data points (gray stars) represent patients; each of them is associated with the diagnosis (the value of the response variable, either A or B).

The left panel illustrates an unsupervised approach – principal component analysis. To visualize the data in a lower-dimensional space (here, the reduction would be from two to a single dimension), PCA, as always, finds the direction of the most variation in the data; being unsupervised, PCA ignores the diagnosis. The first – and here, the only – principal component direction (PC1) is represented by the blue line. If the original two variables are replaced by PC1, that is, if the patient data points are projected onto the blue line, all discriminatory information (which, for the original data, would allow us to separate the two classes) would be lost.

The right panel illustrates the way in which a supervised method perceives the same training data – the crucial information for each patient is the diagnosis (class). The goal is to find a direction that maximally separates the classes – this direction is represented by the red line. If the dimensionality of the data is now reduced by projecting the data points onto the red line, the classes would be perfectly separated in the one-dimensional space identified by the supervised approach.

One may say that "discriminatory information" is associated with classification projects, but what about regression projects? The situation is very similar for regression. In biomarker discovery where the response variable is continuous, our goal is to explain as much of the *variation in the response variable* as possible. If we imagine a $(p + 1)$-dimensional cloud of our training data points (p independent variables plus the response variable y), this would be the variation in the single direction of our response variable y. However, PCA is totally blind to y values, as it uses only the independent variables and tries to find the directions of most *data variation* in a p-dimensional space of p independent variables – the space that does *not* include our response variable. Thus, the goals of PCA are very different from the goals of biomarker discovery, irrespective of whether the response variable is categorical (classification) or continuous (regression).

To conclude: since unsupervised methods are blind to the response variable, using them for reducing the dimensionality of a biomarker discovery problem should be seen as equivalent to blinding ourselves to the target and throwing darts in random directions, in a situation when the information about the target (the response variable) is not only available but also paramount for supervised modeling.

3 Predictive Modeling for Biomarker Discovery

3.1 Regression versus Classification

As mentioned earlier, the ultimate goal of predictive-modeling data science projects (including biomarker discovery ones) is to create a model that can accurately predict values of the response variable for new observations. If the response variable is qualitative (categorical), we build a classification model, which will assign new observations to one of the differentiated classes. If the response variable is quantitative, we build a regression model that will estimate the numerical value of the response variable. Sometimes supervised learning algorithms are categorized as classification methods or as regression methods. One might, for example, say that Discriminant Analysis and Support Vector Machines are examples of supervised learning algorithms for classification. Similarly, we may describe Multiple Regression, Ridge Regression, and Lasso as examples of supervised learning algorithms for regression. The distinction, however, is not that clear. Some learning algorithms – Random Forests, for example – can be quite naturally used for classification as well as for regression. Even support vector machines, the algorithm that was specifically designed for classification, can also be adapted for regression.

3.2 Parametric and Nonparametric Methods

Please be advised that definitions of parametric and nonparametric methods may be quite ambiguous. Not much has really changed in this area since John E. Walsh wrote "*A precise and universally acceptable definition of the term 'nonparametric' is not presently available*" (Walsh 1962).[1] Consequently, in the literature of the topic, one may encounter points of view that seem to be reasonable, even if they do not necessarily converge to the same definition of – and distinction between – parametric and nonparametric methods. One can also find "definitions" that are completely wrong (deeming, for example, that the Random Forests algorithm is parametric because to build a random

[1] It is also worth noting that the term "nonparametric", if taken literally, may be misleading – most nonparametric methods are not parameter-free. In fact, they may have many parameters; however, those methods are completely flexible in the sense that the number and type of their parameters are not associated with any specific functional form assumed in advance.

forest model we need to specify two parameters of the learning algorithm). Among the reasonable definitions are those that focus on the *functional relationship* between the independent variables and the response variable, and those that focus mainly on *distributional assumptions*.[2]

Let's discuss first the definitions of parametric and nonparametric approaches that focus on distributional assumptions. In the view of such definitions, *parametric* methods are those that make some distributional – and eventually other – assumptions about the target population. For example, a parametric method may assume a particular distribution of independent variables, such as a multivariate normal distribution or Poisson distribution. It may also assume that observations are independent of one another, or that class covariance matrices are equal. Some of the other common assumptions may be about the absence of multicollinearities among independent variables,[3] or the absence of outlier observations. *Nonparametric* methods would then be those that make no specific distributional assumptions. For parametric methods that are robust to some violations of their assumptions, it may be advantageous to make the assumptions even if they are (slightly) violated, as it may lead to much simpler predictive models. For example, the homoscedasticity (that is, the homogeneity of class covariance matrices) assumption of linear discriminant analysis leads to models with linear boundaries between classes, which are not only simpler, but may be – at least in some situations – more robust than models with nonlinear boundaries.

For those definitions that focus on the functional relationship, *parametric* methods are characterized as those that make an assumption about the functional relationship between independent variables and the response. For example, multiple regression (see Chapter 6) assumes a linear relationship between y (the dependent, or response, variable) and p independent variables, $E(y) = \beta_0 + \beta_1 x_1 + \beta_2 x_2 + \cdots + \beta_p x_p$. Thus, instead of searching for any pattern that would maximize the predictive abilities of the biomarker, the search is limited to only such patterns that fit the assumed functional relationship. For this regression example, this would mean estimating from the training data only the values of the $p + 1$ *parameters* of the linear relationship, $\beta_0, \beta_1, \ldots, \beta_p$. The advantage of this approach is the simplification of the process of predictive modeling. However, if the real relationship between the independent variables and the response variable is far from that which was assumed, the results would be far from optimal (due to high bias; see Section 3.4). And, of course, from the point of view of this type of definition, *nonparametric* methods are those that make no assumptions about such a functional relationship. Consequently, nonparametric methods are free to search for any pattern (any functional relationship) that would be optimal when based

[2] Since there are overlaps between these two approaches, they could be consolidated; however, this would lead to more vague definitions.

[3] Multicollinearity refers to high correlation between two or more variables (singularity is a special case of multicollinearity, when the variables are totally redundant). It is worth noting that pairwise correlations and multicollinearity are different – although overlapping – concepts. While screening for pairwise correlations may be sufficient for some low-dimensional data, it is insufficient for high-dimensional data. For example, in high-dimensional data, it is quite likely that a variable may be strongly correlated with a group of variables while not strongly correlated with any of them individually.

on particular training data; however, this usually requires a larger set of training data than what would be needed for training a parametric model.

3.3 Predictive Modeling for Multivariate Biomarker Discovery

The placement of the *multivariate biomarker discovery* step in the context of biomarker studies[4] has been described in Section 1.3. This step typically uses the experimental data in the form of a data matrix, with variables as the rows and observations as the columns of the matrix. Observations may represent patients, tissues, cell lines, etc. Variables represent specific measurements for the observations, such as gene expression levels or metabolite concentrations. For each observation, we also have the value of the target variable, the variable that the predictive model – based on an optimal multivariate biomarker – will try to predict. If the target variable is categorical (for example, when its values are labels corresponding to different disease states), we will identify a classification biomarker and build a predictive model using classification learning algorithms. If the target variable is continuous (for example, the probability of disease recurrence), our goal will be an estimation biomarker and predictive modeling will use regression learning algorithms.

The core elements of predictive modeling for multivariate biomarker discovery are presented in Figure 3.1. However, before discussing them we will look first at some issues related to data preparation.

3.3.1 Additional Data Evaluation and Preparation

If the samples were collected and preprocessed specifically for our biomarker study, the quality control of the data should have already been completed during the previous steps of the study. If, however, we are using data that were generated for a different project, or if we acquired the data from one of the public repositories, it is good practice to re-evaluate the data quality (unless the data were acquired from a reliable source, and we are convinced that proper quality control has already been performed). The details of how it should be done depend on the type of the training data and also – to some extent – on the kind of supervised learning algorithms we will use in our predictive modeling. For example, if we are going to use algorithms that are sensitive to outliers, it would be a good idea to check for outlier observations. We may also want to visualize observations in a low-dimensional space by either clustering them or by performing principal component analysis.[5] Such visualizations may provide some information about grouping of the observations, and thus may identify problems with the quality

[4] In the context of applying data science methods to any data, this step would be called *predictive modeling*.

[5] Remember, however, that PCA (an *unsupervised* method) would be used here only for visualization. Principal components may not be used as variables for our supervised analysis; the original data (or components identified via a *supervised* dimensionality reduction method) have to be used there.

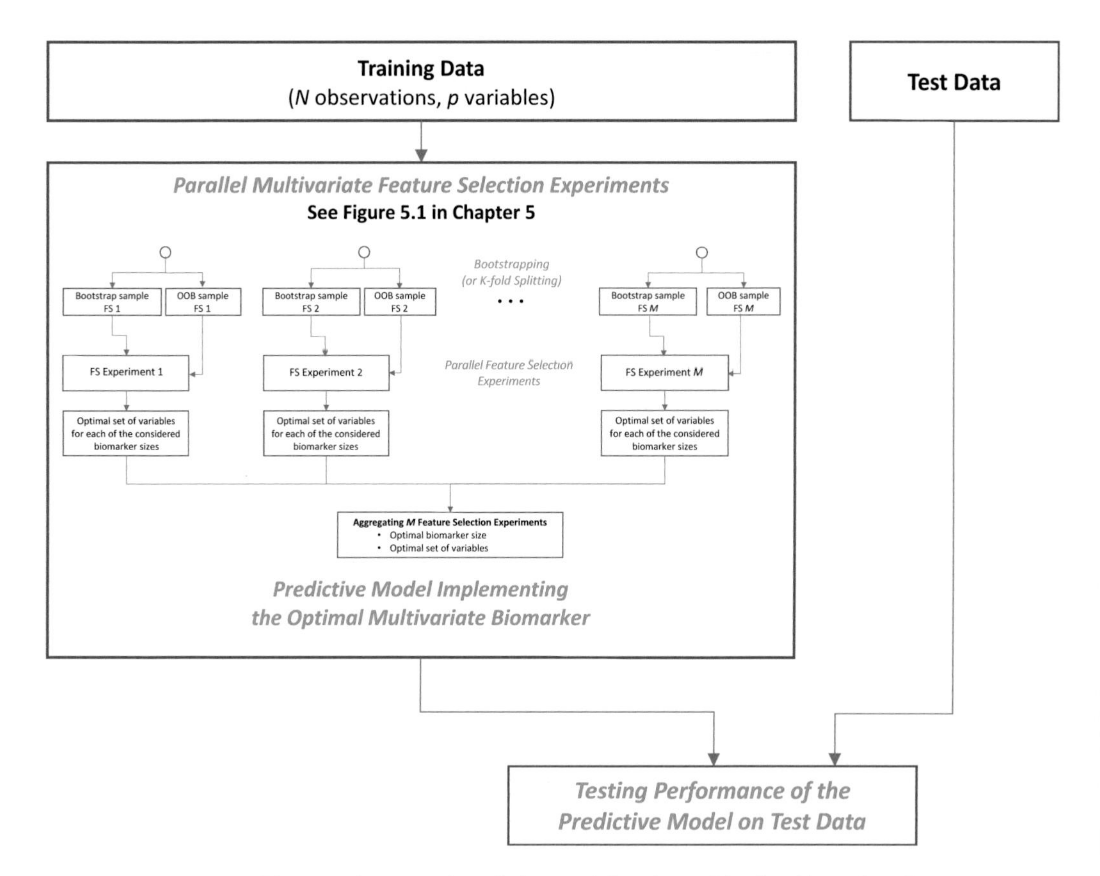

Figure 3.1 The core elements of predictive modeling for multivariate biomarker discovery. Training data are used for feature selection and model building. Test data are used only for testing the final predictive model. The training and test data sets may be the result of splitting our biomarker study data, for example, by stratified random sampling. In this case, the test set is a holdout set that is set aside and not used during the analysis. However, the test data may be – and this is preferable – a data set independently selected from the target population. The training data are used first for multivariate feature selection experiments. Many parallel feature selection experiments are performed, each based on a random selection of observations from the training data (using either bootstrapping or K-fold cross-validation). Each of the parallel experiments would use the same feature selection method, would consider the same set of biomarker sizes, and would estimate the predictive abilities of each of them. If necessary, and if the feature selection process is utilizing a supervised learning algorithm, tuning may also be incorporated into the feature selection experiments. Aggregating results of many parallel feature selection experiments allows for the identification of an optimal multivariate biomarker that has the best chance of being robust. The final evaluation of the predictive model implementing the optimal multivariate biomarker is based on testing its performance on the test data.

of the data (for example, if observations are grouped by their batch numbers, or by the date they were processed in the lab).

Once we have good quality training data, we may consider additional data preparation procedures, such as transformation (for instance, it is a common practice to apply a log-base-2 transformation to gene expression data in order to stabilize the variance of the expression level across different signal ranges). The most important step of high-dimensional data preparation is, however, the removal of the variables that represent noise (such as those, for which the amplitude of their values is below the experimental noise level determined for the data), as well as those variables whose values are not reliably determined. For example, for some gene expression data sets, each expression level value (that is, the value for a particular observation and particular gene, or a probe representing the gene) is associated with a p-value indicating whether this expression value was reliably determined. Usually, the p-values are translated into *detection calls* (such as Present or Absent), and the expression level measurement is considered *Present* if its p-value is below some significance level (for example, 0.05). In such a case, removing variables with unreliable measurements may be done by filtering out the variables with the proportion of *Present* calls below a particular threshold level.

There is no One-Size-Fits-All *Approach in Biomarker Discovery*

Some texts recommend standardizing the data before predictive modeling; however, no such blanket recommendation should be followed blindly. Whether it is or is not advantageous to standardize the independent variables depends on the data and the methods used in the project. If we deal with continuous variables measured in different units, then standardization is, usually, advantageous. However, when our continuous variables use the same unit (which is common for high-throughput biomedical data), there is no need to standardize them; furthermore, standardizing such variables may elevate noise (that is, if there are variables that represent noise, they will be treated in the same way as meaningful ones), and in extreme situations may result in useless models that fit noise.

3.3.2 Training and Test Sets

For predictive modeling, we need to have two data sets – the training set and the test set.[6] The training set will be used to perform all steps of biomarker discovery analysis. The test set has to be unavailable during this entire analysis; it will be used – only after the analysis is completed – to test the performance of the final predictive model (or models) on the unseen data. In an ideal situation, we would have two sufficiently large data sets, and the test set would be independent from the training one. For example, samples for the training and test sets could be collected in different countries and processed in different laboratories. With the continually increasing number of data sets submitted to public repositories, this is not an entirely unlikely scenario. Often, however, we have only one data set available for a biomarker study, and thus we need to split it into training and test sets.

There are, however, two contradictory aspects of this splitting: (i) the larger the training set, the higher the quality of the resulting biomarker expected; and (ii) the larger

[6] Sometimes a third set, the *validation* set, is also used (for example, for selecting one of the considered predictive models, or for tuning; see Section 3.3.6).

the test set, the better the estimate of the predictive abilities of the biomarker expected. A proper compromise between these two aspects depends primarily on the number of observations in the data – we may use 80% of the data for training and 20% for testing if the number of observations is relatively small (say, a hundred or so), we may opt for 70/30 split for larger data sets, and even for two-thirds to one-third for large data sets (say, with thousands of observations). However, it may also depend on other factors specific to a particular data set and project,[7] so – at least to some extent – this decision is usually quite arbitrary.

For classification biomarker studies, we usually perform *stratified random sampling* driven by the proportion of observations in each of the differentiated classes. What this means is that if we decided, for example, to select 80% of the observations into the training set, then we would randomly select 80% of the observations from each of the classes into the training set, with the unselected observations constituting the test set. For estimation biomarker studies, we may simply randomly split the entire data into the training and test sets. Another option (rather infrequently used) would be to match the proportions of observations in different ranges of the value of the continuous response variable (a kind of pseudo-stratified split).

Training Data

The training data can be represented by a $p \times N$ matrix $\mathbf{X}$ and a $N \times 1$ vector $\mathbf{y}$,

$$\mathbf{X} = \begin{bmatrix} x_{11} & x_{12} & \cdots & x_{1N} \\ x_{21} & x_{22} & \cdots & x_{2N} \\ \vdots & \vdots & \ddots & \vdots \\ x_{p1} & x_{p2} & \cdots & x_{pN} \end{bmatrix}, \quad \mathbf{y} = \begin{bmatrix} y_1 \\ y_2 \\ \vdots \\ y_N \end{bmatrix}, \tag{3.1}$$

where N is the number of observations (say, patients or biological samples) and p is the number of independent variables. Each column of matrix $\mathbf{X}$ corresponds to one of the N observations, and $\mathbf{y}$ is a vector of N values of the response (dependent) variable. Hence, each observation $i, i = 1, \ldots, N$, is represented by a $p \times 1$ vector $\mathbf{x}_i = \left[x_{1i}, \ldots, x_{pi}\right]^T$, and a value of the response variable, y_i, associated with the observation. Recall that for classification biomarkers, the response variable is categorical (and its values are labels of the differentiated classes), and that for estimation (regression) biomarkers, the dependent variable is continuous.

3.3.3 Parallel Multivariate Feature Selection Experiments

Feature selection is the single most important step of multivariate biomarker discovery based on high-dimensional data. During data preparation (see Section 3.3.1) we removed those variables that represented experimental noise, or whose measurements

[7] For example, if – for a classification project – class proportions in the training data are not balanced, then these considerations should not be based on the size of the training data, but on the size of its smallest class.

were not reliably determined. This was the first step in the multi-step approach of removing some noise and elevating signal. Now, multivariate feature selection will try to identify such a small set of variables that is estimated to have high predictive ability (sometimes it may be advantageous to identify more than one parsimonious multivariate biomarker, if such sets have similar size and predictive power). Thus, feature selection may be seen as a step that removes some more noise; in this case, removed are those variables that are either noninformative or in other ways unimportant (for example, redundant) for predicting the value of the response variable. Recall that parsimony of a multivariate biomarker is crucial for its robustness (generalizability to unseen data) – the larger the size of a biomarker, the higher the chance of overfitting the training data; this is especially true for biomarkers including noninformative variables.

There are many heuristic methods that can be used for feature selection. Chapter 5 is dedicated to multivariate feature selection methods and approaches; here we will mention the most common ones – recursive feature elimination and stepwise hybrid selection. The former starts with all of the variables in the training data and at each step eliminates one or more variables that are – at this step – deemed least important for predicting the response variable. The latter starts with an empty set, and at each step (except the first one) considers both adding and removing variables in a way that maximizes some measure of the estimated predictive power of a set. This way, each method follows some path of evaluating only a relatively small proportion of all possible subsets of variables, and provides results corresponding to some local optima. A complete search that would find the global optimum by evaluating all possible subsets of variables would be infeasible not only for data with thousands of variables but even for data with several dozen variables. However, even if we could find the global optimum, we would not be interested in it, as it would overfit the training data.[8] Thus, our goal for multivariate feature selection is to find a local optimum associated with a parsimonious and generalizable biomarker with high predictive power. This means that the optimization is based on *a combined criterion* that is a compromise between minimizing the size of the biomarker and simultaneously maximizing its predictive power.

Each considered subset of variables may be evaluated either by calculating some measure of predictive power based on intrinsic characteristics of the data (such as the ratio of the variation between classes to the variation within classes; here, the predictive power would be represented by the set's ability to separate the differentiated classes), or by using a supervised learning algorithm to build a predictive model for each considered subset and thereby evaluating its predictive abilities (for example, by sensitivity, specificity, and accuracy for classification, or by the mean squared error for regression).

As indicated earlier, the process of finding a multivariate biomarker will remove a lot of noise from the data. However, when we deal with high-dimensional and sparse training data (for example, data with several thousands of variables and a thousand or even fewer observations), we may expect that the data include a huge number of random

[8] As it would be the global optimum for the training data set, not for the target population.

patterns that could be found in the training data, but which do not correspond to real patterns that exist in the population (or the populations) from which our training data were selected. If we perform feature selection only once, it is quite likely that we will end up with such a spurious pattern that exists in the training data but not in the target population.

To minimize the chance of identifying a biomarker representing such a spurious pattern (and thus overfitting the training data), the feature selection process should be performed many times, each time using a different subsample randomly selected from the training data (see Figure 3.1). Hence, the feature selection step – for properly performed biomarker discovery – should include a reasonably large number (say, a hundred, or a thousand) of individual multivariate feature selection experiments. Those experiments are performed in parallel,[9] with each of them starting with a different randomized version of the training data – the same set of original variables, but a different subset of observations. The most common way of randomizing the training data is to use bootstrapping or *K*-fold cross-validation.

For bootstrapping, we randomly select some number of observations to include in the randomized training data (a bootstrap sample), with the remaining observations constituting an out-of-bag (OOB) sample, which can be used to test predictive models built during the feature selection process. If selection to the bootstrap sample is performed with replacement, we are implementing Efron's nonparametric bootstrap method, in which the bootstrap sample is of the same size as the original training data (though some observations are included in the bootstrap sample more than once, and some not at all). If the feature selection process utilizes a learning algorithm that requires independence of observations (meaning that observations cannot be included in the bootstrap sample more than once), we would generate the bootstrap sample by selecting observations without replacement, and the bootstrap sample would be smaller than the original training data. When using *K*-fold cross-validation, the training data observations would be randomly split into *K* approximately equal subsets (folds), $K - 1$ of them will be used as the randomized training data and the remaining one for testing.[10] By selecting different $K - 1$ folds, we can create *K* different training sets; thus, this would translate to *K* parallel feature selection experiments. However, common values for *K* are rather small (say, 5 or 10), so in order to have a reasonably large number of parallel experiments, such *K*-fold cross-validation should be repeated many times (M/K times for *M* parallel experiments).[11]

Although each of the parallel feature selection experiments will utilize a different version of randomized training data, all of them would evaluate the same sequence of the considered subset sizes. Therefore, if we perform *M* parallel experiments, for each of the considered biomarker sizes, we will have *M* different subsets of variables, and

[9] This means that those *parallel* feature selection experiments are independent from one another, it does not necessarily mean that they are performed as parallel computer threads (though this would, of course, result in decreased processing time).

[10] Note that this "testing" refers to evaluating models built during the feature selection process, it has nothing to do with the final testing of an optimal multivariate biomarker.

[11] For more information on resampling techniques, see Chapter 4.

associated with them M values of each metric of their predictive power. By aggregating the results of the M parallel experiments, we can select a reasonably small biomarker size with reasonably high estimates of predictive power. The composition of the biomarker may be based on the distribution of variables in the M subsets of the selected size; for example, we may decide to select those variables that were most frequently selected into the M subsets – they are most likely to be associated with a real, rather than a spurious, pattern. However, since some of those variables may be correlated – or even highly correlated – we may opt to perform one more multivariate feature selection experiment that would select an optimal set only from among those variables. Consequently, the aggregation of the results of parallel feature selection experiments may be seen as another (third) step of removing more noise and elevating the signal even further.

3.3.4 Selecting an Optimal Multivariate Biomarker

The selection of an optimal multivariate biomarker (that is, its optimal size and composition) may be conducted as a part of the procedure aggregating the M parallel feature selection experiments, or it may be undertaken thereafter. In any case, the selection is based on an – to some extent – arbitrary decision about what is an optimal compromise between minimizing the size of the biomarker and maximizing its estimated predictive power. Common approaches include the one-standard-error method, the tolerance method, and simultaneous evaluation of several performance metrics (for example, for classification, we may evaluate the values of sensitivity, specificity, and accuracy for different set sizes, and select a reasonable compromise between maximizing those values and minimizing the number of variables in the biomarker).

3.3.4.1 *One-Standard-Error Method*

The one-standard-error method (Breiman et al. 1984) is used to select the smallest subset of variables whose performance is within one standard error from that of the subset deemed optimal according to the used metric of performance. Assume that in a search for a classification biomarker, we performed $M = 100$ parallel feature selection experiments and decided to use the area under the ROC curve (AUC) as the measure of performance of models built during the experiments (also an arbitrary decision).[12] Assume also that the average AUC (over M experiments for each of the considered subset sizes) is maximized for sets with $v = 30$ variables. A biomarker consisting of 30 variables is hardly a parsimonious one. Since we have M subsets of 30 variables, we may, in addition to their average AUC, calculate the standard deviation of their AUC values. The standard error is the standard deviation of the *mean* AUC, and is calculated as the standard deviation of the M AUC values divided by the square root of M (with M representing in this case the sample size for calculating the mean AUC). Assume that

[12] AUC and other metrics of predictive model performance are discussed in Chapter 4.

the smallest subset with its average AUC not further than such calculated standard error (from the average AUC of the sets with v variables) includes nine variables – this would be the optimal size of our multivariate biomarker. By decreasing the size, while still retaining rather high performance, we would select a simpler model and thus decrease the chance of overfitting.

3.3.4.2 *Tolerance Method*

The tolerance method is very similar to the one-standard-error approach, but instead of considering one-standard-error range around the best average performance measure, we use a tolerance range defined by the percent decrease in performance. For example, we may choose to select the smallest subset size whose average performance is no more than one or two percent below the maximal average performance. This would be, again, an arbitrary decision about implementing the compromise between a reasonably small (and thus, likely, parsimonious) size of the biomarker and its reasonably high predictive power.

3.3.4.3 *Simultaneous Evaluation of Several Performance Metrics*

The above two methods of selecting an optimal size of multivariate biomarker have a significant weakness – they are based on the evaluation of a single performance metric. Consider, for example, a classification biomarker. Evaluating its performance using any single basic classification performance metric (that is, accuracy, sensitivity, or specificity) is not recommended, especially when the differentiated classes are unbalanced in the training data (see discussion on this topic in Chapter 4). Even using the area under the ROC curve (AUC) – although more appropriate – may not be the best way to go. A much better approach is to simultaneously evaluate several relevant performance metrics. For this classification example, we may evaluate either three (sensitivity, specificity, and accuracy) or all four of the mentioned metrics, and base our decision about the optimal size of a multivariate biomarker on a reasonable compromise between maximizing their values and minimizing the number of variables in the biomarker (see Figure 4.4, Chapter 4, for an example of visualization of these metrics).

3.3.5 Hyperparameters

Although the term *parameter* may refer to a predictive model's parameter as well as to its *hyperparameter* (quite understandably, since any hyperparameter is a parameter), a distinction between the two is usually made. Parameters are such characteristics of a predictive model that can be calculated or estimated directly from the training data; thus, they are *internal* to the model. Hyperparameters, however, are either set manually, or require tuning. Hence, for the sake of simplicity, we may consider hyperparameters as synonymous with *tunable* model parameters. They are important for the optimization of a predictive model, yet are *external* to the model. For example, for support vector machines, the number of support vectors is a parameter, but the regularization parameter

C (the cost of violating the model constraints) is a hyperparameter that requires tuning. When a predictive model has more than one hyperparameter (for example, an SVM classifier using a nonlinear kernel function), their tuning is typically performed via a grid search. The grid defines a set of values to be considered for each of the hyperparameters, and then cross-validation is performed for models that are built for each combination of the hyperparameter values from the grid.

3.3.6 Building, Tuning, and Validating Predictive Models Based on the Optimal Multivariate Biomarker

Although not all supervised learning algorithms have parameters that require tuning, many do. For example, methods implementing regularization have at least one tuning parameter – the penalty coefficient whose value should be tuned in order to find a proper balance between overfitting and underfitting (the bias-variance tradeoff; see Section 3.4).

Tuning a predictive model is often described in terms of methods that resample the training data. For each of the considered values of the tuning parameter (or for each combination of values when multiple parameters are simultaneously tuned), a model would be built on randomly selected training observations and its performance validated on the remaining observations. This would only work if the training data were not already used for feature selection (which could be the case if we were dealing with low-dimensional data).

However, when we start with high-dimensional data, and then, *after* feature selection, want to tune a predictive model built on the resulting multivariate biomarker, using the same training data for tuning the model is not recommended. The reason is that if – during the tuning – each of the models is built on a part of the training data and validated using the remaining observations (whether cross-validation, bootstrapping, or any other method is used for resampling), those remaining observations cannot be treated as unseen by the model – we need to remember that the entire training data set was used for feature selection; thus, any model utilizing the resulting biomarker is using characteristics based on all observations from the training set. We would have a similar situation if – instead of tuning – we now wanted to use a resampling method to validate any model based on the optimal biomarker. In both cases, this would be an ***internal validation*** (see Chapter 4), which would use the same observations for defining the model (basing it on the biomarker resulting from the feature selection step) and then for validating it.

Proper methods for tuning and validation depend on the search model used for feature selection, and – more precisely – on whether each of the parallel feature selection experiments uses a learning algorithm (for building predictive models and for estimating their performance) or not. If a supervised learning algorithm was incorporated into the feature selection experiments, then tuning can also be incorporated into those experiments (see Figure 3.1). This means that instead of building and evaluating a single predictive model for each considered subset size and each parallel experiment, many models will be built for various values of the tuning parameter (or parameters). Their performance (as measured on an OOB sample for bootstrapping, or Kth-fold

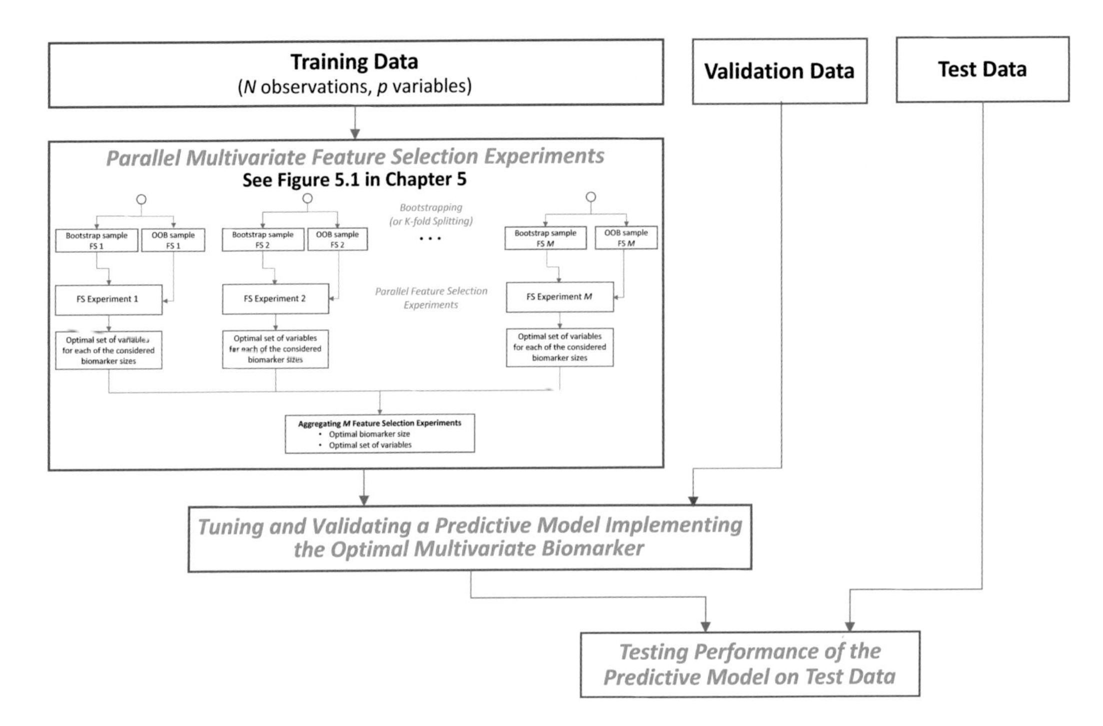

Figure 3.2 Using a separate validations data set. If feature selection experiments do not incorporate a supervised learning algorithm, or if – after the feature selection step – we want to build predictive models utilizing other learning algorithms, tuning and validation would require a separate validation data set.

sample for cross-validation) will be saved along with the values of the tuning parameters. Then, when the results of the *M* parallel feature selection experiments are aggregated, considerations about optimal values of the tuning parameters will also be included. This way, the tuning of the model based on an optimal multivariate biomarker will be incorporated into the parallel feature selection step.

However, if no learning algorithm was used during feature selection (or if one was used, but after feature selection we wanted to build and validate more than one predictive model, utilizing, in addition, learning algorithms different from that incorporated into the feature selection process), we would need a separate validation data set. This means that during the step of splitting the available data (Section 3.3.2), we would create not only the training and test sets, but also the *validation* data set (see Figure 3.2).

3.3.7 Testing Predictive Models on the Test Data

After an optimal multivariate biomarker has been identified and a predictive model implementing this biomarker has been built, tuned (if necessary) and validated, we have some – more or less reliable – estimates of how the model should perform on new data. Even if we are quite sure that all steps of our analysis were properly performed, the final evaluation of the model's predictive abilities has to be performed on the test data that were unseen during the analysis. As indicated earlier, this could be either a totally independent test data set (for example, coming from a different research group and processed at a different lab) – this would be the best option, or this could be a holdout set that was randomly selected from our available data, set aside, and never used during the analysis.

What is important to understand is that the test data set may be used only for the evaluation of the final predictive model. It may not be used for tuning the model or for selecting one of several models. Sometimes, we may have good reasons to build more than one predictive model implementing our optimal biomarker; for example, one model may be expected to have better performance, while another may be easier to interpret, or yet another may be expected to be less expensive in clinical implementation. In such a situation, we would use the test data set to test the performance of all of our final models, but the results of such testing should *not* be used for selecting among them. Since they are our *final* models, selecting one of them should be done independently from our biomarker discovery project, for instance, during their clinical evaluation. If we wanted to select one of the models within our project, this would have to be done before involving the test data, and then only the selected, single *final* model would be tested on the test data.

3.4 Bias-Variance Tradeoff

Although predictive modeling for biomarker discovery is always based on training data, it is not our goal for the predictive model to perfectly fit the training data. Our goal is for the model to be *generalizable*, that is, to have the ability to provide highly accurate predictions for new or future observations. A model that is so complex that it perfectly fits all training observations (and thus has zero training error) but poorly predicts new observations (and thus has a large test error), would have zero bias and high variance – a case of severe *overfitting* and poor generalization (the high variance here means that the model is highly sensitive to changes in the training data). In another extreme, when a model ignores the training data (for example, when it always predicts the mean value of the response variable), it would have no variance but high bias – a case of *underfitting*, that is, poor prediction for both training and new observations. Hence, our goal is to build a model that is not too complex to overfit, but complex enough to provide good generalization capabilities allowing for accurate predictions for new observations. Figure 3.3 shows illustrations of severe overfitting and underfitting.

To decompose the expected error of the model, we will consider the regression setting. Assume that the relationship between the dependent and independent variables is represented by $y = f(\mathbf{x}) + \varepsilon$, where ε is the error term with an expected value of zero, $E(\varepsilon) = 0$, and constant variance $Var(\varepsilon) = \sigma^2$. Let the model prediction for a particular test observation $\mathbf{x}$ be $\hat{y}(\mathbf{x})$; and for simplicity denote it as $\hat{y} \equiv \hat{y}(\mathbf{x})$. Then, the bias of model prediction (averaged over all possible training data sets of the same size *N*) would be:[13]

[13] Note that the model prediction here, $\hat{y}$, is a random variable, as different training data sets would lead to different model parameters and thus different predictions $\hat{y}$.

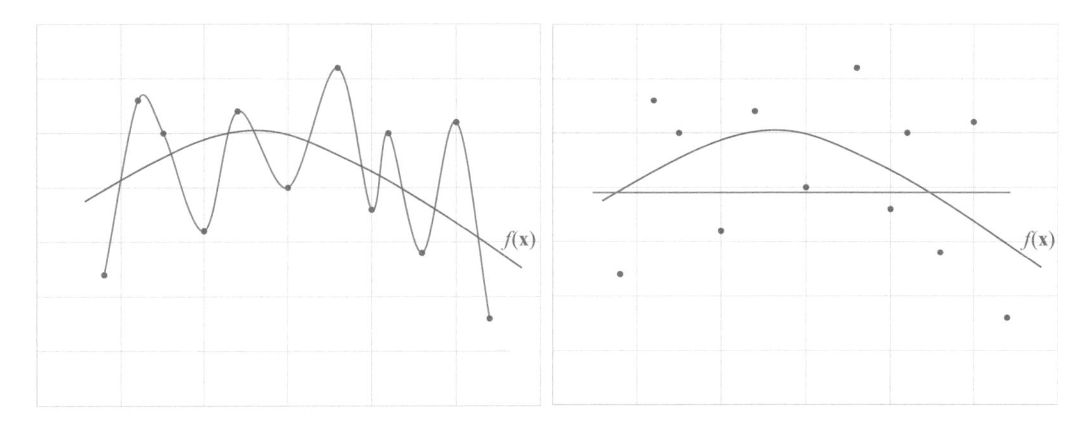

Figure 3.3 Extreme overfitting and underfitting of training data. The points correspond to the training observations, and the green line represents the modeled function $f(\mathbf{x})$. **Left panel:** the predictive model (the orange line) perfectly fits all training observations and thus, has zero training error. It overfits the training data, has low bias, and high variance; it would poorly predict new observations, and would be very sensitive to changes in the training data. **Right panel:** the predictive model (the orange line) seems to ignore the training observations and thus, has no variance but high bias. It underfits the training data and would poorly predict both training and new observations.

$$Bias = E[\hat{y}] - f(\mathbf{x}), \tag{3.2}$$

and it would represent the difference between the expected value of the model prediction and the true (unknown) mean value of y, that is, $E[y] = f(\mathbf{x})$.

The variance of the model prediction would represent the spread of the predicted values around the expected value of the model prediction (again, averaged over potential training sets):

$$Variance = E\left[(\hat{y} - E(\hat{y}))^2\right]. \tag{3.3}$$

It can be shown that the squared-error of model prediction may be then decomposed into the following three elements:

$$\begin{aligned} E\left[(y - \hat{y})^2\right] = \sigma^2 & \\ + (E[\hat{y}] - f(\mathbf{x}))^2 & \\ + E\left[(\hat{y} - E(\hat{y}))^2\right]. & \end{aligned} \tag{3.4}$$

This represents the bias-variance decomposition, that is,

$$\text{Expected prediction error} = Noise + Bias^2 + Variance. \tag{3.5}$$

The noise term is an irreducible error, intrinsic to the data; thus, for particular training data, we have no control over this term.[14] Bias represents a systematic error of

[14] Although this noise may be reduced during data collection and preprocessing.

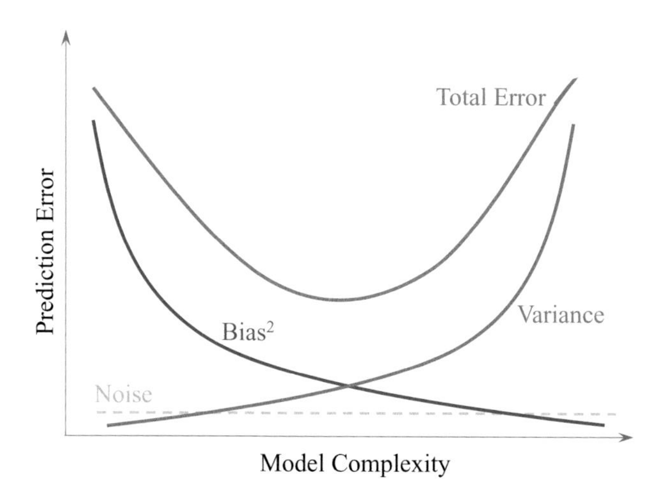

Figure 3.4 Illustration of the bias-variance tradeoff. Models that are too complex have low bias and high variance; they overfit the training data and show poor generalizability (large error when predicting unseen test observations). Models that are too simple have low variance and high bias, they underfit the training data and can neither accurately predict the training data nor the test data. Our goal in predictive modeling is to find the "sweet spot", that is, a model that neither overfits nor underfits the training data and can be well generalized to new observations. In predictive modeling for biomarker discovery based on high-dimensional data, the primary technique to deal with the bias-variance tradeoff is to properly use multivariate feature selection or regularization (which penalizes models that are too complex).

prediction, that is, how far – on average – our prediction is from the true value of the dependent variable. Variance reflects the sensitivity of the predictive model to changes in the training data set. Complex models that overfit training data will have high variance, while models that are too simple and underfit the data will have high bias. This tradeoff between bias and variance is illustrated in Figure 3.4, while an intuitive illustration of the terms is shown in Figure 3.5.

3.4.1 Notes on Overparameterization

The bias-variance tradeoff is a central tenet of predictive modeling. Finding an optimal balance between overfitting and underfitting the training data is considered the best approach to building predictive models that are well generalizable and have optimal performance.

Recently, however, there have been suggestions that overparameterized models may – in some situations – achieve a comparable, or even better, performance than those following the bias-variance tradeoff (Belkin et al. 2019; Hastie et al. 2022; Rocks and Mehta 2022). Overparameterized models can be characterized by a large number of parameters and a complexity so high that it is beyond the point at which the model perfectly fits the training data (that is, where the model *interpolates* the data). There are suggestions that beyond the interpolation limit – at which the training error is zero and the total error very high – there is an overparameterized region, in which the total error

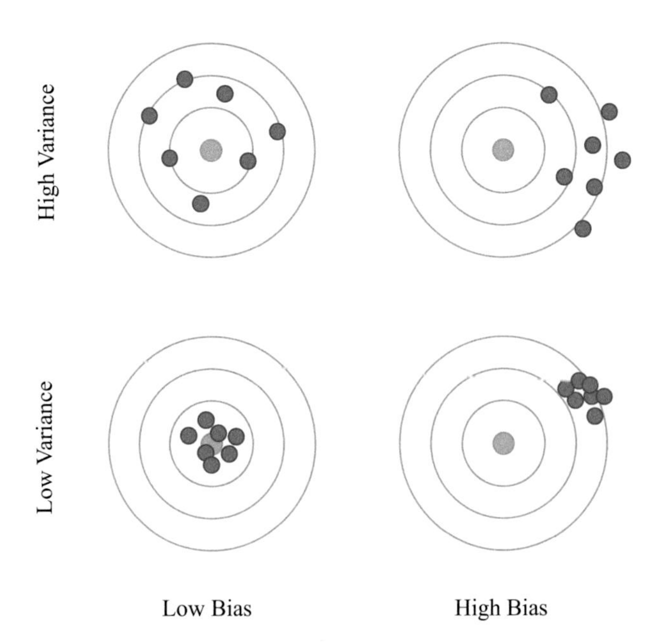

Figure 3.5 Intuitive illustration of bias and variance in predictive modeling. Our target (the true value of the dependent variable for test observations) is the orange bulls-eye. Bias is a systematic error in our predictions while variance represents the spread of our prediction values over the expected value of the model prediction.

decreases. However, no consensus has been achieved regarding the behavior of variance and bias in this overparameterized region. Therefore, while discussions around potential benefits of interpolation and overparameterization continue, we should follow the bias-variance tradeoff's approach to predictive modeling and, for example, use regularization to penalize models that are too complex.

3.5 Screening Biomarkers, Segmentation Models, and Biomarkers for Personalized Medicine

Different multivariate biomarker discovery studies may have significantly different goals. For example, we may be interested in identifying a parsimonious biomarker capable of the early detection of multiple cancer types. In such a case, our goal would be a *screening* biomarker whose implementation would require a relatively noninvasive sample collection (for example, peripheral blood), so it can be, ideally, included as a part of routine testing. For such screening biomarkers, we would assume (or bet) that there are some molecular profiles, as well as biological processes associated therewith, that are common not only for the considered multiple disease states, but also for the target heterogeneous population.

Another – more common – goal could be to identify a multivariate biomarker for a specific disease state or condition. Such a biomarker would target a specific population of patients. However, if the predictive abilities of the biomarker are not satisfactory, we may consider splitting the target population into more homogeneous subpopulations (or segments), and identify separate biomarkers for each of them. In this case, we would be building separate *segmentation models* whose predictive abilities might be substantially better when compared to the model targeting the entire population.

The latter can be seen as the very first step toward personalized medicine. Although segmentation models would typically be built for subpopulations defined by such factors as age groups, gender, or ethnicity, including more factors (such as specific molecular profiles) and thus splitting the target population into even more homogeneous strata would move us toward *stratified* medicine (see Section 1.2). Continuing this process by adding more and more individualized signatures (omic or otherwise) could eventually achieve a flavor of *personalized* medicine.

3.6 Committees of Predictive Models

Ensemble-based predictive models (such as the random forests models described in Chapters 8 and 10), which are ensembles of many *weak* learners, are sometimes called committee machines. However, in this section we will discuss committees that are composed of a relatively small number of independent (or, at least, relatively independent) predictive models, which are *not* weak learners; on the contrary, each of them should have at least satisfactory predictive power as well as good generalization abilities. Hence, each of the base models of the committee could be used individually, and the reason for combining them into a committee is to achieve – possibly – even better predictions and generalization.[15] To increase the likelihood of achieving this goal, the individual models should sufficiently differ among themselves.[16] This can be accomplished by developing the models on different training data, using different types of data, using different learning algorithms, using different parameters of the same learning algorithm, or – eventually – by combining some of these approaches (Cichosz 2015).

Obtaining different and independent training data sets (that is, sufficiently large and independent statistical samples from the target population) is usually not very realistic. Basing prediction on different types of omic data (genomic, transcriptomic, proteomic, etc.) may seem prudent; however, using a committee of models that require the gathering of multiple data types may be challenging in a clinical environment (Azuaje 2010). Nevertheless, this may be attainable if a high-throughput technology allows the generation of data of diverse types from a single biological sample. To some extent, this is already possible with next generation sequencing. Using different learning algorithms, or the same algorithm with different values of its parameters (if this yields

[15] This is analogous to making a decision based on the opinions of several experts on a subject.

[16] If the base models make similar mistakes, the committee's predictions will not provide much (or any) improvement over the individual models. However, if the base models make wrong predictions for different patients, aggregating their results should increase the overall accuracy of predictions.

sufficiently different models), currently seems to be the most straightforward approach to committee-based modeling.

The prediction provided by the committee is determined by aggregating the predictions of the base models. The ways of doing this are different for continuous response variables (regression modeling) and for categorical ones (classification modeling).

3.6.1 Committees of Regression Models

Assume that the committee consists of K base models $\mathcal{M}_k, k = 1, \ldots K$, and denote by x the observation (such as a patient), for which we want to make a prediction. Let then $\widehat{y}(\mathcal{M}_k, x)$ represent the value – of the continuous response variable – predicted for x by model $\mathcal{M}_k$, and $\widehat{y}(x)$ the committee prediction. The latter can be calculated by averaging the individual predictions. When we use simple averaging, the committee prediction is

$$\widehat{y}(x) = \frac{1}{K} \sum_{k=1}^{K} \widehat{y}(\mathcal{M}_k, x). \tag{3.6}$$

If we want to assign different importance to different base models (and thus vary their influence on the committee prediction), we may use weighted averaging,

$$\widehat{y}(x) = \sum_{k=1}^{K} w_k \widehat{y}(\mathcal{M}_k, x), \tag{3.7}$$

where $w_k \geq 0$ represents the weight assigned to model $\mathcal{M}_k$, and the weights sum up to 1.

3.6.2 Committees of Classification Models

Assume that each base classification model (of a committee of K models) differentiates among J classes, say disease states labeled $\mathcal{D}_1, \mathcal{D}_2, \ldots, \mathcal{D}_J$. Let $\widehat{y}(\mathcal{M}_k, x)$ represent class predicted for observation x by model $\mathcal{M}_k$, where

$$\widehat{y}(\mathcal{M}_k, x) = \{\mathcal{D}_1, \mathcal{D}_2, \ldots, \mathcal{D}_J\}. \tag{3.8}$$

The most common method of aggregating the base model results into a committee prediction is plurality voting, in which committee prediction corresponds to the class that received the largest number of votes,[17]

$$\begin{aligned} \widehat{y}(x) &= \text{plurality vote } \{\widehat{y}(\mathcal{M}_k, x)\}_{k=1}^{K} \\ &= \arg\max_{j} \sum_{k=1}^{K} \mathbf{1}_{\widehat{y}(\mathcal{M}_k, x) = \mathcal{D}_j}, \end{aligned} \tag{3.9}$$

[17] If more than one class receives the largest number of votes, one of them is chosen via some arbitrary decision.

where $\mathbf{1}_{\widehat{y}(\mathcal{M}_k,x)=\mathcal{D}_j}$ is the function that equals one if the class predicted by model $\mathcal{M}_k$ is $\mathcal{D}_j$, and equals zero otherwise. Observe that for binary classification ($J = 2$), plurality voting is equivalent to majority voting.

Similar to regression, for classification committees we may also assign different weights to different base classifiers, and then aggregate their results via weighted voting,

$$\widehat{y}(x) = \arg\max_j \sum_{k=1}^{K} w_k \mathbf{1}_{\widehat{y}(\mathcal{M}_k,x)=\mathcal{D}_j}. \tag{3.10}$$

When the base models implement classification learning algorithms that provide class probabilities, we may – instead of aggregating their class predictions – aggregate their class probabilities. Let $P_j(\mathcal{M}_k, x), j = 1, \ldots, J, k = 1, \ldots, K$ represent the probability of class j returned by model $\mathcal{M}_k$ for observation x. Then the committee-based probability of class j may be determined by simple averaging of the base model probabilities,

$$P_j(x) = \frac{1}{K}\sum_{k=1}^{K} P_j(\mathcal{M}_k, x), \quad j = 1, \ldots, J, \tag{3.11}$$

or we may use weighted averaging, if weights $\left(w_k \geq 0; \sum w_k = 1\right)$ are assigned to the base classifiers,

$$P_j(x) = \sum_{k=1}^{K} w_k P_j(\mathcal{M}_k, x), \quad j = 1, \ldots, J. \tag{3.12}$$

4 Evaluation of Predictive Models

4.1 Methods of Model Evaluation

Although methods for evaluation of predictive models are algorithmically the same for models developed from low- and high-dimensional data, the ways and contexts of their utilization may be different (this has been briefly discussed in Section 2.2). We have to remember about the curse of dimensionality and hence about impending disasters if the approaches that are commonly used for low-dimensional data are blindly applied to high-dimensional data. Hence, even if the names of the evaluation and validation techniques (for example, cross-validation, bootstrapping), as well as their basic intrinsic structures, are the same, their implementation within biomarker discovery projects that are based on high-dimensional data may be conceptually different. The decision about which evaluation methods to use also depends on whether we are evaluating the *final* predictive model (or models) built on the already identified optimal multivariate biomarker, or whether we want to evaluate *intermediate* models considered during the biomarker discovery process (for example, possibly hundreds or thousands of models built during the parallel feature selection experiments).[1]

4.1.1 Testing the Final Predictive Model

4.1.1.1 *Independent Test Data Set*

For biomarker discovery studies, the best approach to evaluate the optimal multivariate biomarker – and the performance of the predictive model implementing this biomarker – is to utilize an *independent test data set*. Such a data set should be selected from the target population independently from the training data set used to identify the biomarker. Moreover, it would be particularly desirable for the test set to be another statistical sample from the same target population, but collected from a different geographical area (say, a different country), by a different group of researchers, and then processed in a different laboratory. If the training and test sets are sufficiently large (of course, the

[1] Note that, in some situations, these two approaches may have an overlap. For example, if parallel feature selection experiments are performed with the use of a learning algorithm, then the final optimal biomarker will most likely be already associated with resampling-based metrics of its performance (calculated by aggregating results of the parallel feature selection experiments), before it is tested on a test data set.

larger the better), this approach would provide the most reliable estimate of the generalization abilities of our optimal multivariate biomarker.

If we are, however, not collecting the data for our biomarker study, but rather using data previously collected for another project, then it is still possible that another data set – from the same target population and utilizing the same or compatible technology, but independent from our data set – is available in public repositories. If that is the case, we could use this other set as an independent test set.[2]

4.1.1.2 *Holdout Set*

When an independent test set is not available, and we only have one data set (that is, only one statistical sample collected from the target population) available for a biomarker discovery study, then we need to split these available data into a training set and a test set; in this case the test set is a *holdout set*.[3] This holdout set will be set aside and not used, and not seen, during the biomarker discovery analysis. It will be used only – after the analysis is completed – for testing the final predictive model implementing the identified optimal multivariate biomarker. Such a holdout set can be seen as independent only in a limited sense – its observations had no influence on the identification of the optimal biomarker and no influence on any characteristics of the predictive model; they were, however, part of the same – rather than an independent – statistical sample selected from the target population.

4.1.2 Evaluating Intermediate Models

During the biomarker discovery process, we are building and evaluating (and possibly also tuning) many intermediate predictive models. For example, if the multivariate feature selection step (recall that this is the most important step in multivariate biomarker discovery based on high-dimensional data) consists of only a hundred parallel feature selection experiments, and if each of them utilizes a supervised learning algorithm to build and evaluate predictive models for 50 considered biomarker sizes, then this step alone would involve evaluation of 5,000 intermediate predictive models. If model tuning was also included within the feature selection step, then this number would be much higher – for each of the considered biomarker sizes and for each of the parallel experiments, not one but many models (with different values of the tuning parameters) would be built and evaluated.

Furthermore, depending on the design of the biomarker discovery process, we may also want or need to validate or tune predictive models *after* the feature selection experiments are completed (see Section 3.3.6). Additionally, we may want to build several predictive models implementing the optimal biomarker (for example, using

[2] Even if the two data sets were generated with the use of the same technology, it is quite likely that their preprocessing protocols were not identical. Therefore, in such situations, we should consider obtaining the raw data (rather than the already processed data) and preprocess the test and training data in the same way.

[3] Recall from Chapter 3 that sometimes it may be convenient, or even necessary, to split the available data into training, validation, and test sets.

different supervised learning algorithms) and then evaluate them in order to select the final model (or maybe even more than one final model).

Consequently, depending on the situation, we may – to evaluate intermediate models – use either one of the resampling techniques or a separate validation set. However, it is very important to realize that no model should be evaluated on observations that have in any way influenced the creation of the model. Such evaluation would amount to an *internal validation*, which is as unreliable as re-predicting observations from the training data set. Therefore, before describing resampling techniques (such as cross-validation and bootstrapping), we focus first on their proper and improper use for model validation, that is, on external and internal validations.

4.1.2.1 *Internal Validation (Improper)*

To illustrate such an improper validation, we will consider two examples related to feature selection experiments. For the first example, assume that feature selection experiments were using training data of 500 patients and 10,000 variables and that, as a result of these experiments, an optimal multivariate biomarker of 10 variables has been identified. Now, if we attempt to validate or tune this biomarker by randomly splitting the 500 training observations into a subset that would be used to build a predictive model (on the 10 variables of the optimal biomarker), with the remaining observations used to evaluate the model, then this would amount to an inappropriate evaluation, with results that would be quite irrelevant.[4] The reason for this is that those remaining observations (being previously a part of the training data) were already used for the identification of the variables selected into the optimal biomarker, and thus – by influencing the model in this way – cannot be treated as unseen by the model.

For the second example, let us just extend this consideration to the feature selection experiments themselves. We would have a similar improper-validation situation if, at any iteration of a feature selection experiment, we would try to validate a current model on the observations that were already used in a previous iteration to decide about selection of the current set of variables. In both cases, this would be an *internal validation* – evaluating the model on the observations that had already influenced what variables are used by the model, and thus are, in a sense, internal to the model.[5]

4.1.2.2 *Proper, that is, External Validation*

To properly evaluate each of the *intermediate* models built during the parallel feature selection experiments, each parallel experiment should not only be based on a different subsample randomly selected from the original training data (this subsample will constitute a second-order training set that will be used for all iterations – that is, all considered

[4] Using a resampling technique, such random splits and evaluations would be performed multiple times, but the fallacy of the approach would be exactly the same.

[5] It can be shown that using internal validation may result in perfect or near perfect performance metrics even for data that are random noise.

subset sizes – of this particular experiment), but should also have its own holdout set (that is, the remaining observations from the original training data) that will be used exclusively to evaluate intermediate models built during this particular experiment. Hence, at each iteration of the experiment, a predictive model is built from the experiment's training set, but is evaluated on its holdout set, which had no influence on the characteristics of the model, and no influence on calculating the variable importance metrics used to decide which variables should be considered at the next iteration. Observe that the performance of the intermediate models is not used in any way during the feature selection process; it will be used – after the process is completed – to aggregate the results and suggest an optimal size and composition of the multivariate biomarker.

In the other situation, that is, when we want to evaluate or tune models that would be built *after* the feature selection process is completed, we would need a separate ***validation set*** that has not been used during any of the previous steps of the biomarker discovery process. Such necessity should be identified during the design of the biomarker discovery process, and the available data should be split not only into the training and test sets, but also into a validation set (see Section 3.3.6).

Nonetheless, a question may arise as to whether there is any proper way of evaluating predictive models that are built *after* the feature selection step has been completed (for example, models utilizing different learning algorithms) if the biomarker discovery process was designed without selecting a validation set. Before answering this question, let's first reiterate that all steps of the entire biomarker discovery process should be properly designed before starting the analysis. Consequently, if we found ourselves in such a situation, then we should consider redesigning the process and starting the analysis anew, with the available data split into training, test, and validation sets.[6] If, however, for whatever reason, redesigning the entire process is not an option, then yes, there is another way out of this situation. Let us assume again that, during the parallel feature selection experiments, an optimal biomarker of 10 variables has been identified. What we can do now is *not* evaluate a model built on this particular biomarker (at least not directly, as this would amount to internal validation), but evaluate many models built on biomarkers composed of different "optimal" sets of 10 variables, and then use their *average* performance as an estimate of the performance of our optimal biomarker. To implement this, we would again use a resampling technique (say, bootstrapping) and set up many new parallel feature selection experiments, and for each of them select (from the original training data) a random subsample to be used for training and a holdout set (remaining observations) for validation. For each of these parallel experiments, we would perform feature selection that would be stopped when a ten-variable biomarker is identified. Then, a predictive model would be built for each of those different ten-variable biomarkers, evaluated using its own experiment's holdout set, and we would average performance metrics over all of those models.[7]

[6] Unless the test set is sufficiently large that splitting it now into a validation set and a new (and smaller) test set would not compromise the reliability of the ultimate testing step.

[7] Be advised, however, that in some situations we may actually *prefer* to have the average performance of many "optimal" biomarkers (i.e., each of them optimal for a particular feature selection experiment) of the same size, in addition to – or even instead of – the metrics for our biomarker that is deemed to be optimal

4.1.3 Resampling Techniques

Resampling techniques are used in situations where we want to aggregate results over many experiments based on different random subsamples of the training data. They may be used at different levels of biomarker discovery analysis; for example, to aggregate results over many feature selection experiments, as well as at the level of individual ensemble-based classifiers implementing such methods as the random forests learning algorithm. One of the main uses of resampling during multivariate biomarker discovery is in the design of parallel multivariate feature selection experiments. The most commonly used resampling approaches are bootstrapping and cross-validation.

4.1.3.1 *Bootstrapping*

Bootstrapping refers to the repeated random selection of subsamples (from the original training data) that would be used, in parallel, as the second-order training sets. They are called bootstrap samples. Furthermore, the observations that are not selected into a particular bootstrap sample constitute this bootstrap sample's out-of-bag (OOB) sample, which can be used as a second-order holdout set. There are several variations of bootstrap methods, but the most commonly used is *Efron's nonparametric bootstrap* (Efron 1979; Efron and Tibshirani 1993). This bootstrap method performs N random selections of observations from training data of size N. However, the selections are *with replacement*, that is, after an observation is selected, it is "replaced" and may be selected again. Thus, even if each bootstrap sample will include exactly N observations, some of the observations will be selected into a bootstrap sample more than once, and some not at all. Those that are not in the bootstrap sample are automatically included into the OOB sample associated with this bootstrap sample. With this method of selection, different bootstrap samples may have a different number of unique observations; on average, however, they include about $0.632*N$ unique observations (hence, OOB samples have, on average, $0.368*N$ observations).

Some learning algorithms (for example, discriminant analysis) assume that training observations are independent of each other. In such situations, the nonparametric bootstrap with replacement (which allows for duplicated observations in the bootstrap samples) is inappropriate, and we should use a bootstrap method selecting observations without replacement. In this case, fewer than N samples will be randomly selected from the training

overall. For example, this may happen when we expect that many parsimonious biomarkers of the same size tap into the same set of biological processes and thus, may represent similar predictive information. Our optimal biomarker could be thus seen as being selected to represent predictive information associated with those underlying biological processes. Therefore, even if our biomarker was deemed optimal based on particular training data, its performance estimate based on averaging many similar biomarkers may, possibly, better represent its true performance for the target population. Hence, in such a situation, using the average performance to represent the performance of our optimal biomarker, would be *by the design* of the biomarker discovery process. Please remember, however, that we are discussing here the evaluation or tuning of intermediate predictive models. The final model (built on the optimal biomarker) has to be ultimately evaluated on the test data. Recall also that averaging performance metrics over many parallel experiments of the feature selection step is – also by design – a way to decide about the size and composition of the optimal multivariate biomarker.

data, but each of them may be selected only once into a particular bootstrap sample. The selection is driven by the proportion of training observations that need to be selected into each bootstrap sample. As for any bootstrap method, observations that were not selected into a particular bootstrap sample will constitute its corresponding OOB sample.

Regardless of the bootstrap method used, after bootstrapping we will have B bootstrap samples and associated with them B OOB samples (B is usually between 100 and 1,000). Then, using the bootstrap samples as their training sets and the OOB samples as the holdout sets, B parallel analyses of the same kind are performed. One of the major goals of approaches utilizing bootstrapping is to build an ensemble of predictive models (see, for example, the random forests algorithms described in Chapters 8 and 10); another is to perform many parallel feature selection experiments in order to aggregate their results to identify an optimal size and composition of a multivariate biomarker. A diagram of a bootstrap-based analysis is presented in Figure 4.1).

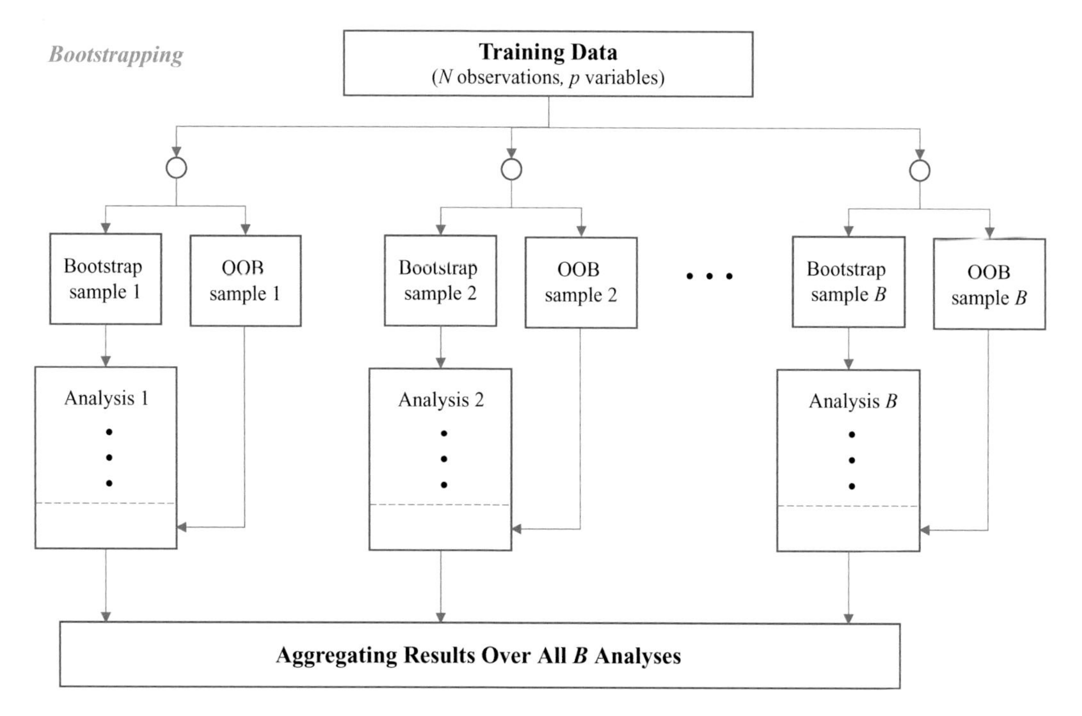

Figure 4.1 A schematic illustration of a bootstrap-based analysis. B bootstrap samples are selected from the training data. Depending on the bootstrap method, random selection of observations from the training data may be performed either with or without replacement. If the selection is with replacement, each bootstrap sample is of the same size as the training data (N). For the selection without replacement, bootstrap samples include fewer than N observations. In either case, observations not selected into a bootstrap sample constitute the out-of-bag (OOB) sample associated with the bootstrap sample (OOB samples may be used as holdout sets to evaluate results of each analysis). B parallel analyses of the same kind are performed, each of them is using its own bootstrap sample. Then, the results of all of the B analyses are aggregated. Any kind of analysis for which it would be advantageous to aggregate results over many parallel runs can be considered; typical kinds are ensemble-based predictive modeling and multiple feature selection experiments.

4.1.3.2 *Cross-Validation*

In cross-validation, observations from the training data are randomly assigned into K nonoverlapping groups (which are called *folds*) of roughly the same size. Traditionally, a distinction is made between the K-fold cross-validation and the Leave-one-out cross-validation; however, the latter (for which $K = N$) is just a special case of the former. Once the observations to be randomly allocated into each of the K folds are selected, we may perform K parallel experiments. For each of them, $K - 1$ folds will be used as a second-order training set, and the remaining fold as a second-order holdout set; a different fold will be used as the holdout set for each of the K experiments. The most commonly used values of K are 5 and 10, which means that we would have only 5 or 10 parallel experiments. Usually, for more meaningful aggregation of the results, we want to perform many more parallel experiments. To achieve this, we would repeat random splitting of the training data many times and run many K-fold cross-validation-based experiments. Figure 4.2 illustrates 5-fold cross-validation.

For the leave-one-out cross-validation (LOOCV), each holdout set includes only one observation, which means that the number of parallel experiments is the same as the number of observations N in the original training data. Since each training set consists of $N - 1$ observations, those training sets are almost identical – in their size and composition – to the original training data. This poses a danger that the results of the N LOOCV experiments may be highly correlated, and thus aggregating over them may not add much value. This also means that estimates based on LOOCV tend to have

Figure 4.2 A schematic illustration of K-fold cross-validation for $K = 5$. Each observation from the original training data is randomly assigned to one of the five folds of approximately equal size (for the purpose of illustrating this random selection, observations selected into a particular fold are marked in the original training data diagram with the color representing this fold). For each of the possible K experiments, one of the folds is used as the holdout set and the remaining $K - 1$ folds constitute the training set. If we want to aggregate over more than K experiments, the entire K-fold cross-validation schema would be repeated an appropriate number of times.

lower bias but substantially higher variance (that is, are more sensitive to changes in the original training data) than those based on a typical K-fold cross-validation. Hence, the typical K-fold cross-validations that use K equal to 5 or 10 may be seen as the ones providing a reasonable tradeoff between bias and variance.[8]

4.2 Evaluating Regression Models

The goal of an estimation biomarker is to predict (estimate) the value of a continuous response (dependent) variable. Predictive models implementing such biomarkers are using regression learning algorithms to estimate the parameters of the following functional relationship between the dependent variable y and p independent variables,

$$y = \beta_0 + \beta_1 x_1 + \beta_2 x_2 + \cdots + \beta_p x_p + \varepsilon, \tag{4.1}$$

where β_0 and β_k, $k = 1, \ldots, p$ are the parameters to be estimated, and ε is the error term (for details, see Chapter 6).

It is crucial to realize that some of the metrics used to evaluate how well a predictive model fits the training data are different from the metrics that evaluate the performance of a predictive regression model on test data. There are texts and software implementations that use the same name (for example, *MSE*) for those metrics that are calculated differently when quantifying the fit to the training data and when evaluating performance on the test data. This may be misleading, if the difference in their calculations – in the two situations – is not explained. To clarify this issue, we will first look at the metrics of the fit, and then focus on the performance metrics.

4.2.1 Metrics of Fit to the Training Data

Fitting a regression model to the training data while using the ordinary least squares method is described in Section 6.1. Discussed there are metrics used to evaluate how well the optimal hyperplane, which represents the estimated regression equation,

$$\hat{y} = \hat{\beta}_0 + \hat{\beta}_1 x_1 + \hat{\beta}_2 x_2 + \cdots + \hat{\beta}_p x_p, \tag{4.2}$$

fits the training data (the data that include N observations and p independent variables). Those metrics are based on the following sums of squares (variations):

- $SST = \sum_{i=1}^{N} (y_i - \bar{y})^2$ is the total variation in the dependent variable,
- $SSR = \sum_{i=1}^{N} (\hat{y}_i - \bar{y})^2$ is the variation due to regression, which represents the part of the total variation that can be explained by the estimated regression equation,

[8] When compared to bootstrapping, K-fold cross-validation has higher variance, but lower bias.

- $SSE = \sum_{i=1}^{N} (y_i - \hat{y}_i)^2$ is the sum of squared errors (where the error is defined as the difference between the observed and the predicted value of the response variable for a particular training observation), representing the remaining, unexplained part of the total variation,

where:

- y_i is the observed value of the response variable for training observation i, $i = 1, \ldots, N$,
- $\hat{y}_i$ is the value predicted by the estimated regression equation (4.2) for training observation i, $i = 1, \ldots, N$,
- $\bar{y}$ is the mean of all y_i values.

The metrics described in Section 6.1 are:

- Coefficient of determination $R^2 = \frac{SSR}{SST}$, which represents the proportion of the total variation in the dependent variable that can be explained by the estimated regression equation.
- Adjusted coefficient of determination $R_a^2 = 1 - \left(1 - R^2\right) \frac{N-1}{N-p-1}$, which better estimates this proportion for regression problems with many independent variables.[9]
- Mean squared error $MSE = \frac{SSE}{N-p-1}$.

However, it is very important to understand that these metrics are used to determine how well our predictive regression model ***fits the training data***. They should not be blindly applied when evaluating the performance of the predictive model on the test data. This is because the degrees of freedom associated with regression problems are different during training the model and when we are using it to predict the response for any other data. When we are fitting the model, degrees of freedom for a particular variation term depend on the number of population parameters that have to be estimated in order to calculate the term. For example, degrees of freedom for the error term are $N - p - 1$, and for the total variation $N - 1$.[10] However, those degrees of freedom have no meaning when we are testing our predictive model on the test data, for when we are doing this, we are not estimating any population parameters. Hence, neither MSE nor

[9] Observe, however, that the adjusted coefficient of determination is meaningless for $p \geq N - 1$. This means that, for example, it is not applicable when we evaluate regression models (built with such methods like support vector regression or random forests) created during recursive feature elimination performed on high-dimensional data.

[10] Degrees of freedom for a particular term are defined as the number of observations in the training data, N, minus the number of population parameters that have to be estimated before we can perform calculations for the term. To calculate prediction errors for training observations, we have to have estimates for all of the $p + 1$ parameters of Function 4.1; hence, for the error term we have $N - p - 1$ degrees of freedom. To calculate the total variation in the dependent variable, we need to estimate only one parameter – the population mean of the dependent variable.

R_a^2 – as defined above – may be used for evaluating performance of regression models implementing estimation biomarkers.[11]

4.2.2 Metrics of Performance on Test Data

Assume that we want to evaluate the performance of an estimation biomarker – and of a regression model implementing the biomarker – on test data consisting of n observations.[12] The test data here may be an independent test set, a holdout set, a validation set, or – more generally – any data different from the training data used to develop the biomarker and the predictive model. Since the formulas for the variations (SST, SSR, SSE) are not using any degrees of freedom, they are calculated in the same way as when fitting the model (except that the size of the test data, n, will be used instead of N). The same is true for the coefficient of determination,

$$\begin{aligned} R^2 &= \frac{SSR}{SST} = \frac{\sum_{i=1}^{n}(\hat{y}_i - \bar{y})^2}{\sum_{i=1}^{n}(y_i - \bar{y})^2} \\ &= \frac{SST - SSE}{SST} = 1 - \frac{SSE}{SST} = 1 - \frac{\sum_{i=1}^{n}(y_i - \hat{y}_i)^2}{\sum_{i=1}^{n}(y_i - \bar{y})^2}, \end{aligned} \tag{4.3}$$

where

- y_i is the known value of the response variable for test observation i, $i = 1, \ldots, n$,
- $\hat{y}_i$ is the value predicted by our model for test observation i, $i = 1, \ldots, n$,
- $\bar{y}$ is the mean of all y_i values in the test data set.

R^2 now represents the proportion of the total variation (in the dependent variable) in the test data that is explained by the regression. However, observe that the *adjusted* coefficient of determination R_a^2 now has no meaning.

When we were fitting the model, the mean squared error, MSE, was used to estimate the population variance δ^2 of the error term ε. In order to have an unbiased estimate of this variance, we were dividing the sum of squared errors, SSE, by the error term degrees of freedom, $N - p - 1$. Now, during the testing, we are not interested in

[11] It may be worth mentioning that – although model evaluation should be performed on a test (or validation) data set not used during training the model – sometimes MSE or R^2 are used to select from among several models (under the assumption that the relative size of the training error may be related to the unknown test error). While this could work for models using the same number of independent variables, it would be inappropriate for comparing models with a different number of variables. In such situations, criteria that estimate test error by adjusting training error for the number of variables may be used. Examples of such criteria are: adjusted coefficient of determination R_a^2, AIC – Akaike information criterion, and BIC – Bayesian information criterion. However, it should be noted that these criteria are not suitable for high-dimensional data (Hastie et al. 2009; James et al. 2014).

[12] Recall that the size of the *training* data was denoted by N.

estimating any population parameters, but only in finding how well our predictive model works on the test data. Hence, once the sum of squared errors is calculated for all of the test observations, we are interested in its simple average value. Though this quantity will still have a general meaning of a mean squared error, in order to make an explicit distinction between how it is calculated when fitting the model and when the model is tested, we call it *the mean squared error for prediction* (*MSEP*),

$$MSEP = \frac{SSE}{n} = \frac{1}{n}\sum_{i=1}^{n}(y_i - \hat{y}_i)^2. \tag{4.4}$$

Of course, *MSEP* still has a flavor of variance, and thus is measured in the squared units of the response variable. Therefore, it may be convenient to define also *the root mean squared error for prediction* (*RMSEP*),

$$RMSEP = \sqrt{MSEP} = \sqrt{\frac{1}{n}\sum_{i=1}^{n}(y_i - \hat{y}_i)^2}, \tag{4.5}$$

which is measured in the units of the response variable.

Another measure that is often reported among the results of testing regression models is *the mean absolute error* (*MAE*),

$$MAE = \frac{1}{n}\sum_{i=1}^{n}|y_i - \hat{y}_i|, \tag{4.6}$$

which is the average of the absolute values of the prediction errors (residuals).[13] Since large residuals do not increase the value of *MAE* as much as they increase *MSEP*, in typical situations when large residuals are not desired, *MSEP* may be preferred over *MAE*.

4.3 Evaluating Classifiers Differentiating between Two Classes

The goal of classification biomarkers is to accurately assign new observations to one of the differentiated classes (such as disease states). The response variable is categorical, and its values can be seen as labels identifying the differentiated classes. Since the basic token of information underlying classification metrics of performance is a binary outcome of classifying any individual observation (say, *TRUE* when the observation is assigned to its true class, and *FALSE* otherwise; irrespective of the number of differentiated classes), the same metrics may be used when predictive models are built on the training data, as well as when they are tested on the test data. Although the basic metrics have the same general meaning when we differentiate between two classes and

[13] Since we were not using the mean absolute error during model development and fitting it to the training data, we denote it here as *MAE*. Nonetheless, one may argue that for consistency it could (or should) be denoted as *MAEP*, for the *mean absolute error for prediction*.

Table 4.1 Confusion matrix summarizing results of testing a predictive classification model differentiating between two classes (Disease and Healthy). True Positive (TP) and True Negative (TN) provide the numbers of test observations that were correctly classified, with their true class being $\mathcal{D}$ and $\mathcal{H}$, respectively (they are marked with a green background). Accordingly, False Positive (FP) and False Negative (FN) represent the numbers of test observations that were misclassified (pink background).

		Predicted Class	
		Disease ($\mathcal{D}$)	Healthy ($\mathcal{H}$)
True Class	Disease ($\mathcal{D}$)	**TP** (True Positive)	**FN** (False Negative)
	Healthy ($\mathcal{H}$)	**FP** (False Positive)	**TN** (True Negative)

when we consider more than two classes, their calculations may differ. In this section, we focus on evaluating predictive models that differentiate between two classes.

A multivariate classification biomarker – as well as a classification model implementing the biomarker – may be built to differentiate between any two populations (classes) as long as such differentiation makes sense from a biomedical point of view. To focus our attention, we will consider differentiating between the following two classes: *Disease* and *Healthy*, which will be labeled by $\mathcal{D}$ and $\mathcal{H}$, respectively. When a predictive model for classification (a classifier) is tested on a test data set, it assigns each of the observations (in this case, patients) from the test set to one of the two classes. Since we know the true class of the test observations, we will know, for each of these assignments, whether it is correct or not. This information is typically summarized in a confusion matrix (see Table 4.1).

4.3.1 Confusion Matrix

A confusion matrix (sometimes called a contingency table) provides the results of testing a classifier on a particular test data set. Multiple metrics of classifier performance are calculated using information from the confusion matrix. When the confusion matrix is created, the numbers of *correct* classifications are tabulated separately for those test observations whose true class is $\mathcal{D}$ and for those whose true class is $\mathcal{H}$. They are called *True Positive* (TP) and *True Negative* (TN), respectively, and are shown on the main diagonal of the confusion matrix. Likewise, the *misclassifications* are also summarized separately – *False Negative* (FN) provides the number of test observations whose true class is $\mathcal{D}$, but the classifier assigned them to $\mathcal{H}$; *False Positive* (FP) refers to the number of test observations whose true class is $\mathcal{H}$ and predicted class is $\mathcal{D}$. Observe also that the number of "positive" test observations, say $\|P\|$, that is, the number of test patients

Table 4.2 Basic performance metrics for predictive models for binary classification. The true class for test patients with the disease is $\mathcal{D}$, and the true class for healthy test patients is $\mathcal{H}$. A test result refers to the class to which the predictive model assigns the patient. A positive test result ($test+$) means that the patient is classified as having the disease; a negative test results ($test-$) means that the patient is classified as healthy.

Sensitivity	$\frac{\text{TP}}{\text{TP}+\text{FN}} = \frac{\text{TP}}{\lVert\text{P}\rVert}$	Proportion of test patients with disease ($\mathcal{D}$) who get a positive test result ($test+$)
Specificity	$\frac{\text{TN}}{\text{TN}+\text{FP}} = \frac{\text{TN}}{\lVert\text{N}\rVert}$	Proportion of healthy test patients ($\mathcal{H}$) who get a negative test result ($test-$)
Accuracy	$\frac{\text{TP}+\text{TN}}{\lVert\text{P}\rVert+\lVert\text{N}\rVert} = \frac{\text{TP}+\text{TN}}{n}$	Proportion of test patients for whom the predicted class is the same as their true class (correct classifications)
Misclassification rate	1 – Accuracy	Proportion of test patients for whom the predicted class is different from their true class (incorrect classifications)

with the disease (their true class is $\mathcal{D}$) is $\lVert\text{P}\rVert = \text{TP} + \text{FN}$. Similarly, the number of the "negative" test observations, $\lVert\text{N}\rVert$, for whom the true class is $\mathcal{H}$, is $\lVert\text{N}\rVert = \text{FP} + \text{TN}$. Of course, with n being the size of the test data set, we have $n = \text{TP} + \text{FN} + \text{FP} + \text{TN} = \lVert\text{P}\rVert + \lVert\text{N}\rVert$.

4.3.2 Basic Performance Metrics for Binary Classifiers

The most basic metrics calculated from the information provided in the confusion matrix are: *sensitivity*, *specificity*, and *accuracy*. Furthermore, from accuracy, we can directly calculate the *misclassification rate* (the overall test error rate). Descriptions of these metrics, as well as how are they calculated, are presented in Table 4.2. Sensitivity and specificity refer to the proportions of test patients who are correctly classified into their true class (which is $\mathcal{D}$ for sensitivity, and $\mathcal{H}$ for specificity). Hence, sensitivity[14] can be seen as the ability of the predictive model to correctly identify patients who do have the disease (that is, how sensitive the classifier is). Similarly, specificity[15] refers to the ability of the model to correctly identify healthy patients (that is, how specific the classifier is). Accuracy can be interpreted as the weighted average of sensitivity and specificity (weighted by the number of positive and negative test observations); using information from Table 4.2, we can express this by:

$$\text{Accuracy} = \frac{\text{Sensitivity} \times \lVert\text{P}\rVert + \text{Specificity} \times \lVert\text{N}\rVert}{\lVert\text{P}\rVert + \lVert\text{N}\rVert}. \tag{4.7}$$

It is important to understand that none of these basic performance metrics should be used *alone* to evaluate a predictive model. For example, an accuracy of 90% may sound good, but it may be very misleading when the classes are unbalanced – if 90% of the test patients do not have the disease, then a dummy classifier that would assign each and every

[14] Sensitivity is also called *true positive rate* or *recall*.

[15] Specificity is also called *true negative rate*.

patient to class $\mathcal{H}$ would also have a 90% accuracy while being completely useless. In this case, the accuracy of the dummy classifier represents the *no-information* rate[16] (a baseline rate), to which the accuracy of the evaluated classifier should be compared. Observe that this useless dummy classifier would have a specificity of 100% (and, of course, a sensitivity of 0%). Therefore, when evaluating a classifier, class-specific accuracies, that is, sensitivity and specificity, should be considered together, while accuracy (and the associated misclassification rate) would summarize them for the entire test data.

4.3.3 Proper and Improper Interpretation of Sensitivity and Specificity

Sensitivity and specificity are important metrics of the performance of a biomarker (and a classifier implementing the biomarker), and they should be interpreted in this context. To illustrate their potential misinterpretation in a clinical context, consider the following situation.

> A patient had blood work done. It included a screening test for early detection of a particular CNS disease.[17] After the test came back positive, the patient asks the physician: *Having a positive result of this test, what is the chance that I have this disease?*
>
> Assuming that the only information the physician has about the test is that it has 99.5% sensitivity and 99.5% specificity (and thus, 99.5% accuracy), what should be the physician's answer to this question?

Let us hope that no physician would answer: *Since the test has 99.5% sensitivity and 99.5% specificity, the probability that you have this disease is 99.5%.* The proper answer – in this situation – should be: *I do not have enough information about the test to answer this question.* So, what information is missing? To answer the patient's question, we must first know the *prevalence* of this CNS disease in the population targeted by the test (with prevalence being defined as the proportion of the population that has the disease in a given time period).

Assume that the prevalence is 0.1% and that this screening test was targeting 1,000,000 people.[18] Hence:

- With a prevalence of 0.001, we expect that 1,000 of the 1,000,000 people have this disease, and 999,000 do not have the disease
- With a sensitivity of 0.995, of the 1,000 people with the disease, 995 would test positive (true positive)
- With a specificity of 0.995, of the 999,000 people without the disease, 994,005 would test negative (true negative)
- This leaves 4,995 people without the disease who would test positive (false positive)

[16] For data with unbalanced classes, the no-information rate is defined as the proportion of the largest class.

[17] CNS: Central Nervous System.

[18] This number by itself is arbitrary; since only the proportions are important, we could pick any number and the final result would be the same.

This means that out of a million people, 5,990 patients would test positive (true positive plus false positive). Since only 995 among them have the disease, the probability that the patient with the positive test result has this CNS disease is 995/5,990 = 0.166, or 16.6%. This proportion is called the *positive predictive value* (PPV), and it is among other important metrics that can be estimated from the confusion matrix. This result may raise questions about the usability of this screening test, which may be answered only in a clinical context. For example, if the follow up testing (to differentiate between the true positive and the false positive results of our screening test) would be invasive and/or associated with possibly serious adverse reactions, then the fact that it would not be needed for 83.4% of the patients that tested positive would make the original screening test unsuitable for clinical practice.[19]

Observe that, with the low prevalence of the disease, specificity has much greater influence on the positive predictive value than sensitivity. If the sensitivity – in this example – decreased to 95% (keeping the prevalence and specificity unchanged), then the PPV would decrease only to 16%; however, if the specificity decreased to 95%, the PPV would drop to 2%. Consequently, high specificity is extremely important for screening biomarkers targeting populations of asymptomatic individuals with low prevalence of the disease.

Let us also observe that our PPV calculations are equivalent to determining the conditional probability of having the disease given a positive test result. If we denote this conditional probability $P(\mathcal{D}|test+)$, then from Bayes' rule on conditional probability, we have:

$$P(\mathcal{D}|test+) = \frac{P(test+|\mathcal{D})P(\mathcal{D})}{P(test+|\mathcal{D})P(\mathcal{D}) + P(test+|\mathcal{H})P(\mathcal{H})}. \tag{4.8}$$

Observing that $P(test+|\mathcal{D})$ corresponds to sensitivity, $P(\mathcal{D})$ to prevalence, $P(test+|\mathcal{H})$ to (1 – specificity),[20] and $P(\mathcal{H})$ to (1 – prevalence), we may rewrite this conditional probability as:

$$P(\mathcal{D}|test+) = \frac{\text{Sensitivity} \times \text{Prevalence}}{\text{Sensitivity} \times \text{Prevalence} + (1 - \text{Specificity}) \times (1 - \text{Prevalence})}. \tag{4.9}$$

Using the values from our screening test case, we would have for this example:

$$P(\mathcal{D}|test+) = \frac{0.995 \times 0.001}{0.995 \times 0.001 + (1 - 0.995) \times (1 - 0.001)} = 0.166. \tag{4.10}$$

4.3.4 Other Important Performance Metrics for Binary Classifiers

The positive predictive value[21] (discussed earlier) as well as other important performance metrics that are also based on the information from the confusion matrix are summarized in Table 4.3. Although prevalence (also discussed in the screening test

[19] Furthermore, for diseases with low survival rates, a false positive test result could – by itself – have a psychologically devastating impact on the patient.

[20] Since $P(test-|\mathcal{H})$ corresponds to specificity.

[21] Positive predictive value is also called *precision*.

Table 4.3 Other important performance metrics for predictive models for binary classification. For each of these metrics, in addition to its description and the method of calculating it from the confusion matrix, the probability estimated by the metric is provided (again, under the assumption that the test data set is a random and representative sample from the target population).

Positive predictive value (PPV)	$\frac{TP}{TP+FP}$	$P(\mathcal{D}\vert test+)$	Probability that a patient has the disease given a positive test result
Negative predictive value (NPV)	$\frac{TN}{TN+FN}$	$P(\mathcal{H}\vert test-)$	Probability that a patient does not have the disease given a negative test result
False discovery rate (FDR)	$\frac{FP}{TP+FP}$	$P(\mathcal{H}\vert test+)$	Proportion of patients who tested positive but who do not have the disease
False positive rate	1 – Specificity	$P(test+\vert\mathcal{H})$	Proportion of healthy test patients ($\mathcal{H}$) who get a positive test result
False negative rate	1 – Sensitivity	$P(test-\vert\mathcal{D})$	Proportion of test patients with disease ($\mathcal{D}$) who get a negative test result

example) is an epidemiological variable and thus, not explicitly available in the confusion matrix, it may be – but only under the assumption that the test data set is a random and representative sample from the target population – estimated from its entries as (TP + FN)/n. As mentioned before, the *positive predictive value* is one of the most important metrics for clinical implementation of the biomarker. It answers the question of what is the probability that a patient has the disease if the patient tested positive. Similarly, *negative predictive value* estimates the probability that a patient with a negative test result does indeed not have the disease. *False discovery rate* (FDR) describes the proportion of false positives among all of the positive test results (thus, it corresponds to 1 – positive predictive value), and in our example would be equal to 83.4%.

4.3.5 ROC Curves and the Area Under the ROC Curve

As illustrated in the example of the screening test for early detection of a particular CNS disease, even relatively high sensitivity and specificity (they both were 99.5% in this hypothetical example) may be insufficient for a predictive model to be clinically useful. For this particular screening test, a specificity of 99.5% was too low for the prevalence of 0.001, as it resulted in an unacceptable 83.4% false discovery rate. What if we wanted to trade a bit of sensitivity for a possibly substantial increase in specificity (in the hope of a more acceptable clinical test)?

A typical cost-insensitive classifier estimates the probability that a tested patient belongs to class $\mathcal{D}$, and then assigns the patient to class $\mathcal{D}$ if this probability is greater than or equal to 0.5, and to class $\mathcal{H}$ otherwise. This is the probability threshold t, for which the misclassification cost is equal for both classes (that is, the costs for false positives and for false negatives are the same). By changing this threshold, we may try to find such a tradeoff between sensitivity and specificity that is optimal for our particular situation. Increasing the probability threshold would increase specificity but decrease sensitivity, and vice versa. The ROC curve is a common way of illustrating this tradeoff.

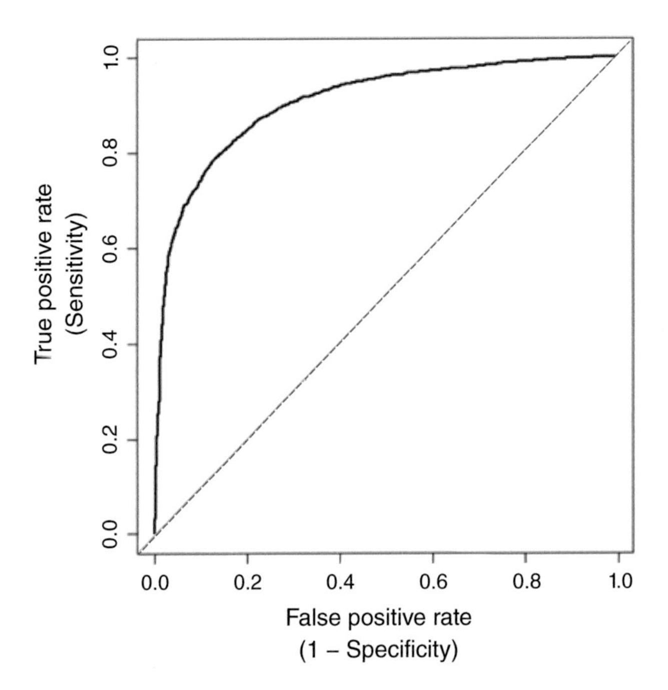

Figure 4.3 An example of the ROC curve illustrating the tradeoff between sensitivity and specificity for the full range of values of the probability threshold. The southwest point of (0, 0), for which sensitivity is zero and specificity one, corresponds to the probability threshold of 100%; this represents a classifier that would assign all test patients to class $\mathcal{H}$. On the other end, the northeast point of (1, 1), for which sensitivity is one and specificity zero, corresponds to a probability threshold of 0%, and to a classifier that would assign all test patients to class $\mathcal{D}$. The red dashed line corresponds to a classifier that is no better than a random choice. The area under the ROC curve (AUC) is another metric of the overall performance of a classifier; the larger the AUC (which usually corresponds to the ROC curve being closer to the northwest corner of the plot) the better the classifier's performance.

The ROC curve[22] is a plot of sensitivity (aka the true positive rate) versus the false positive rate (equal to 1 – specificity). An example of the ROC curve is shown in Figure 4.3. An ideal classifier would have both sensitivity and specificity equal to 1; its ROC curve would touch the northwest corner of (0, 1) of the plot (true positive rate of one, false positive rate of zero). The diagonal line – from the southwest corner of (0, 0) to the northeast corner of (1, 1) of the plot – represents a classifier performing no better than making random decisions about class membership.

The *area under the ROC curve* (AUC) is another metric of a classifier's performance. If we randomly select two patients, one from class $\mathcal{D}$, and one from class $\mathcal{H}$, then the AUC can be interpreted as the probability of the test patient from class $\mathcal{D}$ having a higher chance of being assigned to class $\mathcal{D}$ than the test patient from class $\mathcal{H}$. A perfect

[22] ROC stands for *receiver operating characteristics*; ROC curves were a part of the signal detection method designed during WWII for the interpretation of radar images.

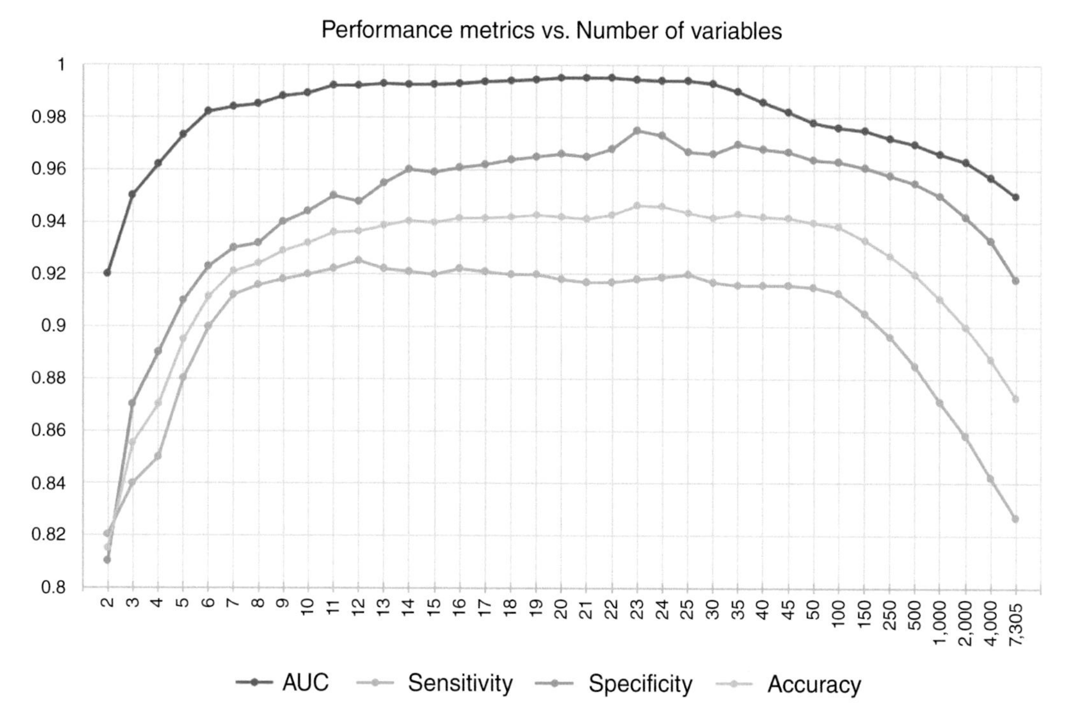

Figure 4.4 An example of using the AUC – as well as the basic performance metrics – for evaluating the results of multivariate feature selection. AUC, sensitivity, specificity, and accuracy are plotted against the subset sizes considered during the feature selection experiments. For each of the subset sizes, the plotted value of each metric is its average across all parallel feature selection experiments.

classifier would have an AUC equal to 1.[23] As a classifier's performance metric, AUC has some advantages over accuracy. Unlike accuracy, AUC is insensitive to class imbalance, and it does not depend on a particular probability threshold underlying classification decisions.

ROC curves may also be used to visualize the sensitivity–specificity tradeoff when varying the values of other parameters of a classifier (for example, its tuning parameters), in order to decide about their optimal values. During multivariate feature selection, AUC (averaged over many parallel feature selection experiments) may be plotted against the size of the considered subsets of variables in order to decide on the optimal size of a multivariate biomarker. However, such decisions would be better informed if we also plot – and consider – other performance metrics, such as accuracy, sensitivity and specificity (see an example in Figure 4.4).

Plotting and comparing ROC curves for different predictive models may help in selecting an optimal model (that is, a model that is best suited for a particular

[23] A classifier represented by the dashed red line would have an AUC of 0.5. This means that for a randomly selected pair of test patients, one from class $\mathcal{D}$ and one from class $\mathcal{H}$, each of them would have the same chance of being classified into class $\mathcal{D}$; hence, the classifier's decisions are as reliable as a coin toss.

application). Models with their ROC curves closer to the northwest point of (0, 1) would usually have a higher AUC, and would generally be considered better. However, in some situations, using the AUC alone may be misleading – if the ROC curves cross, then instead of maximizing the AUC, we may want to focus on a particular area of the ROC curves that would be more relevant for a specific tradeoff that we are evaluating.

4.3.6 A Few More Metrics of Performance

4.3.6.1 *Balanced Accuracy*

Recall that accuracy represents the weighted average of sensitivity and specificity (weighted by the number of positive and negative test observations). The metric that takes their simple average (and is therefore insensitive to class imbalance) is called *Balanced Accuracy*,

$$\text{Balanced Accuracy} = \frac{1}{2}(\text{Sensitivity} + \text{Specificity}). \tag{4.11}$$

Recall the example where 90% of test patients did not have the disease, and thus a dummy classifier assigning each patient to class $\mathcal{H}$ would have a 90% accuracy. Since such a useless classifier would have specificity of 100% and sensitivity of 0%, its balanced accuracy would be only 50%.

4.3.6.2 *F_1-Measure*

The F_1-measure aggregates sensitivity and positive predictive value (PPV) by calculating their harmonic mean,

$$\begin{aligned} F_1 &= \frac{2}{\frac{1}{\text{Sensitivity}} + \frac{1}{\text{PPV}}} \\ &= 2 \times \frac{\text{Sensitivity} \times \text{PPV}}{\text{Sensitivity} + \text{PPV}}. \end{aligned} \tag{4.12}$$

If true positive (TP) is greater than zero, then F_1 may also be calculated as[24]

$$F_1 = 2 \times \frac{\text{TP}}{n + \text{TP} - \text{TN}}. \tag{4.13}$$

Under the assumption that sensitivity is greater than zero, the F_1-measure assumes values in (0, 1]; having the value of one only when both sensitivity and PPV are at 100%. As a harmonic mean, the value of F_1 is between the values of sensitivity and PPV, but is closer to the smaller of them. This implies that a large F_1 indicates large values of both sensitivity and PPV. In our example of the screening test for early detection of a CNS disease (where we had sensitivity, specificity, and accuracy all equal

[24] Observe that if TP = 0, then both sensitivity and PPV are equal to zero, and F_1 is undefined.

to 99.5%, and PPV of 16.6%), the F_1-measure would be 28.5%. Since both sensitivity and positive predictive value equally contribute to the F_1-measure, this metric may be used to search for an optimal tradeoff between them (similar to how the ROC curve is used for a tradeoff between sensitivity and specificity).

It may be worth noting that the F_1-measure is a special case of the family of F_β metrics,

$$F_\beta = \left(1+\beta^2\right)\frac{\text{Sensitivity} \times \text{PPV}}{\text{Sensitivity} + \beta^2 \times \text{PPV}}, \tag{4.14}$$

where β is a non-negative parameter representing the relative importance of sensitivity and positive predictive value, with $\beta > 1$ corresponding to a higher importance of sensitivity, and $\beta < 1$ to a higher importance being assigned to PPV.

4.3.6.3 *Kappa*

The kappa metric was originally designed (Cohen 1960) as a coefficient of agreement between the decisions of two judges (for example, clinical psychologists) who are independently assigning subjects to one of the mutually exclusive and exhaustive categories (such as psychological disorders). This metric is defined as

$$\kappa = \frac{p_o - p_c}{1 - p_c}, \tag{4.15}$$

where:

- p_o is the proportion of cases for which the judges agreed, that is, the proportion of observed agreements
- p_c is the proportion of cases for which agreement is expected purely by chance

Thus, kappa may be interpreted as the proportion of agreements above what could be expected by chance.

When applied to the evaluation of predictive models for classification, kappa measures the agreement between the predicted classes and the true classes of the test observations. In this case, p_o is equal to the classifier's accuracy,

$$p_o = \text{Accuracy} = \frac{\text{TP} + \text{TN}}{n}. \tag{4.16}$$

Since the maximal value of accuracy is 1, the denominator of (4.15) represents the maximum value of the difference $p_o - p_c$, and, hence, scales the largest possible value of kappa to 1. If accuracy (p_o) is equal to that obtained by chance (p_c), the numerator of (4.15) and thus, the value of kappa are equal to zero. Theoretically, kappa may be negative (with the minimum value of -1) when accuracy is below that which would be obtained by chance; however, we are not interested in such classifiers.[25]

[25] In practice, such underperforming classifiers would have kappa values only slightly below zero; larger negative values are rather unlikely to be observed.

To find out p_c, assume first that $P_{True}(*)$ denotes the probability that the true class of an observation is *, and $P_{Pred}(*)$ is the probability that the predicted class is *. From the confusion matrix, we may estimate these probabilities for both classes as:

$$\begin{aligned} P_{True}(\mathcal{D}) &= \frac{\text{TP}+\text{FN}}{n}, \\ P_{Pred}(\mathcal{D}) &= \frac{\text{TP}+\text{FP}}{n}, \\ P_{True}(\mathcal{H}) &= \frac{\text{TN}+\text{FP}}{n}, \\ P_{Pred}(\mathcal{H}) &= \frac{\text{TN}+\text{FN}}{n}. \end{aligned} \tag{4.17}$$

Consequently, under the assumption that the model's prediction and the true class are independent variables, the proportion of agreement between them by chance is:

$$\begin{aligned} p_c &= P_{True}(\mathcal{D}) \times P_{Pred}(\mathcal{D}) + P_{True}(\mathcal{H}) \times P_{Pred}(\mathcal{H}) \\ &= \frac{\text{TP}+\text{FN}}{n} \times \frac{\text{TP}+\text{FP}}{n} + \frac{\text{TN}+\text{FP}}{n} \times \frac{\text{TN}+\text{FN}}{n} \\ &= \frac{(\text{TP}+\text{FN}) \times (\text{TP}+\text{FP}) + (\text{TN}+\text{FP}) \times (\text{TN}+\text{FN})}{n^2}. \end{aligned} \tag{4.18}$$

Observe, that the numerators in Formulas (4.17) represent either row or column totals in the confusion matrix. Hence, for the binary classification, when the number of classes is $J = 2$, we may rewrite (4.18) in a more general form[26] as

$$p_c = \sum_{j=1}^{J} \frac{rowTotal(j) \times columnTotal(j)}{n^2}. \tag{4.19}$$

As mentioned earlier, $\kappa = 1$ indicates perfect agreement between the model's predictions and the true classes of the test observations, while $\kappa = 0$ corresponds to agreement by chance. However, the interpretation of other positive values of the metric is somewhat arbitrary, as arbitrary as deciding on what should be considered a good agreement. Different sources suggest various ranges of positive kappa values to correspond to agreements considered, for example, as "slight", "moderate", "reasonable", "substantial", etc. Usually, values above 0.5 or 0.6 are considered to indicate substantial agreement. In our earlier example involving the test for a CNS disease (with the accuracy of 99.5%, and the prevalence of 0.001 indicating very unbalanced classes), we would have the probability of agreement by chance (p_c) about 0.993 and kappa of 0.283.

4.4 Evaluating Multiclass Classifiers

Although most of the time we want to differentiate only between two classes (such as disease versus healthy), there are situations that call for differentiating among more than

[26] This form may be extended to the multiclass classification.

two classes; for example, differentiating among several subtypes of the same disease, or – more generally – among disease states that may have a similar clinical presentation, but require different treatments. Some supervised learning algorithms (such as discriminant analysis or random forests) may naturally be used for multiclass problems. Some others, however (like support vector machines), may have been designed specifically for two-class problems; therefore, their application to the multiclass situation may require either extensions to their algorithms (if at all possible), or the decomposition of a multiclass problem into a number of binary comparisons. Furthermore, when using algorithms capable of multiclass classification, we may either classify all classes simultaneously or – especially with larger numbers of classes – opt for the multistage approach to a multiclass problem. Hence, we will discuss four approaches to dealing with multiclass classifications:

- Classifying all classes simultaneously
- Implementing the multistage approach to multiclass classification
- Decomposing the problem into a number of *one-versus-one* two-class problems
- Decomposing the problem into a number of *one-versus-rest* two-class problems

4.4.1 Classifying All Classes Simultaneously

Assume that we want to evaluate a predictive model that simultaneously differentiates among J disease states when $J > 2$. Let us label those disease states $\mathcal{D}_1, \mathcal{D}_2, \ldots, \mathcal{D}_J$. The confusion matrix for such a multiclass problem is shown in Table 4.4. This is now a $J \times J$ matrix, with rows representing the true classes of the test patients, and columns representing the classes predicted by the classifier. The summary numbers for correct classifications (that is, where the true and the predicted class are the same) are shown on the main diagonal (marked with a green background). They are called True Positives and are denoted as TP(j) for each class j, $j = 1, \ldots, J$. Other cells of the confusion matrix summarize all possible misclassifications; they are denoted as $\mathrm{M}(m, k)$, where $m \neq k$; and where m is the index of the true class and k is the index of the predicted one.

We may add to the confusion matrix row totals $n(j)$, and column totals $p(j)$, $j = 1, \ldots, J$, where $n(j)$ is the number of test patients whose true class is $\mathcal{D}_j$, and $p(j)$ is the number of test patients whose predicted class is $\mathcal{D}_j$. If the size of the test data set is n, then of course $\sum n(j) = \sum p(j) = n$.

Directly from the confusion matrix, we can calculate the *Overall Accuracy* of the multiclass classifier as:

$$\textit{Overall Accuracy} = \frac{1}{n} \sum_{j=1}^{J} \mathrm{TP}(j). \tag{4.20}$$

However, for the multiclass confusion matrix there is no notion of overall sensitivity or overall specificity as those metrics are associated with individual classes. Similarly, at the level of the entire multiclass confusion matrix, there is no notion of true negative, false positive, or false negative. However, we can focus on one class at a time, and – in addition to TP(j) directly available from the confusion matrix – calculate these three other characteristics separately for each class.

Table 4.4 Confusion matrix summarizing the results of testing a predictive model for multiclass classification. J classes, $\mathcal{D}_1, \mathcal{D}_2, \ldots, \mathcal{D}_J$, are simultaneously differentiated. Entries on the main diagonal (green background) provide – for each class $j, j = 1, \ldots, J$ – the number of test patients that are correctly classified. All other entries represent the numbers of test patients who were misclassified (pink background). Also included are *row totals* and *column totals*, which provide the numbers of test patients, $n(j)$, whose true class is $\mathcal{D}_j$, and the number of patients, $p(j)$, whose predicted class is $\mathcal{D}_j$.

True Class \ Predicted Class	$\mathcal{D}_1$	$\mathcal{D}_2$	...	$\mathcal{D}_J$	Row Total
$\mathcal{D}_1$	**TP(1)** True Positive for class $\mathcal{D}_1$	**M(1, 2)**	...	**M(1, *J*)**	$n(1)$
$\mathcal{D}_2$	**M(2, 1)**	**TP(2)** True Positive for class $\mathcal{D}_2$	...	**M(2, *J*)**	$n(2)$
...	...	...	...	...	...
$\mathcal{D}_J$	**M(*J*, 1)**	**M(*J*, 2)**	...	**TP(*J*)** True Positive for class $\mathcal{D}_J$	$n(J)$
Column Total	$p(1)$	$p(2)$	...	$p(J)$	n

False Positive for class j, FP(j), will represent the number of test patients who were assigned by the classifier into disease state $\mathcal{D}_j$ while their true class is different from $\mathcal{D}_j$. Hence, FP(j) can be calculated by summing up the numbers from the misclassification cells in column j (corresponding to predicted class $\mathcal{D}_j$),

$$\begin{aligned} \mathrm{FP}(j) &= \sum \mathrm{M}(*, j) \\ &= p(j) - \mathrm{TP}(j), \end{aligned} \tag{4.21}$$

where $\mathrm{M}(*, j)$ represents any misclassification cell from column j.

Similarly, *False Negative for class j*, FN(j), summarizes misclassifications for the test patients whose true class is $\mathcal{D}_j$ and predicted class different than $\mathcal{D}_j$; thus, FN(j) is the sum of the misclassification numbers from row j,

$$\begin{aligned} \mathrm{FN}(j) &= \sum \mathrm{M}(j, *) \\ &= n(j) - \mathrm{TP}(j), \end{aligned} \tag{4.22}$$

and $\mathrm{M}(j, *)$ corresponds to any misclassification cell from row j.

Finally, *True Negative for class j* can be calculated as the sum of all entries in the confusion matrix except those in column j and row j:

Table 4.5 J binary confusion matrices showing the results of the decomposition of the information from the multiclass confusion matrix into class-specific characteristics. Each of these binary confusion matrices is focusing on an individual class while all other classes are treated as one class representing "all other $J-1$ classes".

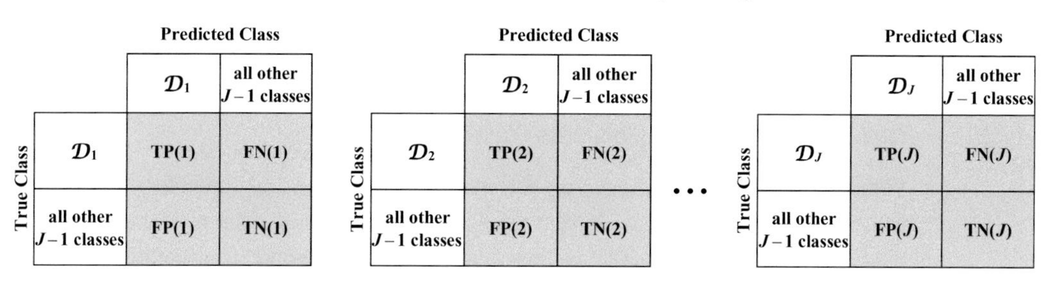

Table 4.6 Class-specific performance metrics for multiclass classification.

Metric	Formula	Description
Sensitivity for class j	$\frac{\mathrm{TP}(j)}{\mathrm{TP}(j)+\mathrm{FN}(j)}=\frac{\mathrm{TP}(j)}{n(j)}$	Proportion of test patients for whom both true and predicted class is $\mathcal{D}_j$
Specificity for class j	$\frac{\mathrm{TN}(j)}{\mathrm{TN}(j)+\mathrm{FP}(j)}=\frac{n-p(j)-n(j)+\mathrm{TP}(j)}{n-n(j)}$	Proportion of test patients for whom both true and predicted class is different from $\mathcal{D}_j$
Positive predictive value for class j	$\frac{\mathrm{TP}(j)}{\mathrm{TP}(j)+\mathrm{FP}(j)}=\frac{\mathrm{TP}(j)}{p(j)}$	Probability that a patient has disease $\mathcal{D}_j$ given that predicted class is $\mathcal{D}_j$
Negative predictive value for class j	$\frac{\mathrm{TN}(j)}{\mathrm{TN}(j)+\mathrm{FN}(j)}=\frac{n-p(j)-n(j)+\mathrm{TP}(j)}{n-p(j)}$	Probability that a patient does not have disease $\mathcal{D}_j$ given that predicted class is different from $\mathcal{D}_j$
False discovery rate (FDR) for class j	$\frac{\mathrm{FP}(j)}{\mathrm{TP}(j)+\mathrm{FP}(j)}=\frac{p(j)-\mathrm{TP}(j)}{p(j)}$	Proportion of patients for whom predicted class is $\mathcal{D}_j$ but their true class is different from $\mathcal{D}_j$

$$
\begin{aligned}
\mathrm{TN}(j) &= \sum_{\substack{m=1\\ m\neq j}}^{J}\sum_{\substack{k=1\\ k\neq j}}^{J}\mathrm{M}(m,k)+\sum_{\substack{m=1\\ m\neq j}}^{J}\mathrm{TP}(m)\\
&= n-\left[\sum \mathrm{M}(*,j)+\sum \mathrm{M}(j,*)+\mathrm{TP}(j)\right]\\
&= n-p(j)-n(j)+TP(j).
\end{aligned}
\tag{4.23}
$$

This represents the number of test patients for whom neither the true nor predicted class is $\mathcal{D}_j$. Please note that these classes may be different from each other, so those patients may be misclassified. However, from the point of view of class j, they are from a "different class" and are assigned to a "different class".

Once we have the true positive, true negative, false positive, and false negative calculated for each of the J classes, we may present them in J binary confusion matrices (see Table 4.5), as well as use them to calculate class-specific metrics (such as sensitivity and specificity). Recall that we have here only one multiclass classifier, which assigns each of the test patients to a class of the highest probability. We are *not* decomposing this multiclass classifier into a number of binary classifiers. What we are decomposing is the general information from the multiclass confusion matrix – which shows the results of testing our multiclass classifier on the test data set – into class-specific information, which illustrates how class-specific performance metrics may be calculated. Table 4.6 shows formulas and descriptions of class-specific performance metrics.

4.4.2 Multistage Approach to Multiclass Classification

Simultaneous differentiation among too many classes is not recommended. It is a quite common situation – when we have more than, say, five classes – that one or two of them dominate the classification, that is, they are more or less reasonably separated from other classes, as well as from themselves. Often, in such situations, the other classes are nearly indistinguishable; hence, it would be better to treat them as a single class. This would call for a *multistage classification schema* (Dziuda 2010). Once such a situation is recognized, then, at the first stage, we would identify a multivariate biomarker – and build a classifier – that differentiates among those well separated classes, as well as the class combining all other (overlapping) classes. Then, at the next stage, only the previously overlapping classes will be considered. If we still have several classes, we may again observe that some of them are well separated, with the rest overlapping and thus, we will again treat the currently overlapping classes as a single class (and identify a biomarker and build a classifier for this stage). Depending on the original number of classes, we will continue with the identification of subsequent multivariate biomarkers and the building of classifiers until, at the last stage, all of the remaining classes may be reasonably differentiated. Note that once we build such a multistage classification schema, we do not need to use all of its classifiers to classify a particular new observation – once the new observation is assigned to one of the individual classes (not a class combining overlapping classes), we do not need to continue with the remaining stages.

To illustrate the multistage approach to multiclass classification, assume that six classes are differentiated (say, disease states $\mathcal{D}_1, \ldots, \mathcal{D}_6$). When an initial multivariate biomarker for simultaneous classification of all six classes is identified, and a predictive model built, we may observe that two of the classes are well separated, while the other four are quite indistinguishable (see panel A in Figure 4.5).[27] This is a situation in which multistage classification should be considered. As a result, the predictive modeling could be performed at the following stages:

Stage 1

- Panel A in Figure 4.5 shows a two-dimensional projection of the five-dimensional discriminatory space for a classifier simultaneously differentiating all of the six

[27] In such situations, it is very convenient to use a supervised learning algorithm capable of visualizing the classification (and the training data) in a low-dimensional space. One such algorithm is discriminant analysis, which can present high-dimensional data in a discriminatory space with no more than $J - 1$ dimensions (number of classes minus one). For algorithms that cannot provide such visualizations, we could identify the well separated and the overlapping classes by investigating their class-specific performance metrics.

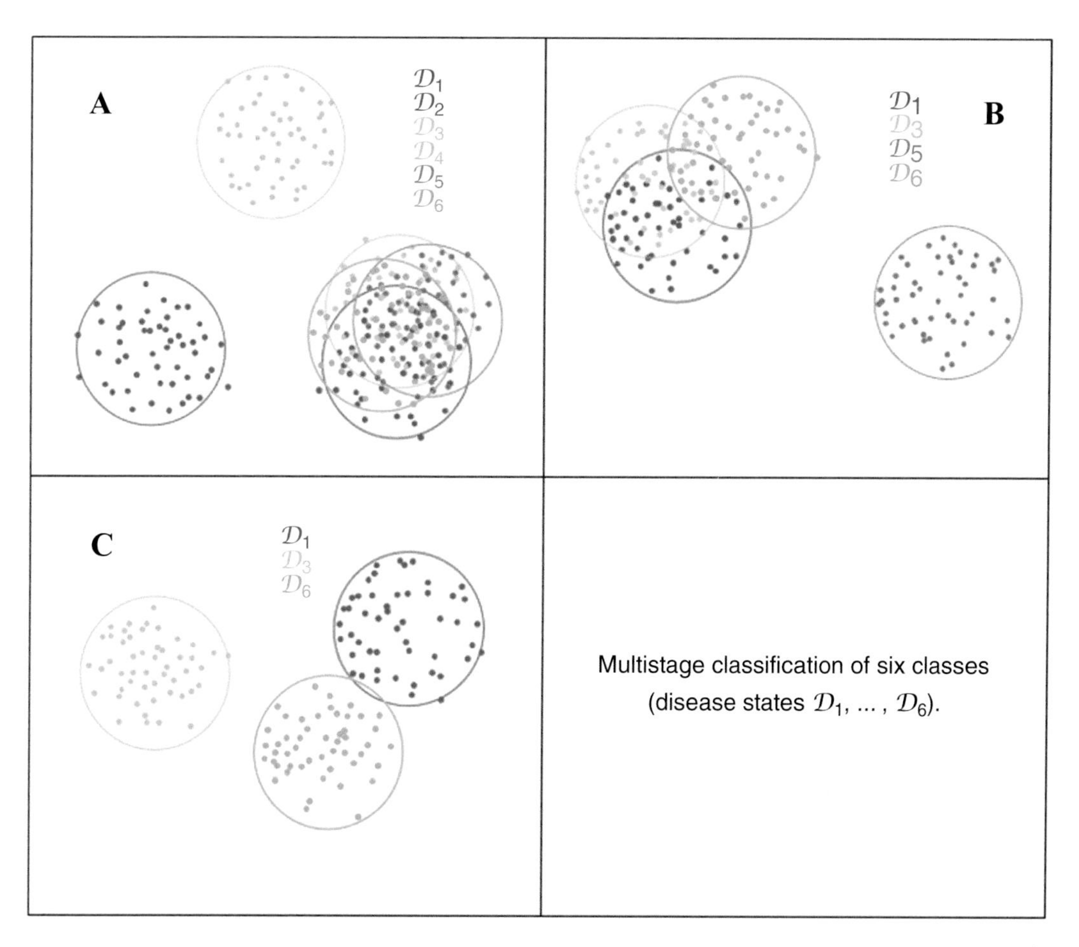

Figure 4.5 Illustration of multiclass classification using the multistage approach. Panel A shows a projection of the five-dimensional discriminatory space of a predictive model simultaneously differentiating among all of the considered six classes ($\mathcal{D}_1, \ldots, \mathcal{D}_6$). The first-stage classifier would differentiate three classes – the well separated $\mathcal{D}_2$ and $\mathcal{D}_4$, and a single class combining the other four classes. Panel B shows a projection of the three-dimensional discriminatory space of a model simultaneously differentiating the remaining four classes. The second-stage classifier would differentiate two classes – the well separated $\mathcal{D}_5$ versus a class combining the other three classes. Finally, since the remaining three classes are now reasonably separated (panel C), the third-stage classifier would differentiate $\mathcal{D}_1$, $\mathcal{D}_3$, and $\mathcal{D}_6$. The class circles represent a high probability (usually, 95%) area of each class; the points correspond to training observations.

classes.[28] Two of the classes ($\mathcal{D}_2$ and $\mathcal{D}_4$) are well separated, while the others are heavily overlapping. Thus,

- The overlapping classes are combined into a single class (this is not shown in Figure 4.5). Multivariate feature selection experiments are performed, an optimal

[28] The two-dimensional projection is defined here by the first two linear discriminant functions; such projections usually represent a majority of the discriminatory information (discriminant analysis is described in Chapter 12).

biomarker is identified, and a three-class classifier is built (and tested) for this stage. It will be assigning new patients into either $\mathcal{D}_2$ or $\mathcal{D}_4$, or "other classes".

Stage 2

- All steps of predictive modeling are performed again, while considering only the previously overlapping four classes. Panel B in Figure 4.5 shows a two-dimensional projection of the three-dimensional discriminatory space for a classifier simultaneously differentiating these four classes. $\mathcal{D}_5$ is now well separated from the other three classes, which still have substantial overlaps.
- Again, the currently overlapping classes are treated as a single class. All steps of predictive modeling are performed for two classes ($\mathcal{D}_5$ versus the remaining three), and an optimal multivariate biomarker is identified, and a two-class classifier is built (and tested) for this stage.

Stage 3

- All steps of predictive modeling are performed now for the three remaining classes ($\mathcal{D}_1$, $\mathcal{D}_3$, and $\mathcal{D}_6$). Panel C in Figure 4.5 shows the two-dimensional discriminatory space for a classifier simultaneously differentiating these three classes. Since the classes are reasonably separated, this will be the classifier for the third and last stage of this multistage classification.

Classification of each new patient would start with the first stage classifier. If the patient is assigned to class $\mathcal{D}_2$ or $\mathcal{D}_4$, this is the patient's diagnosis. Otherwise, the second stage classifier is used, and if the patient is assigned to class $\mathcal{D}_5$, that's the diagnosis. If not, the third stage classifier is invoked, which assigns the patient to one of the remaining three classes ($\mathcal{D}_1$, or $\mathcal{D}_3$, or $\mathcal{D}_6$).

Observe that – like for the simultaneous multiclass classification – each patient classified via the multistage approach is assigned to only one of the individual classes. This is not the case for the later described "one-versus-one" or "one-versus-rest" approaches, which generate many independent classifiers whose results need to be somehow aggregated in order to arrive at the final decision. A description of a study implementing the multistage approach to the classification of seven subgroups[29] of pediatric acute lymphoblastic leukemia (ALL) may be found in Dziuda (2010).

4.4.3 One-Versus-One Approach

Although predictive models that perform simultaneous multiclass classification (including those implementing the multistage approach) have the advantage of directly assigning any classified observation to one of the differentiated classes, sometimes we may have a reason[30] to decompose a multiclass problem into a number of binary classifications. One such approach consists of identifying a separate multivariate

[29] Six subtypes of ALL plus a class including ALL patients not assigned to any of the subtypes.

[30] For example, one may perhaps argue that sometimes there may be a need to use – for a multiclass problem – a learning algorithm that does not support multiclass classification.

biomarker – and building on it a classifier – for each pair of the considered classes. Thus, if our multiclass problem deals with J classes, we would need to build $\binom{J}{2}$ binary classifiers.[31] Then, after an observation is classified by all of the binary classifiers, we would need to aggregate the votes for each class and decide about the final assignment of the observation (usually by assigning the observation into the class that received most votes in the pairwise comparisons).

4.4.4 One-Versus-Rest Approach

In this approach, we decompose multiclass classification of J classes into J binary classifications. Each of them compares one of the classes with the remaining $J - 1$ classes combined into the "rest" class. Since each individual class could receive only one "vote" (from its classifier), a classified observation is assigned to the class with the largest value of some objective function (for example, the class probability).

4.5 More on Incorporating Misclassification Costs

Classification learning algorithms are typically designed to minimize the misclassification error rate, and work under the implicit assumption that all misclassifications have the same cost. However, when developing diagnostic biomarkers, we may want – in some situations – to assign different costs to different kinds of misdiagnoses, and then have the predictive model to minimize the expected misclassification *cost* (rather than its error rate). Some aspects of incorporating misclassification costs have already been mentioned earlier in this chapter (for example, using the ROC curve to find an optimal tradeoff between sensitivity and specificity); here, we will take a closer look at this topic.

In general data science literature, misclassification costs are often associated with class imbalances, and resampling methods may be offered as a solution. For example, if one class in a binary classification problem is severely underrepresented in the training data, the training data may be balanced either by randomly undersampling the majority class or by oversampling the minority class. This can be done multiple times and would result either in an ensemble of classifiers, or in a classifier founded on the aggregation of the results of the multiple experiments. Although such an entirely data-driven "technical" balancing tactic may be appropriate and useful in some situations, especially when using learning algorithms that are sensitive to class imbalances in training data (random forests and support vector machines are examples of such algorithms), it does not necessarily properly reflect real misclassification costs. Therefore, if we are interested in incorporating misclassification costs into a biomarker discovery study, we should have those costs provided by a domain expert, who would take into

[31] Where $\binom{J}{2} = \frac{J!}{2!(J-2)!}$ is the number of combinations of J classes taken two at a time. For example, for six classes (the number of classes considered in the example illustrating the multistage approach), we would need 15 binary one-versus-one classifiers.

consideration the specific context and goals of the biomarker. Whether such costs are aligned with class imbalances in the training data would then be quite irrelevant.

Among other cost-sensitive learning approaches (Cichosz 2015; Ling and Sheng 2017; Fernandez et al. 2018) are:

- Designing a cost-sensitive learning algorithm (usually by modifying a cost-insensitive one). This is an active research and development area, which has already resulted in implementations of cost-sensitive versions of such learning algorithms as, for example, support vector machines, decision trees, and neural networks. This option is usually out of the scope of a typical biomarker discovery study, unless the study includes the creation of a specialized machine learning software as a part of the project.
- For cost-insensitive classifiers that can accept observation weights, we may provide class-wide weights that are proportional to misclassification costs of the differentiated classes, and thus convert such weight-sensitive classifiers into cost-sensitive ones.
- For cost-insensitive classifiers that provide class probabilities for the classified observations, we may design a wrapper that will modify the output from the classifier by changing its decision threshold. For example, for a binary classifier with a typical probability threshold of $t = 0.5$, we would change this decision threshold to reflect different misclassification costs. This can be interpreted as moving class boundaries away from the class with a higher misclassification cost.[32]

In the literature on misclassification costs, the most commonly discussed medical examples are related to the cost of misdiagnosing a cancer (or any potentially fatal illness). The reasoning provided for increasing the cost of cancer misdiagnosis is that such misclassification is likely to result in serious negative consequences for cancer patients misdiagnosed as healthy. Thus, a usual conclusion is that it is necessary to increase the test sensitivity to cancer, even if this results in a decreased test specificity. Although such reasoning and its ensuing conclusion are valid in some (possibly many) situations, they should not be generalized.

Consider first a situation when such a conclusion is valid. If a diagnostic test targets the population of patients who are already (for other medical reasons) suspected of having the cancer, that is, if the prevalence of this cancer in the target population is large, then assigning a high misclassification cost to the cancer misdiagnosis – and thus increasing the test sensitivity to cancer – will be a proper and advisable approach. Although the decreased test specificity will increase the proportion of healthy patients diagnosed as having cancer, this increase will be relatively small (due to the large prevalence of the cancer among the tested), and will definitely be outweighed by the benefits to the properly diagnosed cancer patients who otherwise would be misclassified as healthy.

[32] There are also wrappers for cost-insensitive classifiers that do not provide class probabilities, for example the *MetaCost* algorithm (Domingos 1999).

Now, consider a screening test aimed at the early detection of the same cancer in a large population, in which the prevalence of this cancer is very low. In this situation, it is very important for the test to have a very high specificity (see the example in Section 4.3.3, in which a specificity of 99.5% was too low for the screening test to be acceptable). Hence, increasing the test sensitivity to the cancer and thus, decreasing its specificity, would be improper in such situations – if the proportion of false positives (healthy patients misclassified into the cancer class) greatly overwhelms the proportion of true positives, the screening test would most likely be rendered unsuitable for practical use. Hence, the alternative here would be either to have a screening test that misses some early-cancer patients (more than we would like), or to have no screening test at all. Consequently, when designing such a screening test, we may even consider increasing the cost of false positives – even if this would slightly decrease the test sensitivity to the cancer – in order to have an acceptable screening test, rather than not having any.

4.5.1 Cost Matrix

Misclassification costs may be specified in the form of a $J \times J$ matrix $\mathbf{C}$, whose elements $C(j, m)$, where $j, m = 1, \ldots, J$, represent the cost of classifying a patient whose true class is j into class m. Recall, however, that even when the number of differentiated classes (J) is large, a properly designed biomarker discovery study would usually decompose such a problem into ones dealing with a smaller number of classes (see, for example, the multistage approach described in Section 4.4.2). This means that there should be no need for domain experts to provide misclassification costs for large cost matrices with possibly complex interrelations among classes.[33] Furthermore, most of the typical biomedical tests that require the incorporation of misclassification costs are based on biomarkers for binary classification. A cost matrix for binary classification is presented in Table 4.7. The differentiated classes are called *Disease* ($\mathcal{D}$) and *Healthy* ($\mathcal{H}$).

Although the cost matrix may incorporate various costs that are associated with particular predictions, without losing generality we may focus only on the misclassification costs and assume zero costs for diagonal elements $C(j, j)$, for which the predicted and true class is the same. Thus, we have

$$C(\mathcal{D}, \mathcal{D}) = C(\mathcal{H}, \mathcal{H}) = 0, \tag{4.24}$$

and a domain expert only needs to provide misclassification costs for the false negatives – $C(\mathcal{D}, \mathcal{H})$, and the false positives – $C(\mathcal{H}, \mathcal{D})$.

[33] If, however, misclassification costs need to be provided for a multiclass problem, we may make this more manageable for domain experts by replacing a $J \times J$ cost matrix $\mathbf{C}$ with a $J \times 1$ cost vector $\mathbf{c}$, in which each element $c(j)$, $j = 1, \ldots, J$, represents the cost of misclassifying patients with true class j into any other class (Cichosz 2015).

Table 4.7 Cost matrix for a classifier differentiating between two classes (Disease and Healthy). Matrix elements $C(j,m)$, where $j,m = 1,2$ (and where $\mathcal{D}$ corresponds to 1, and $\mathcal{H}$ to 2) represent the costs of classifying a patient whose true class is j into class m. The correct classifications are marked with a green background and their costs $C(\mathcal{D},\mathcal{D})$ and $C(\mathcal{H},\mathcal{H})$ are assumed to be zero. Misclassification costs, $C(\mathcal{D},\mathcal{H})$ and $C(\mathcal{H},\mathcal{D})$ – pink background, are provided by a domain expert and should reflect the level of danger that a particular misclassification poses to a patient.

		Predicted Class	
		Disease ($\mathcal{D}$)	Healthy ($\mathcal{H}$)
True Class	Disease ($\mathcal{D}$)	$C(\mathcal{D},\mathcal{D})$ True Positive Cost	$C(\mathcal{D},\mathcal{H})$ False Negative Cost
	Healthy ($\mathcal{H}$)	$C(\mathcal{H},\mathcal{D})$ False Positive Cost	$C(\mathcal{H},\mathcal{H})$ True Negative Cost

4.5.2 Cost-Sensitive Classification

Assume that we want to classify a patient represented by a $p \times 1$ vector $\mathbf{x} = [x_1, \ldots, x_p]^T$ where p is the number of variables included in the multivariate biomarker. Assume further that a cost-insensitive classifier implementing this biomarker provides for this patient the following posterior class probabilities:

- $P(\mathcal{D}|\mathbf{x})$ – the classifier's estimate of the probability that the patient belongs to the *Disease* class
- $P(\mathcal{H}|\mathbf{x})$ – the classifier's estimate of the probability that the patient belongs to the *Healthy* class

Consequently, the *cost-insensitive* diagnosis proposed by the classifier would be:

$$\underset{\text{(cost-insensitive)}}{\text{Predicted Class}} = \begin{cases} \textit{Disease} & \text{if } P(\mathcal{D}|\mathbf{x}) \geq P(\mathcal{H}|\mathbf{x}) \\ \textit{Healthy} & \text{otherwise.} \end{cases} \tag{4.25}$$

To incorporate misclassification costs into the classifier's prediction, we will first calculate the expected cost of classifying the patient into the *Disease* class, $\text{EC}(\mathcal{D})$, as well as the expected cost of predicting *Healthy*, $\text{EC}(\mathcal{H})$. Based on the cost matrix, we would have:

$$\begin{aligned} \text{EC}(\mathcal{D}) &= P(\mathcal{D}|\mathbf{x}) \cdot C(\mathcal{D},\mathcal{D}) + P(\mathcal{H}|\mathbf{x}) \cdot C(\mathcal{H},\mathcal{D}), \\ \text{EC}(\mathcal{H}) &= P(\mathcal{H}|\mathbf{x}) \cdot C(\mathcal{H},\mathcal{H}) + P(\mathcal{D}|\mathbf{x}) \cdot C(\mathcal{D},\mathcal{H}). \end{aligned} \tag{4.26}$$

However, due to the assumption expressed in (4.24), the expected costs can be simplified to:

$$\begin{aligned} \text{EC}(\mathcal{D}) &= P(\mathcal{H}|\mathbf{x}) \cdot C(\mathcal{H},\mathcal{D}), \\ \text{EC}(\mathcal{H}) &= P(\mathcal{D}|\mathbf{x}) \cdot C(\mathcal{D},\mathcal{H}). \end{aligned} \tag{4.27}$$

Cost-sensitive classification will now assign a patient to the class with the minimum expected cost. For example, patient $\mathbf{x}$ will be assigned to the *Disease* class if and only if $\mathrm{EC}(\mathcal{D}) \leq \mathrm{EC}(\mathcal{H})$. Consequently, the *cost-sensitive* prediction will be:

$$\underset{\text{(cost-sensitive)}}{\text{Predicted Class}} = \begin{cases} \textit{Disease} & \text{if } P(\mathcal{H}|\mathbf{x}) \cdot C(\mathcal{H}, \mathcal{D}) \leq P(\mathcal{D}|\mathbf{x}) \cdot C(\mathcal{D}, \mathcal{H}) \\ \textit{Healthy} & \text{otherwise.} \end{cases} \tag{4.28}$$

Since $P(\mathcal{H}|\mathbf{x}) = 1 - P(\mathcal{D}|\mathbf{x})$, the condition – in Formula (4.28) – for predicting *Disease* can be replaced by,

$$P(\mathcal{D}|\mathbf{x}) \geq \frac{C(\mathcal{H}, \mathcal{D})}{C(\mathcal{H}, \mathcal{D}) + C(\mathcal{D}, \mathcal{H})}, \tag{4.29}$$

with the right side of inequality (4.29) representing a new and cost-sensitive probability threshold t^* for classifying patient $\mathbf{x}$ into the *Disease* class. Summing up, the original cost-insensitive probability threshold $t = 0.5$, would be replaced by a cost-sensitive threshold,

$$t^* = \frac{C(\mathcal{H}, \mathcal{D})}{C(\mathcal{H}, \mathcal{D}) + C(\mathcal{D}, \mathcal{H})}, \tag{4.30}$$

which considers the misclassification costs from the cost matrix. Consequently, the cost-sensitive diagnosis provided by the classifier, which would now minimize the expected cost of misclassification, would be:

$$\underset{\text{(cost-sensitive)}}{\text{Predicted Class}} = \begin{cases} \textit{Disease} & \text{if } P(\mathcal{D}|\mathbf{x}) \geq t^*) \\ \textit{Healthy} & \text{otherwise.} \end{cases} \tag{4.31}$$

5 Multivariate Feature Selection

5.1 Introduction

Multivariate feature selection is the most important part of predictive modeling for biomarker discovery based on high-dimensional data (the data for which the number of variables p is greater – and often much greater – than the number of observations N). After an optimal, robust, and parsimonious multivariate biomarker with satisfactory predictive abilities is identified, many learning algorithms[1] can be used for creating efficient predictive models for classification or regression. Thus, properly designing the feature selection step is even more important than selecting the learning algorithm.

The commonly used "feature selection" term is quite general; it may mean *variable selection*, which refers to considering and selecting only subsets of the original variables included in the training data, but it may also mean selecting subsets of *features*, where features – in addition to, or instead of, the original variables – include variables that are derived from the original variables (for example, *engineered* features). However, in the context of biomarker discovery, we are almost always interested in biomarkers that consist of only some of the original variables (to be able to interpret biomarkers and predictive models in terms of the original variables). Although we will use the common term *feature selection*, in the context of this book it will be considered synonymous with ***variable selection***.

Furthermore, in the context of multivariate biomarker discovery, the term *feature selection* is understood to be synonymous with ***multivariate feature selection*** (and thus with *multivariate variable selection*). This means that when we are searching for an optimal subset of variables, each variable has to be considered in the context of other variables.

As mentioned in Chapter 2 (recall the discussion of two scenarios for predictive modeling), the typical methods and habits that may work well for low-dimensional data will virtually guarantee overfitting when applied to high-dimensional data (especially when $p \gg N$). Imagine, for example, a regression problem with a toy data set including three variables (two independent variables plus response) and two observations, that is, two data points in a three-dimensional space. We can easily find a line (or, actually, a plane) that will perfectly fit these two points, producing a model with zero training error. However, such a model will, in all likelihood, perform very poorly in predictions for new observations. Now, imagine data with 10,000 variables and 100 observations (yes,

[1] In addition to the learning algorithm used during the feature selection process (if any was used).

I know, our brains can imagine this only via analogy) – fitting a model[2] in such a high-dimensional space, without performing feature selection first, will virtually guarantee overfitting (that is, producing a model that is useless for prediction). Recall that such useless models would be associated with perfect (or near perfect) *MSE* (zero) and R^2 (one). All this means is that the *training-data-based* metrics of fit that one may have been successfully using for $p < N$ data are useless when dealing with high-dimensional data. This is equally true for regression as well as for classification modeling. To have a reliable estimate of performance, each biomarker (and a predictive model implementing the biomarker) considered during the feature selection process should be evaluated on data that were not used for the identification of the biomarker (for example, via *external* validation utilizing bootstrapping or cross-validation: see Chapter 4).

The main goal of feature selection for biomarker discovery is to identify a small subset of the original independent variables (representing, for example, gene or protein expression levels) whose combined pattern of values allows for a highly accurate prediction of the value of the response variable (such as disease state, or the probability of a specific response to a treatment) for new observations. Hence, we are searching for biomarkers that are: **parsimonious, multivariate, and robust**. Furthermore, it would be preferable if such a biomarker is associated with additional information that may facilitate a plausible biological interpretation of "why or how does it work?". Let us summarize these characteristics:

Multivariate

Multivariate biomarker means not only that a biomarker is *a set* of variables, but also, and most importantly, that such a set has been identified via *multivariate approaches*, that is, the methods that consider each independent variable in the context of other independent variables. Consequently, sets composed of variables that were selected via univariate methods (such as calculating correlation of each independent variable with the response variable, or by using univariate filter methods) are *not* considered multivariate biomarkers. This means that feature selection algorithms for biomarker discovery must implement multivariate methods that evaluate the association between each considered *set of independent variables* and the response variable; whether variables included in a multivariate biomarker are individually associated with the response is quite irrelevant.[3] Furthermore, it is important to understand that a multivariate biomarker is ***a biomarker***, not a "set of biomarkers". A properly performed feature selection identifies a set of variables, which together – as a set – constitute *a single biomarker*.

Parsimonious

An optimal multivariate biomarker should consist of as few variables as possible; such parsimony is very important for the identification of a biomarker that is well generalizable to unseen data. Adding uninformative or redundant variables increases

[2] Assuming that we are using a learning algorithm for regression that is not performing feature selection, and is capable of working when $p > N$.

[3] Usually, some of them are and some are not.

the chance for overfitting the training data.[4] Parsimony is thus one of the crucial aspects of selecting an optimal biomarker by using a combined criterion of minimizing its size and maximizing its predictive abilities.

Robust

The robustness (that is, the generalizability to unseen data) of a multivariate biomarker is associated with its parsimony, with the proper approach to feature selection, and with the proper ways of evaluating predictive abilities of the biomarker. For example, it is unlikely for a biomarker identified via a single feature selection experiment to be robust. Similarly, if feature selection is driven by improper methods of subset evaluation (such as internal validation), a biomarker deemed – by such methods – as highly predictive may have no predictive abilities when used for new data.

Facilitating plausible biological interpretation

The meaningful biological interpretation of a multivariate biomarker is still an open, and quite challenging, problem. Variables (such as genes, proteins, or metabolites) included in the biomarker should be interpreted as a set, as their individual interpretations do not translate into the biological interpretation of the set. Ideally, the multivariate interpretation of a biomarker should tap into interactions among biological processes (or molecular networks) represented by the variables that constitute the biomarker. A first step toward such interpretation may consist of linking the biomarker variables to groups or clusters of other variables with similar characteristics (such as similar gene expression pattern across training observations), under the assumption that associating such clusters with underlying biological processes may be easier than in the case of individual variables.

5.2 General Characteristics of Feature Selection Methods

Multivariate feature selection methods are characterized by their approach to searching for an optimal subset of variables. Traditionally, they are categorized by their search model (such as filter, wrapper, or embedded), and by their search strategy (for example, heuristic, or complete). Of course, for biomarker discovery we are interested only in feature selection methods that are based on supervised approaches (see Section 2.4).

5.2.1 Feature Selection Search Models

We will start with discussing the search model, in which feature selection is performed ***independently of any learning algorithm***. This search model is commonly called the *filter* model, which is a very unfortunate name, and for more than one reason: (i) it does

[4] There are additional advantages of the biomarker parsimony, such as the possibly lower cost of its clinical implementation as well as more efficient software applications.

not properly describe the essence of this model (that it is independent of any learning algorithm), and (ii) it is misleading in that it allows the inclusion of methods that perform univariate filtering, which is just that, *filtering*, and has nothing to do with feature selection.

As discussed earlier (see Section 2.3), the univariate significance of variables is irrelevant for multivariate biomarker discovery. Univariate methods (such as a t-test or the ANOVA F-test) ignore any and all interactions among the variables and thus, would be appropriate *only* when the variables are independent of each other, which is definitely not the case in biomarker discovery based on high-dimensional data.[5] Hence, ordering variables by any univariate criterion and then limiting the analysis to those deemed significant would – most likely – remove from consideration important subsets of variables and thus, prevent the identification of multivariate biomarkers with the best predictive abilities. Recall that variables that are individually insignificant may be very important for prediction when combined with some other variables, and no univariate approach is capable of identifying such combinations. Consequently, we would not consider – for feature selection – methods that are either univariate or univariately-biased (for example, methods that apply a multivariate approach only *after* the considered variables are limited to those that are individually correlated with the response). This means that into the category of search models that are independent of any learning algorithm, we place only those multivariate feature selection methods that evaluate sets of variables by using some multivariate metrics of their predictive abilities (for example, a metric of class separation for classification). They are usually heuristic methods that, in the search for an optimal biomarker, add and remove variables (to and from a current set) in a way that maximizes some intrinsic metric of the set's predictive ability (for example, the ratio of the between-to-within class variation).

In another search model, the ***wrapper*** model, the feature selection procedure is wrapped around a supervised learning algorithm. The search process is driven by whatever feature selection method is implemented, and the learning algorithm is used to build (and evaluate) a predictive model for each of the considered subsets of variables. Some methods (like the popular recursive feature elimination) may also require that the learning algorithm has the ability to provide a measure of multivariate importance of the variables (which would then be used to decide on the next subset of variables). Incorporating a learning algorithm within the feature selection process may result in biomarkers that are better tailored to this particular learning algorithm.

The third commonly distinguished search model, the ***embedded*** model, is not – *sensu stricto* – a feature selection search model; rather, it refers to those supervised learning algorithms that have the intrinsic ability to perform feature selection (such as the *lasso*; see Chapter 7).

[5] With possibly rare exceptions, the same is true for any data analysis project based on high-dimensional data, whether the data are biomedical or otherwise.

5.2.2 Feature Selection Search Strategies

Theoretically, search strategies for feature selection include *exhaustive* and *complete* searches, which would guarantee finding the global optimum. The former would evaluate all possible combinations of the independent variables, while the latter would achieve the goal without considering all of such possible subsets. Neither of these approaches is, however, feasible for high-dimensional data. Furthermore, as indicated in Chapter 3, even if we could find the global optimum for the training data, we would not be interested in it; such a global optimum would overfit the training data and – most likely – represent a spurious pattern existing in the training data, but not in the target population.

Efficient search strategies for high-dimensional data are ***heuristic searches***, which evaluate subsets of variables that are on their search path, the path that is dynamically determined by a process that optimizes a specific objective criterion. The simplest of such searches are *stepwise forward selection* and *stepwise backward selection*. The former starts with an empty set, and then at each step adds to the current set the variable whose addition maximally improves a metric that estimates the set's predictive power. The latter acts similarly, but starts with all variables and at each step removes one of them. These procedures are known as *greedy* searches as they do not revisit already evaluated subsets (and thus do not reevaluate the search path). Forward selection would usually be stopped either when some predetermined size of the subset is reached, or when the measure of predictive power is high enough. Consequently, it may use any relevant metric of predictive power,[6] even those that cannot be calculated for a subset with $p > N$ (more variables than observations in the training data). Backward selection, however, is limited to metrics that can be calculated for $p > N$, as it always starts with the subset including all of the p independent variables. Both of these greedy approaches produce nested subsets, that is, variables that are included in a smaller set are also included in any larger one.

Better results may be achieved by ***stepwise hybrid selection***, which is a non-greedy approach that dynamically reevaluates already considered subset sizes. In the hybrid search, variables may be added as well as removed from a currently evaluated set until no further improvement is, at the moment, possible. This means that the current subset size may increase as well as decrease, depending on which path maximizes the metric of predictive power. This also means that the variables included in an optimal subset for a particular size do not have to be included in the larger optimal subsets.

Whichever search model or strategy we utilize, a feature selection experiment would provide results consisting of one optimal subset for each of the considered subset sizes. However, it is very important – in order to avoid a spurious solution – to perform multiple feature selection experiments using a resampling of the training data, and then aggregate results over all such parallel experiments, in order to decide about the size and composition of the optimal multivariate biomarker.

[6] It has to be a metric estimating predictive abilities of a set of variables, *not* a p-value or any other measure of statistical significance; recall (from Chapter 2) that the statistical significance of a biomarker is irrelevant in the context of predictive modeling.

5.3 Multiple Feature Selection Experiments

Performing a single feature selection experiment is not recommended in any situation; however, when dealing with $p > N$ (and especially $p \gg N$) high-dimensional data, a single search will – in most situations – guarantee overfitting, that is, finding a spurious pattern that exists in the training data, but does not exist in the target population, from which the training data were selected. Basically, this would most likely be a random pattern similar to that described in the toy example in the Introduction.

When we deal with sparse data whose dimensionality is much higher than the number of training observations, we must overcome *the curse of dimensionality* by using methods capable of elevating signal (predictive information) from the overwhelming noise. To achieve this, we should perform many searches (hundreds or even thousands) that are randomized in some way. We may, for example, use bootstrapping or repeated cross-validation (see Chapter 4) to generate many random subsamples of the training data, and then for each of them perform an independent feature selection experiment.

This way we would perform many parallel feature selection experiments (see Figure 5.1). Since each of them is using only a subsample of the original training data, the observations not selected to the subsample can be used to assess the predictive abilities of each of the considered subsets of variables. Although some of the variables selected into the subsets identified during the individual parallel searches may be selected by random chance, some other variables may be frequently selected into multiple subsets. The latter are more likely to represent real population patterns associated with the predictive information.[7] Thus, by aggregating the results of the parallel feature selection experiments we can identify an optimal subset of variables – a multivariate biomarker – that is most likely to represent a real population pattern. If predictive models are then built on such an optimal biomarker, the models are much more likely to be robust and generalizable than those based on a set of variables identified via a single feature selection experiment.

5.3.1 Design of a Feature Selection Experiment

Each of the parallel feature selection experiments (represented by the colored boxes in Figure 5.1) is designed the same way, and each of them considers the same sequence of sizes of subsets of variables. This way, their results can be meaningfully aggregated. However, there exist various possible designs for the experiments. For example, in each of them, we may use recursive feature elimination (see Section 5.3.2) associated with a

[7] It is important to understand that such a real population pattern corresponds to some biological processes underlying the investigated changes in (or different states of) the response variable, and that our optimal biomarker is only one of the possible subsets of variables associated with those biological processes. Hence, we can assume that there are other multivariate biomarkers of similar size and predictive abilities composed of different sets of variables (either overlapping or not with our biomarker). Therefore, if we repeated our feature selection using different methods and/or parameters, we would most likely find one of those other solutions. However, such apparently different results would (or may) represent a stable solution if those different subsets tap into those same biological processes.

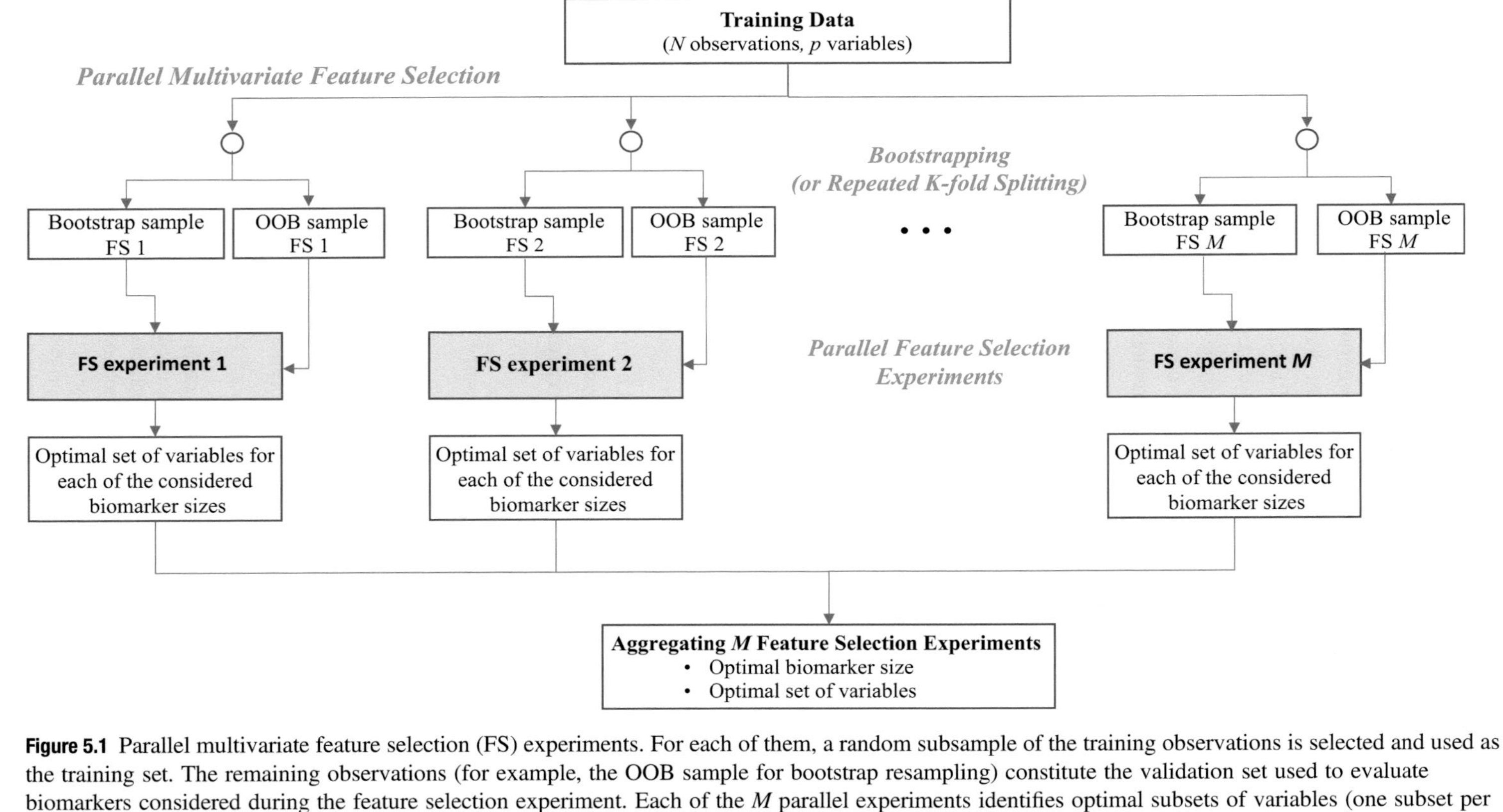

Figure 5.1 Parallel multivariate feature selection (FS) experiments. For each of them, a random subsample of the training observations is selected and used as the training set. The remaining observations (for example, the OOB sample for bootstrap resampling) constitute the validation set used to evaluate biomarkers considered during the feature selection experiment. Each of the *M* parallel experiments identifies optimal subsets of variables (one subset per each of the considered subset sizes). The results of all parallel experiments are then aggregated, and the optimal size and composition of the multivariate biomarker identified.

supervised learning algorithm (such as random forests, or support vector machines), and in each of them a predictive model will be built (and evaluated) for each of the considered subset sizes. In such a design, the learning algorithm has to be capable of calculating the multivariate importance of variables, which would drive recursive variable elimination. Another design may – instead of recursive feature elimination – incorporate a learning algorithm that is intrinsically capable of performing feature selection[8] and thereby identify a subset of variables (as well as build and evaluate a predictive model) for each considered cardinality. Yet another option would be feature selection independent of any learning algorithm. In this case, the evaluation of predictive abilities of the considered subsets of variables is based on some characteristics of the data (for example, for classification it may be a metric of class separation).

When designing parallel feature selection experiments, the computational efficiency of the design has to be considered. For example, a design with 500 parallel recursive feature elimination experiments using the random forests learning algorithm may be quite efficient. Less efficient would be the same design with linear support vector machines, and it would be much less efficient when a nonlinear kernel-based SVM is used.[9] Of course, computational efficiency may be very different for different software implementations of the methods; hence, it is a good idea to first perform small test experiments (e.g. a single feature selection run) to estimate the computation time for a full design.

5.3.2 Feature Selection with Recursive Feature Elimination

Recursive Feature Elimination (RFE) is a popular feature selection method implementing the stepwise backward selection search strategy (Guyon et al. 2002).[10] It is a greedy wrapper method that can work with many supervised learning algorithms. However, such algorithms have to be able to calculate some measure of multivariate variable importance (or, at least, the importance of each variable in the context of some other variables), which would drive the variable elimination process.[11] Another requirement (as indicated in Section 5.2.2) for learning algorithms used with stepwise backward selection is that they have to be able to work when $p > N$. RFE is usually coupled with the random forests or support vector machines learning algorithms.

Figure 5.2 shows an example of one of M parallel feature selection experiments (represented by colored boxes in Figure 5.1) performed with recursive feature elimination. RFE experiment m, $m = 1, \ldots, M$, uses bootstrap sample m as its training data and OOB sample m for evaluation (alternatively, a repeated K-fold cross-validation may

[8] For example, the *lasso*, for regression (see Chapter 7).

[9] The main reason for lower efficiency of the design with SVM used as the learning algorithm is the necessity of tuning hyperparameter(s) of SVM classifiers.

[10] In its original design, RFE was coupled with support vector machines. Predictive models built for each subset size were, however, evaluated using *internal* validation; this error can be corrected by putting RFE within a resampling schema and evaluating models on an OOB sample (or on other data not used for feature selection).

[11] Learning algorithms that would select variables by their univariate metrics should not be used for RFE as such a design would not implement multivariate feature selection.

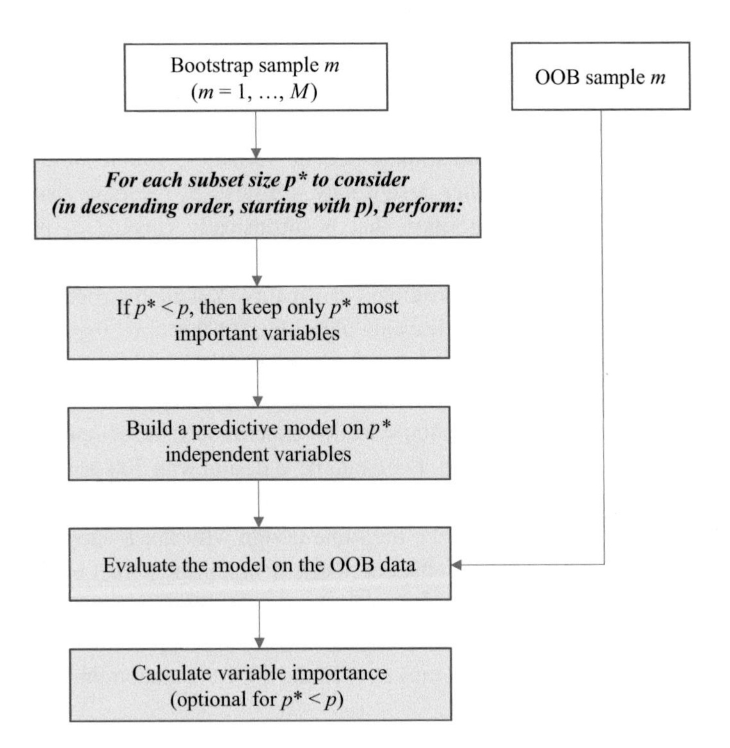

Figure 5.2 A diagram of recursive feature elimination (RFE). For high-dimensional data, each RFE experiment is performed using predetermined subset sizes; p^* corresponds to the current subset size, which starts with all p variables, and then takes the consecutive sizes in descending order. For each of the considered subset sizes, the supervised learning algorithm (such as random forests or support vector machines) is used to: (i) build a predictive model on the bootstrap sample including only the current set of p^* variables; (ii) evaluate the model on the OOB sample (and save the results to be used for aggregation); and (iii) calculate variable importance (which may be made optional for sizes smaller than p). Then, at the next step, only the current (for this step) p^* most important variables are used.

be used). RFE is usually performed for predetermined, project-driven subset sizes. To focus our attention, assume that we are performing a classification experiment based on microarray gene expression data, that the data originally included over 22,000 independent variables (expression levels for probes representing genes), and that after filtering out the variables representing noise and those with unreliable measurements, we are starting our analysis with $p = 7{,}305$ independent variables. Also assume that our biomarker discovery study aims at differentiating between two classes, say $\mathcal{D}$ and $\mathcal{H}$, and that the training data include 250 patients in each of the classes. Since we are interested in a parsimonious multivariate biomarker, our focus is on small subset sizes. This means that when RFE starts with all 7,305 variables, at each of the early iterations we will eliminate many variables[12] (say, up to 50%), and later – after reaching smaller

[12] For high-dimensional data, this approach provides a reasonable compromise between the computational efficiency of the algorithm and the quality of the results.

sizes – we will eliminate at first few and then only one variable per step. Then assume that we will consider the following subset sizes: 7,305, 4,000, 2,000, 1,000, 500, 250, 150, 100, 50, 45, 40, 35, 30, and then every subset size between 25 and 2. Finally, assume that RFE will be coupled with the *random forests* learning algorithm that will be used to build forests consisting of 1,000 trees, and that the other parameter, the number of variables randomly selected for consideration at each node, will be left at its default value (for details on the random forests algorithm for classification, see Chapter 10).

The RFE experiment m starts with building a random forest predictive model using all p independent variables (in this example, $p = 7{,}305$). The model is built on bootstrap sample m (constituting the m-th second-order training set) and is then tested on the corresponding OOB sample m. The results of testing the model (such as AUC, sensitivity, specificity, and accuracy) are saved. Then, using one of the random-forest-specific measures (described in Chapter 10), variable importance is calculated for all of the 7,305 variables, and the variables are ordered by their importance. Since the next of the predetermined subset sizes includes $p^* = 4{,}000$ variables, the 3,305 least important variables are removed. Then, for each of the subsequent subset sizes, only variables included in the then current subset are used, and the entire process is repeated: a new predictive model is built, tested on the OOB data, and the least important variables are removed. There are two approaches to the variable importance calculation during an RFE experiment: (i) variable importance is calculated only once, for the full set of variables, and (ii) variable importance is recalculated at each iteration of the RFE process (that is, for each of the considered subset sizes). There is no compelling evidence that one of these approaches is always better than the other – software implementations of RFE usually provide an option to do it either way. Our preference is to recalculate variable importance for every subset size, as it seems more fitting to evaluate importance of a variable in the context of the *currently* considered variables.

After all of the M parallel feature selection experiments are performed, we have the M values of each performance metric for each of the considered subset sizes. Thus, for each of those sizes, we may consider the average, standard deviation, and standard error values of each performance metric and thereby decide on the optimal size of a biomarker.[13] For example, we could identify the subset size with the largest value of AUC (or accuracy), and then either deem it optimal, or apply the one-standard-error rule (or, alternatively, the tolerance method; see Section 3.3.4) and select the smallest subset of variables whose AUC (or accuracy) is within one standard error range from the maximal value of the metric. However, as discussed in Section 3.3.4, a better approach is to decide on the optimal size of a biomarker by simultaneously evaluating several relevant performance metrics (for example, AUC, accuracy, sensitivity, and specificity – in a case of classification biomarkers).

Once the size of the optimal multivariate biomarker is selected, we will aggregate information about variables included in the M subsets of this size, in order to decide about the composition of the optimal biomarker. We may consider the distribution of

[13] An example of a graphical illustration of the average performance metrics over the considered subset sizes is presented in Figure 4.4.

variables across the M subsets, as well as the distribution of their importance measurements. We may, for example, value much higher those variables that were most often selected into the M subsets, under the assumption that they are more likely to represent a real pattern (rather than a spurious one) than the variables selected only once.[14] Alternatively, we may perform a single feature selection experiment using only the variables that are included in the M subsets of the optimal size. After the optimal biomarker is identified, it is used to build a predictive model on the entire original training data set.

The RFE example described above was a classification example with the random forests learning algorithm. However, the same RFE process is applicable to regression, as well as when a different learning algorithm is used. Some learning algorithms (for example, random forests) do not have parameters that *have to* be tuned.[15] However, there are algorithms that require their parameters to be tuned (such as support vector machines). In such cases, there are two main approaches to tuning: (i) perform tuning *after* the parallel RFE experiments are completed, or (ii) incorporate tuning into the RFE process.

In the first option, once the results of M parallel RFE experiments are aggregated and the optimal multivariate biomarker is identified, the predictive model implementing this biomarker will be tuned on a separate validation set. This means that during the initial splitting of the available data, we would need to create not only the training and test sets, but also a validation set to be used during this tuning. This option has an important advantage – once the optimal biomarker is identified, we may use various learning algorithms to create more than one predictive model implementing this biomarker. Such models would be evaluated and tuned on the validation set, and then the final one (or, possibly, more than one deemed final) will be tested on the test data.

If the tuning is incorporated into the RFE experiments, all models built during each of the RFE experiments would be evaluated as well as tuned on the OOB sample of the experiment.[16] This is a reasonable approach – each of the models is built on the current bootstrap sample, and evaluated (including the versions of the model with different values of its tuning parameters) on the current OOB sample (thus, there is no information leakage between the data used for training the models and the data used for their evaluation and tuning). Nevertheless, if we wanted to be extra cautious about the possibility of overfitting by using the same OOB data for model evaluation as well as for tuning, we may consider adding another level of resampling (Kuhn and Johnson

[14] Although, due to correlations among variables, it may be preferable to include in our considerations information about groups of variables with similar patterns, rather than only looking at the distribution of individual variables (see Chapters 15 and 16).

[15] There are two parameters for random forests – the number of trees to build, and the number of variables to be randomly selected at each node. To decide on the number of trees, we may perform a preliminary run and observe how the OOB error rate changes across different number of trees; often, it stabilizes around 500 or so trees. The number of variables to randomly select at each node may be tuned; however, its default values ($\lfloor p/3 \rfloor$ for regression, and $\lfloor \sqrt{p} \rfloor$ for classification) usually work well (for details, see Chapters 8 and 10).

[16] Using resampling (bootstrapping or cross-validation) of the current bootstrap sample, *after* it was used to train the model, has to be avoided as it would amount to *internal* validation.

2020) by starting the RFE experiment with either resampling or splitting the bootstrap sample (our second-order training data) into third-order training and validation data, the former to be used for training, and the latter for tuning (after such tuning, the models would still be evaluated on the original, second-order, OOB sample). However, since this could significantly increase the computational costs of the method, and since the evaluation and tuning results will be aggregated over all M parallel RFE experiments anyway, this additional level of data splitting would rarely be necessary.

5.3.3 Feature Selection with Stepwise Hybrid Search

As an example of this option, we will discuss the ***stepwise hybrid selection with*** T^2 (Dziuda 2010). This non-greedy feature selection algorithm is driven by the *Lawley–Hotelling trace statistic* T^2, which is used as a metric of class separation (representing the ratio of between-to-within class variations). Since this metric is defined under the assumption that observations are independent of each other, the M parallel feature selection experiments cannot be based on bootstrapping with replacement (that is, on Efron's nonparametric bootstrap), and a bootstrap method without replacement needs to be used (see Section 4.1.3.1).

5.3.3.1 *Basic T^2-Driven Hybrid Feature Selection*

This feature selection method implements a search model independent of any learning algorithm.[17] The search for an optimal biomarker starts either with the empty set, or with a randomly selected variable. Since this is a stepwise hybrid approach (combining both forward and backward selections), at each subsequent step, variables may be added or removed, in order to maximize the T^2 measure of class separation provided by the most current set with p^* variables (where $p^* = 1, 2, \ldots$). This means that a particular subset size may be visited more than once, and that a next step may consider either a larger or a smaller subset size than the previous one (although, once a class separation for a particular cardinality is no longer improving, the overall tendency of the search is toward larger subset sizes). Figure 5.3 shows a diagram of this stepwise hybrid search algorithm.

The search ends when one of the following stopping criteria is met: (i) the maximum subset size (a user-defined parameter) is reached; (ii) the maximum value of the T^2 measure (another user-defined parameter) is reached; or (iii) when the current subset size reaches $N - J - 2$, which is the maximum size for which the T^2 metric of class separation can be calculated (where N is the number of observations in the bootstrap sample m, and J is the number of differentiated classes).[18] The first stopping criterion reflects our goal to search for a parsimonious biomarker, and thus the lack of interest in

[17] Although the T^2 metric is associated with discriminant analysis, no learning algorithm is used, and no predictive models are built, during this feature selection procedure.

[18] Since we are dealing with high-dimensional data with $p > N$ (and, most likely, with $p \gg N$), stopping the search when the current subset size would reach p is impossible.

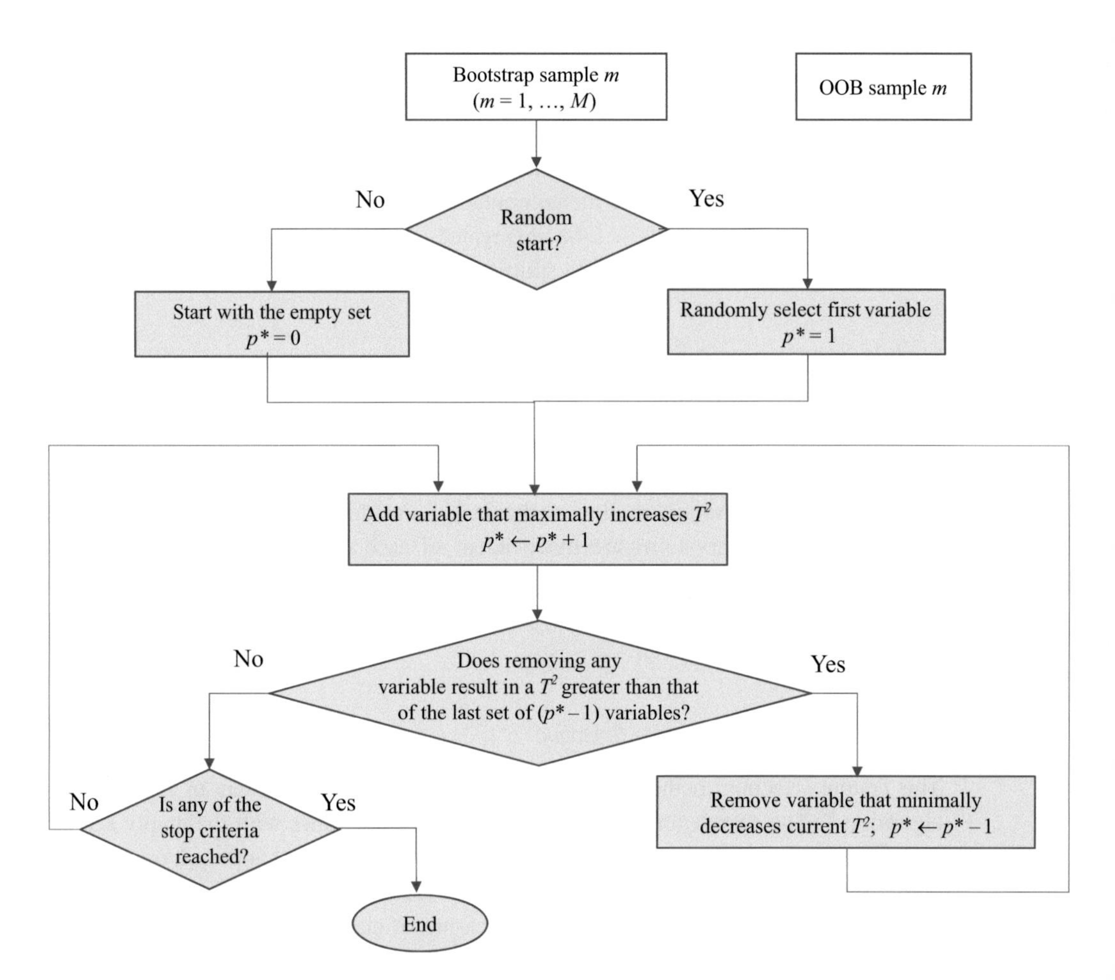

Figure 5.3 A diagram of the stepwise hybrid feature selection with the Lawley–Hotelling trace statistic T^2 as a metric of class separation. Each of the M parallel feature selection experiments is performed for a different bootstrap sample. Since no predictive models are built during the search, OOB samples are not used. The search starts either with the empty set or with a randomly selected first variable. Each forward selection may be followed by a backward one, if this would result in a T^2 measure higher than the best one for the previous cardinality. The feature selection experiment ends when one of the stopping criteria is satisfied.

biomarkers above a particular size. The second criterion stops the search when a large T^2 indicates a strong possibility of overfitting.

When all of the M parallel searches are completed, each of them provides its optimal biomarker for each of the considered sizes, associated with a measure of class separation and a set of selected variables. Since this search is computationally very efficient, we may choose to perform a large number of parallel feature selection experiment (say, M = 1,000); this would give us a better insight into the distribution of variables selected across the M subsets of each size, and especially about which of them are most frequently selected across those subsets. Similarly, we will have more information about the distribution of class separation measures over the subset sizes. We will aggregate this information in order to identify the size of a biomarker providing an optimal compromise between minimizing the size and maximizing the class separation. When deciding on the variables to be included in the optimal biomarker, we may prioritize those that are frequently selected into the M subsets of the selected size.[19] For example, we may take the most frequently selected variables (with the cutoff allowing us to have at least twice as many as the already decided optimal biomarker size, and preferably more), and perform a single feature selection experiment to identify a subset of them (of the selected biomarker size) that maximizes their T^2 measure of class separation.

5.3.3.2 *Extensions to the Basic T^2-Driven Hybrid Feature Selection*

The above described *basic* stepwise hybrid selection with T^2 does not use any learning algorithm, and hence does not use the OOB samples. The search is driven by the intrinsic characteristics of the considered subsets of variables (that is, their capability to separate the differentiated classes, as measured by the T^2 metric). Since this search is computationally very efficient, performing a large number of parallel feature selection experiments allows for a substantial decrease in the variance of the results, and for a more confident selection of the size and composition of the optimal biomarker. Nevertheless, after each of the M parallel searches is completed, we may use its results to build predictive models – one model per each of the final subsets identified during the search for each of the considered cardinalities – and evaluate the models using the OOB sample. Complementing the intrinsic measure of class separation with the estimate of predictive model performance, would result in a more informed identification of the optimal biomarker. Note that even with this extension, the entire process is still computationally efficient – each search is performed exactly as before (without using any learning algorithm), and only after the search is completed, are the final subsets of variables (one per each of the considered sizes) used to build the predictive models. Although any supervised learning algorithm may be used to build and evaluate those models, Fisher's discriminant analysis is directly aligned with the T^2 trace statistic used here as a measure of class separation (see Chapter 12).

This computationally efficient stepwise hybrid search may also be used in a situation when an optimal multivariate biomarker has already been identified, and we now want to estimate its performance via a resampling approach. Resampling the training data and building predictive models implementing the optimal biomarker would constitute a useless *internal* validation, since the training data had already been used for the biomarker identification. However, we may estimate the performance of our biomarker by averaging estimates of many biomarkers of the same size. This means that additional feature selection has to be included within the resampling schema – we will resample the training data, but we will *not* be building models based on the already identified optimal biomarker. Instead, for each resample, a feature selection is performed, and the T^2-driven search stops when the size of our optimal biomarker is reached (recall that the maximum subset size is a user-defined parameter of the T^2 search). If we perform, say, 1,000 such

[19] Although, similar to aggregating RFE results, it would be better to also consider information about groups of correlated variables, rather than considering only the frequency of individual variables (see Chapter 15).

searches, we will have 1,000 biomarkers of the same size as that of our optimal biomarker (though, with different sets of variables) and 1,000 predictive models implementing those biomarkers. Averaging the performance of those 1,000 biomarkers may provide a quite reliable estimate for the performance of our optimal biomarker. When driven by the same objective (such as differentiating the same set of disease states), many parsimonious biomarkers may tap into the same set of biological processes (for example, their variables may belong to the same groups of variables with similar gene or protein expression patterns), and thus averaging their performance over different subsamples of the training data may provide a good estimate of the performance of our optimal biomarker.

5.3.3.3 *Calculating T^2*

The Lawley–Hotelling trace statistic T^2 (Lawley 1938; Hotelling 1951) provides a measure of class separation represented by the ratio of the variation between classes to the variation within classes. Assuming that our classification problem involves J classes (say, disease states labeled $\mathcal{D}_1, \mathcal{D}_2, \ldots, \mathcal{D}_J$), that the current resample (our current second-order training data) includes n_j observations in class j, $j = 1, \ldots, J$, $\sum n_j = N$, that we are currently considering a subset of p^* variables, and that each observation is represented by a $p^* \times 1$ vector $\mathbf{x}_i = [x_{1i}, \ldots, x_{p^*i}]^T$ and a class label y_i, $i = 1, \ldots, N$, the T^2 measure for this subset is calculated as

$$T^2 = tr\left(\mathbf{W}^{-1}\mathbf{B}\right), \tag{5.1}$$

where:

- **B** is a $p^* \times p^*$ matrix representing the *variation between classes* (sum of squares based on the distances between each of the class centers and the overall data center):

$$\mathbf{B} = \sum_{j=1}^{J} n_j \left(\bar{\mathbf{x}}_j - \bar{\bar{\mathbf{x}}}\right)\left(\bar{\mathbf{x}}_j - \bar{\bar{\mathbf{x}}}\right)^T \tag{5.2}$$

- **W** is a $p^* \times p^*$ matrix representing the *variation within classes* (sum of squares based on the distances between each observation and its class center):

$$\mathbf{W} = \sum_{j=1}^{J} \sum_{i:y_i=\mathcal{D}_j} \left(\mathbf{x}_i - \bar{\mathbf{x}}_j\right)\left(\mathbf{x}_i - \bar{\mathbf{x}}_j\right)^T \tag{5.3}$$

- trace *tr*() of a matrix is the sum of its diagonal elements
- $\bar{\mathbf{x}}_j$ is the mean vector for training observations from class j
- $\bar{\bar{\mathbf{x}}}$ is the mean vector for all training observations

5.4 Some Other Feature Selection Algorithms

From among other feature selection algorithms that could eventually be used within multiple feature selection experiments, we will briefly describe the following three evolutionary methods: Simulated Annealing, Genetic Algorithms, and Particle Swarm Optimization. For their efficient use in parallel searches for parsimonious multivariate biomarkers (such searches as recursive feature elimination or hybrid stepwise selection described in Section 5.3), their implementations would have to allow for the algorithms to be driven by the subset size. Currently, such approaches are not routinely supported, and using those evolutionary algorithms within such multiple feature selection schemas may require some further work.

5.4.1 Feature Selection with Simulated Annealing

Simulated annealing (Kirkpatrick et al. 1983) has been designed to be an efficient heuristic optimization technique for finding an optimal value of a given objective (or cost) function of very many variables. It is based on the application of statistical mechanics to the annealing of solids, the process in which metal or glass is first heated to its liquid state (in which particles are distributed randomly), and is then cooled in slow stages (allowing the material to reach thermal equilibrium at each stage) to solidify into a highly-ordered crystalline form (a minimum energy state).

When simulated annealing is applied for feature selection, we start with a random subset of independent variables, and then iteratively try to improve this initial solution. The objective function E – which corresponds to the energy level in annealing – is a performance metric evaluated for each of the considered subsets of variables. It can be a result of testing a predictive model built on a subset of variables, or a value of some intrinsic metric of the set's predictive ability. The goal of the algorithm is to minimize the objective function.[20] In addition to the objective function E, simulated annealing is driven by:

- a control parameter T (representing temperature)
- an *annealing schedule*, which defines how and when an initially high value of T is gradually decreased[21]
- a *neighbor function*, which generates random changes to a current solution

At each iteration, the current solution is randomly perturbed. For example, if the solution is represented by a vector of p binary elements (bits), whose values (0 or 1) indicate which of the p independent variables are currently selected, then perturbing the current solution would correspond to flipping some number of randomly selected bits. This is done by the neighbor function, and the magnitude of the perturbance depends on the current temperature T. At high values of T, a new solution can be taken from a large neighborhood of the current one, that is, the values of many bits may be reversed. When the temperature decreases, fewer and fewer bits may be reversed.

[20] Any maximization problem can be presented as a minimization one, so this will work for any performance metric.

[21] For example, the schedule may define a sequence of T values as well as the number of iterations performed at each temperature.

After the perturbed solution (a new subset of variables) is generated, it is evaluated. For example, a learning algorithm (such as random forests) may be used to build and evaluate a predictive model based on the new subset of variables. If necessary, its performance (measured by a selected metric of the model's performance) may be translated to a value of the objective function E, and if the energy decreases (that is, $\Delta E < 0$), the new, perturbed solution becomes the current one.[22] This corresponds to a downhill move. However, if the performance of the new subset of variables is worse than that of the current one (and thus would represent an uphill move with $\Delta E > 0$), the perturbed solution is not automatically discarded. The probability of accepting the uphill move is calculated as:

$$P(accept) = e^{\frac{-\Delta E}{T}}. \tag{5.4}$$

This probability is compared to $R(0,1)$, a random value generated from the uniform distribution in the interval [0,1]. If $P(accept) > R(0,1)$, the worse solution (the uphill move) is accepted and becomes the current solution. This is the feature of simulated annealing that distinguishes it from greedy, always downhill, algorithms, and allows the simulated annealing algorithm to move away from inefficient local minima.

Depending on its implementation, the algorithm stops either when some predetermined number of iterations is performed, or when the decreasing temperature T reaches its final value (which would be defined in the annealing schedule). Since both the neighborhood (from which a perturbed solution may be selected) and the probability of accepting an uphill move depend on temperature, simulated annealing starts with aggressive random searches allowing the identification of some general patterns (a large neighborhood and acceptance of most uphill moves), and then, when the temperature decreases, focuses on local refinements (a small neighborhood and a very limited acceptance of uphill moves).

Some implementations of simulated annealing for feature selection perform multiple experiments based on resamples of the training data (Kuhn and Johnson 2020). However, such experiments are designed in a different way than those described in Section 5.3. The external resampling is used there to decide on an optimal number of iterations (rather than on an optimal size and composition of a multivariate biomarker). For each of those external resamples, another level of resampling is used to perform a number of simulated annealing experiments, in which second-order (or internal[23]) subsamples are used for the search, and their OOB samples[24] for assessing the values of the energy function. After all of the internal and external resampling experiments are

[22] Recall that, for simplicity of description, we discuss minimization of energy function E. Thus, if our performance metric (such as the mean square error) is to be minimized, it will be the same as the energy function. However, if we use a metric that should be maximized, it needs to be translated into the energy function (for example, by subtracting its value from 1).

[23] Note that "internal" refers here to the placement of the second-order resampling (within the schema of feature selection as described here), and does not refer to the improper internal validation discussed in Chapter 4.

[24] OOB samples here mean either the out-of-bag samples in bootstrapping or the test fold samples in cross-validation.

completed, performance estimates based on the external OOB samples are aggregated for each iteration, in order to determine an optimal number of iterations. Once this is done, the final simulated annealing search is performed – for the optimal number of iterations – on the entire training data, to identify the optimal subset of variables (Kuhn and Johnson 2020). Since such designs are not driven by the subset size (but by the number of iterations), their direct application to high-dimensional data is unlikely to result in a parsimonious multivariate biomarker.[25]

5.4.2 Feature Selection with Genetic Algorithms

Genetic algorithms (Holland 1992; Mitchell 1996) implement stochastic searches inspired by the basic elements of natural evolution. When applied to supervised feature selection, a genetic algorithm starts with some number L of randomly selected subsets of independent variables. Those subsets constitute the initial population. Usually, each subset is represented by a $p \times 1$ binary vector, where p is the number of independent variables. Each bit of the vector corresponds to one variable, and if the value of the bit is 1, the variable is included in the subset.[26] The number of vectors, L, corresponds to the populations size, and each vector corresponds to a chromosome that is treated here as an individual member of the population.[27] Each subset of variables (chromosome) represents a potential solution to a feature selection problem. A general idea of a genetic algorithm is a successive replacement of a current population of chromosomes with a next-generation one (of the same size) in a way that – via maximizing the "fitness" – should lead to the identification of an optimal subset of variables. This evolution-mimicking process involves the following elements: a fitness function, selection of chromosomes for reproduction, and two genetic operators – crossover and mutation. For reference, let's summarize the relation between the feature selection terms and the evolutionary terms:

- Current population (of chromosomes) – all currently considered subsets of variables
- Population size – the number of the subsets
- A chromosome (or an individual) – a subset of variables (represented by a binary vector)

A fitness function is used to evaluate the fitness of each chromosome in the current population (hence, it plays the role of an environment). In feature selection for predictive modeling, a learning algorithm may be used to build a predictive model, for each of the current subsets of variables, and evaluate the performance of the model. The value of a selected performance metric (such as accuracy, AUC, or *RMSE*) may then be

[25] Although we may try to reduce the subset size by forcing the initial solution to include only a small number of variables, it usually would not result in a sparse final solution.

[26] To further the analogy with genetics (though a naive view of it), each independent variable corresponds to a gene, and the location of the bit representing the variable in the vector corresponds to a locus of a chromosome.

[27] Thus, there is an implicit assumption of one-chromosome individuals.

translated into a value of the fitness function, if necessary. Alternatively, performance of a subset may be estimated by some intrinsic characteristics of the subset, without building a predictive model. *Selection* is a process by which some of the chromosomes are selected for reproduction. *Crossover* is an operation of exchanging corresponding segments of two current-generation chromosomes before and after a randomly selected locus;[28] this creates two offspring chromosomes. *Mutation* modifies an offspring chromosome by flipping some of its bits.

Starting with an initial population of *L* randomly generated chromosomes, the genetic algorithm first estimates the fitness of each chromosome. Then a pair of chromosomes is selected for reproduction – the two chromosomes with the currently highest values of the probability of selection, which is calculated as a function of fitness.[29] These two chromosomes are subject to crossover (which is performed with some *crossover probability* and thus, may not take place for some pairs of chromosomes), which results in two offspring chromosomes. Those chromosomes are then mutated, that is, each of their bits may be flipped with the *probability of mutation*. This results in two next-generation chromosomes.

Then, a next pair of chromosomes is selected (since selection is with replacement, a chromosome may be selected more than once), and the process repeats and continues until the next generation includes *L* chromosomes. At this point, the next-generation population becomes the current generation and everything starts all over again. The processing ends either after some predetermined number of generations, or when some other stopping criterion becomes true (for example, when the average fitness exceeds some predefined threshold, or when the best fitness does not improve for some predefined number of consecutive generations).

The parameters of the algorithm (the size of the population, the probability of crossover, the probability of mutation, and the number of generations) may have a significant impact on the quality of the results (Mitchell 1996). To achieve good results, it is important for the search process to strike a right balance between exploration and exploitation. *Exploration* is driven by the randomness introduced via the genetic operators, and – by exploring new regions of the search space – promotes population diversity. *Exploitation*, on the other hand, is driven by the selection mechanism, and focuses on exploiting the neighborhood of the currently best solutions. Some versions of genetic algorithms include *elitism*, a policy of preserving the best chromosomes from the current population. Elitism may be implemented, for example, by selecting some number (or percent) of the best chromosomes from the current generation and replacing with them the worst performing chromosomes of the next generation. This shifts the balance toward exploitation, and quite often improves the algorithm's performance.

[28] Alternatively, two-point, or even multi-point, crossover may be implemented, in which each of the two children will inherit alternate segments whose boundaries are defined by randomly selected crossover loci.

[29] The calculations for this fitness-proportional selection may implement the roulette approach, in which each chromosome is assigned a sector of a *roulette wheel* with the size proportional to its fitness. Each spin of the wheel will select a chromosome for reproduction (Mitchell 1996). Other mechanisms of selection include rank-proportional selection (where chromosomes are sorted by fitness and then selected by rank) and uniform random selection (De Jong 2005).

As it is for simulated annealing, some implementations of genetic algorithms also perform multiple feature selection experiments that are based on different random resamples of the training data. However, analogically to simulated annealing, the goal of such experiments is to determine an optimal number of generations, which is then used as a parameter for the final run of the genetic algorithm on the entire training data (Kuhn and Johnson 2020). However, versions of genetic algorithms that allow for selection of a subset of a fixed, predetermined size have been reported (for example, Wolters 2015; Orestes Cerdeira et al. 2023), and thus it may be possible that a genetic algorithm will eventually be implemented within one of the multiple feature selection experiment schemas described in Section 5.3.

5.4.3 Feature Selection with Particle Swarm Optimization

Swarm intelligence algorithms are optimization methods inspired by the collective behavior of complex self-organizing biological *swarm* systems, such as swarms of flying insects, flocks of birds, schools of fish, or colonies of ants. A swarm is a large population of individuals called *particles*. Although the behavior of an individual particle may seem insignificant, their interactions and adaptations result in *collective swarm intelligence*, which manifests itself in group decisions that are better than individual ones.

Particle swarm optimization (Kennedy and Eberhart 1995) is one of the most popular swarm intelligence algorithms.[30] It is a stochastic optimization technique, in which a population (or a swarm) of particles is moving in a p-dimensional space in search for an optimal solution. Each of the particles is defined by three p-dimensional vector properties – position, velocity, and *personal best* (which is the best position of a particle in its lifetime). Defined is also the *global best* for the entire swarm. Each position of a particle corresponds to a candidate solution to the optimization problem, and the quality of each solution is determined by a fitness function. At each iteration of the algorithm, the position and velocity of each particle are updated based on the current state of the particle as well as on the best historical position of the particle and that of its neighbors. First, the velocity of particle i at iteration $t+1$ is updated as follows (Kennedy et al. 2001; Engelbrecht 2007):

$$\begin{aligned} \mathbf{v}_i(t+1) &= w\mathbf{v}_i(t) \\ &\quad + a_1\mathbf{r}_1[\mathbf{pbest}_i(t) - \mathbf{x}_i(t)] \\ &\quad + a_2\mathbf{r}_2[\mathbf{gbest}(t) - \mathbf{x}_i(t)], \end{aligned} \tag{5.5}$$

where:

- $\mathbf{x}_i(t)$ and $\mathbf{v}_i(t)$ are the current position and velocity of the particle
- w is the inertia weight parameter, which controls the balance between exploration and exploitation by weighing the influence of the previous-iteration velocity on the new one
- a_1 and a_2 are acceleration coefficients (also called trust parameters, or cognitive and social weights), which – in conjunction with $\mathbf{r}_1$ and $\mathbf{r}_2$ – control the balance

[30] Other popular algorithms include ant colony optimization, artificial bee colony, and artificial fish swarm.

between the individual and global influences on the particle's velocity (the tradeoff between local and global learning)

- $\mathbf{r}_1$ and $\mathbf{r}_2$ are two random vectors sampled (independently for each particle) from a uniform distribution on [0, 1][31]
- $\mathbf{pbest}_i(t)$ is the vector representing the best position of the particle (in its history of t iterations)
- $\mathbf{gbest}(t)$ is the global best position for the swarm (across t iterations).[32]

Then, the position of the particle is updated as follows:

$$\mathbf{x}_i(t+1) = \mathbf{x}_i(t) + \mathbf{v}_i(t+1). \tag{5.6}$$

The three velocity components of Formula (5.5) may be interpreted as: (i) the *momentum* component of the particle's movement (inertia bias to keep the current direction); (ii) the *cognitive* component pulling the particle toward the best position saved in its individual memory; and (iii) the *social* component pulling the particle toward the globally best position saved in the swarm's memory (Engelbrecht 2007).

Observe that Formulas (5.5) and (5.6) correspond to situations where particles move in a continuous search space, in which the velocity and position are real-valued vectors. However, when particle swarm optimization is used for feature selection, each dimension k, $k = 1, \ldots, p$, of a p-dimensional search space represents the presence or absence of the kth independent variable, and thus a particle can assume only binary position values (0 or 1) along any dimension. This means that the position of a particle is represented by a $p \times 1$ binary vector (as it is for feature selection with simulated annealing or genetic algorithms). Hence, a continuous velocity described by Formula (5.5) has to be discretized.[33] For example, for each dimension k we may start with calculating $v_{ik}(t+1)$, which now can be interpreted as a particle's predisposition to choose 0 or 1 at iteration $t+1$ (Kennedy et al. 2001; Engelbrecht 2007),

$$\begin{aligned} v_{ik}(t+1) &= w v_{ik}(t) \\ &\quad + a_1 r_{1k}[pbest_{ik}(t) - x_{ik}(t)] \\ &\quad + a_2 r_{2k}[gbest_k(t) - x_{ik}(t)], \end{aligned} \tag{5.7}$$

where:

- $x_{ik}(t)$ is the state of bit k in the $p \times 1$ binary position vector (a subset of variables represented by this particle) of particle i at iteration t

[31] In the original version of the algorithm (Kennedy and Eberhart 1995) there were no acceleration constants, and r_1 and r_2 were random numbers generated from a uniform distribution with an upper limit set often in a way that the two limits summed up to 4.0.

[32] In this version of the algorithm (which is called *global best* particle swarm optimization), the entire swarm is the neighborhood for each of its particles. In the *local best* version, smaller neighborhoods are defined for each of the particles (and the global best position $\mathbf{gbest}(t)$ would be replaced in Formula (5.5) with the local best position $\mathbf{lbest}_i(t)$ defined independently for the neighborhood of each particle).

[33] Such discrete movements can be seen as a particle moving between corners of a hypercube by flipping some bits of its binary position vector. This also means that the distance between any two valid positions of a particle is the Hamming distance, that is, the number of bit positions in which two binary vectors differ.

- $v_{ik}(t)$ is the current particle's predisposition
- $pbest_{ik}(t)$ is the state of bit k in vector $\mathbf{pbest}_i(t)$ representing the best so far position of particle i
- $gbest_k(t)$ is the state of bit k in vector $\mathbf{gbest}(t)$ of the global best position for the swarm

Then the predisposition (the velocity of bit k) is translated into a probability threshold (that is, into the range [0, 1]) via the sigmoid function,

$$S(v_{ik}(t+1)) = \frac{1}{1+\exp(-v_{ik}(t+1))}, \tag{5.8}$$

and the position of particle i along dimension k is updated to

$$x_{ik}(t+1) = \begin{cases} 1 & \text{if } S(v_{ik}(t+1)) > \rho_{ik} \\ 0 & \text{otherwise,} \end{cases} \tag{5.9}$$

where ρ_{ik} is a random number from uniform distribution on [0, 1].[34]

When the binary version of particle swarm optimization is used for feature selection for predictive modeling, it starts with random initialization of the p-dimensional binary vectors, which determine initial positions of the swarm's particles (random subsets of variables).[35] The fitness of each position is determined by building and evaluating a predictive model using the corresponding subset of variables (utilizing, for example, support vector machines or random forests learning algorithm). If necessary, the selected metric of model performance (such as *RMSE* or accuracy) is translated into an appropriate value of the fitness function, which typically is minimized. The initial position of a particle is recorded as its currently best position, and the best of all of them as the global best. Then, at each iteration, each particle's velocity and position are updated according to Formulas (5.7)–(5.9), and evaluated, which may lead to appropriate updates to the best positions of some particles as well as to the global best. The second and third components of Formula (5.7) imply particles' proclivity for stochastic re-visiting of promising areas of the search space. The process ends either after a predetermined number of iterations or when one of other stopping criteria (if defined) is satisfied; one such criterion may represent finding a sufficiently good solution, another – the swarm's convergence, which may be manifested by the lack of significant changes in the swarm's spatial distribution.

As with simulated annealing and genetic algorithms, when particle swarm optimization is applied to high-dimensional data, solutions are typically far from parsimonious. This holds true even for versions of the algorithm that implement multi-objective optimization, and, for example, penalize solutions with a larger number of variables. Although this may limit the number of variables in an optimal subset, the solutions are usually still far from parsimonious.

[34] Note that the only difference in calculations for the continuous and the binary search spaces is the way in which new positions are determined. Applying Formula (5.7) to each of the p dimensions results is the same real-valued velocity that would be calculated from Formula (5.5).

[35] The velocities are usually initialized to zero.

Part II

Regression Methods for Estimation

6 Basic Regression Methods

In Part II of this book we will describe supervised learning algorithms that can be used to build models for predicting the value of a quantitative (and continuous) response variable. The goal of such regression models is to estimate this value from the values of the independent variables. Before we deal with state-of-the-art algorithms (such as those implementing regularization), we will start with basic ones – *Multiple Regression* and *Partial Least Squares Regression*. Even if it is unlikely for multiple regression to be used as the primary method for multivariate biomarker discovery based on high-dimensional data, discussing it here will provide the necessary background for regression analysis, and will highlight the weaknesses of multiple regression, which will be addressed by the subsequently presented methods.

6.1 Multiple Regression

Multiple Regression is a classical regression method taught in elementary statistics classes, and will be discussed here insofar as is necessary to fully understand and appreciate the other regression methods described later on.

In multiple regression problems, we have one response variable y (also called the dependent variable) and p independent variables $x_1, x_2, \ldots, x_p$ (sometimes called predictors).[1] The dependent variable y has to be continuous (or, at least, is assumed to be continuous). The goal of multiple regression is to predict the value of the dependent variable from the values of the independent variables. We assume that the function we will use to predict y is linear and thus, is represented by a p-dimensional hyperplane in a $(p + 1)$-dimensional space (defined by one dependent and p independent variables).[2]

[1] Recall that the term "independent variables" should be understood in the context of independent variables vs. the dependent variable. It does not provide *any* information about relations and correlations among the independent variables. In fact, in biomedical data, we may usually expect a complex correlation structure among the independent variables.

[2] The linearity of the function (and the resulting predictive model) is the linearity in the parameters of this function (that is, $\beta_0, \beta_1, \beta_2, \ldots, \beta_p$). This means that if we include additional variables that are transformations of the original independent variables (for example, x_1^2, or x_2^3, or $x_1 x_2$), then we will still have a linear regression model, even if it would model a nonlinear relationship between the dependent variable and the *original* independent variables.

We will start with the full *multiple regression model* of the relationship between the dependent variable and the p independent variables in the population:

$$y = \beta_0 + \beta_1 x_1 + \beta_2 x_2 + \cdots + \beta_p x_p + \varepsilon, \tag{6.1}$$

where β_0 is the intercept, each β_k, $k = 1, \ldots, p$, is the slope parameter associated with independent variable x_k, and ε is the error term – an unobservable random variable representing the variability in the dependent variable y that cannot be explained by the relationship. It is assumed that the error term is normally distributed with a mean of zero and constant variance δ^2 and that its values are independent.[3]

Since the goal of our predictive modeling is to find an optimal hyperplane that maximizes the amount of variation in the dependent variable that is explained by a set (or a subset) of independent variables, we will neglect the error term in the model (we cannot predict the error anyway). Thus, the hyperplane described by the *multiple regression equation* (6.2)[4] will represent the expected value of y:

$$\begin{aligned} E(y) &= \beta_0 + \beta_1 x_1 + \beta_2 x_2 + \cdots + \beta_p x_p \\ &= \beta_0 + \sum_{k=1}^{p} \beta_k x_k \\ &= \beta_0 + \mathbf{x}^T \boldsymbol{\beta}. \end{aligned} \tag{6.2}$$

Observe that if we have only one independent variable x (*Simple Linear Regression*), then function (6.2) simplifies to a straight line $E(y) = \beta_0 + \beta_1 x$. For two independent variables, the function would be represented by a plane.

To estimate hyperplane (6.2), which describes the relationship in the population, we will use training data (a random sample from the population) to estimate $p + 1$ population parameters – the population intercept β_0 and the p population slopes β_k, $k = 1, \ldots, p$. For each training data point i, $i = 1, \ldots, N$ (represented by a $p \times 1$ vector $\mathbf{x}_i = \left[x_{1i}, \ldots, x_{pi}\right]^T$) and any hypothetical hyperplane $\beta_0 + \mathbf{x}^T \boldsymbol{\beta}$, we can define the error of prediction as the difference between the observed value of the response variable for training point i and the value predicted for this point by the hyperplane,

$$\text{error}_i = y_i - \left(\beta_0 + \mathbf{x}_i^T \boldsymbol{\beta}\right). \tag{6.3}$$

A hyperplane for which the sum of squared errors (over all training data points), SSE,

$$\begin{aligned} SSE &= \sum_{i=1}^{N} \left[y_i - \left(\beta_0 + \sum_{k=1}^{p} \beta_k x_{ki} \right) \right]^2 \\ &= \sum_{i=1}^{N} \left[y_i - \left(\beta_0 + \mathbf{x}_i^T \boldsymbol{\beta} \right) \right]^2 \end{aligned} \tag{6.4}$$

[3] These assumptions are necessary for the validity of the tests for the significance of the relationship and estimates of confidence intervals for the prediction results.

[4] The superscript T denotes vector transposition; that is, transposing a column vector $\mathbf{x}$ into a row vector $\mathbf{x}^T$.

is minimized, is considered to be the best fit to the training data (that is, to a cloud of N points in the $(p + 1)$-dimensional space of the regression problem), as well as an optimal approximation of the relationship between the dependent variable and all independent variables.[5] This approach, which minimizes the sum of squared errors, is called the *Least Squares Criterion*. This is the most common method for estimation of the regression parameters; it is often referred to as the *Ordinary Least Squares* (OLS) regression.[6]

To simplify the analysis implementing the least squares criterion, we may either assume that all variables are mean-centered (in which case the intercept would be zero), or we may treat the intercept the same way as the slope coefficients by adding a dummy independent variable x_0 whose values are set to 1 for all of the N observations. The former approach will be presented in Section 6.2.1; here, we describe the latter. Adding the dummy variable to a $p \times N$ matrix of the training data will define a $(p+1) \times N$ matrix $\mathbf{X}$; we have a $N \times 1$ vector $\mathbf{y}$ of the response values, and we will now seek a $(p+1) \times 1$ parameter vector $\boldsymbol{\beta}$ (of all regression coefficients) that minimizes the sum of squared errors:[7]

$$\mathbf{X} = \begin{bmatrix} 1 & 1 & \cdots & 1 \\ x_{11} & x_{12} & \cdots & x_{1N} \\ x_{21} & x_{22} & \cdots & x_{2N} \\ \vdots & \vdots & \ddots & \vdots \\ x_{p1} & x_{p2} & \cdots & x_{pN} \end{bmatrix}, \quad \mathbf{y} = \begin{bmatrix} y_1 \\ y_2 \\ \vdots \\ y_N \end{bmatrix}, \quad \boldsymbol{\beta} = \begin{bmatrix} \beta_0 \\ \beta_1 \\ \beta_2 \\ \vdots \\ \beta_p \end{bmatrix}. \tag{6.5}$$

The sum of squared errors can be presented now as $SSE(\boldsymbol{\beta}) = (\mathbf{y} - \mathbf{X}^T\boldsymbol{\beta})^T(\mathbf{y} - \mathbf{X}^T\boldsymbol{\beta})$, and the least square estimate of the β parameters can be obtained by solving the following optimization problem:

$$\underset{\boldsymbol{\beta}}{\text{minimize}} \left(\mathbf{y} - \mathbf{X}^T\boldsymbol{\beta}\right)^T\left(\mathbf{y} - \mathbf{X}^T\boldsymbol{\beta}\right). \tag{6.6}$$

This is a quadratic problem that can be solved by setting to zero the partial derivative of SSE with respect to $\boldsymbol{\beta}$,

$$\frac{\delta SSE}{\delta \boldsymbol{\beta}} = -2\mathbf{X}\left(\mathbf{y} - \mathbf{X}^T\boldsymbol{\beta}\right) = 0, \tag{6.7}$$

[5] Minimizing *SSE* means minimizing the amount of variation in the dependent variable that is not accounted for by the regression model.

[6] By minimizing *SSE*, the OLS solution provides the regression parameter estimates that minimize bias (when considered in the context of the bias-variance tradeoff).

[7] Statistical texts usually present matrix $\mathbf{X}$ in the form of $N \times (p+1)$, with observations as rows and variables as columns. However, for high-dimensional biomedical data it is common, as well as more convenient, to have variables represented by rows and biological samples by columns. For example, typical gene expression data have more than 20,000 variables; representing them by columns would prevent saving, viewing, or manipulating such data in Excel, which does not support that many columns.

and the unique solution can be given by

$$\hat{\boldsymbol{\beta}}_{OLS} = \begin{bmatrix} \hat{\beta}_0 \\ \hat{\beta}_1 \\ \hat{\beta}_2 \\ \vdots \\ \hat{\beta}_p \end{bmatrix} = \left(\mathbf{X}\mathbf{X}^T\right)^{-1}\mathbf{X}\mathbf{y}. \tag{6.8}$$

If matrix $\mathbf{X}\mathbf{X}^T$ is invertible (and the error term ε is normally distributed with a mean of zero and constant variance), then $\hat{\boldsymbol{\beta}}_{OLS}$ is a unique and the best linear unbiased[8] estimator of regression parameters, which means that it has the smallest variance among all linear and *unbiased* estimators of $\boldsymbol{\beta}$.

The optimal (sample-based) hyperplane can be described by the following *estimated multiple regression equation*:

$$\begin{aligned} \hat{y} &= \hat{\beta}_0 + \hat{\beta}_1 x_1 + \hat{\beta}_2 x_2 + \cdots + \hat{\beta}_p x_p \\ &= \hat{\beta}_0 + \sum_{k=1}^{p} \hat{\beta}_k x_k \\ &= \hat{\beta}_0 + \mathbf{x}^T \hat{\boldsymbol{\beta}}, \end{aligned} \tag{6.9}$$

where $\hat{y}$ represents the predicted value of y, sample statistics $\hat{\beta}_0, \hat{\beta}_1, \hat{\beta}_2, \ldots, \hat{\beta}_p$ are the point estimators of the population parameters $\beta_0, \beta_1, \beta_2, \ldots, \beta_p$, and $\hat{\boldsymbol{\beta}}$ is a $p \times 1$ vector that is the part of $\hat{\boldsymbol{\beta}}_{OLS}$ representing the slope estimates. Using Formula (6.9), we may now calculate the prediction error for each training data point $\mathbf{x}_i = \left[x_{1i}, \ldots, x_{pi}\right]^T$ as $y_i - \hat{y}_i$ (see Figure 6.1), and the sum of squared errors as $SSE = \sum_{i=1}^{N} (y_i - \hat{y}_i)^2$. To predict the value of the dependent variable y for a new sample $\mathbf{x}_{new}$, we will calculate $\hat{y}(\mathbf{x}_{new}) = \hat{\beta}_0 + \mathbf{x}_{new}^T \hat{\boldsymbol{\beta}}$.

The optimal hyperplane (6.9) is also the one that maximizes the amount of variation in the dependent variable that can be explained by the independent variables. To evaluate how well this hyperplane fits the training data, we can calculate the percentage of the total variation in the dependent variable that can be explained by this hyperplane (or, more precisely, by the relationship represented by the hyperplane):

$$R^2 = \frac{SSR}{SST}, \tag{6.10}$$

where:

- R^2 is the coefficient of determination
- $SSR = \sum_{i=1}^{N} (\hat{y}_i - \bar{y})^2$ is the variation (or sum of squares) due to regression

[8] Unbiasedness of this estimator means that its expected value, $E(\hat{\boldsymbol{\beta}}_{OLS})$, is the same as the true value of $\boldsymbol{\beta}$; $E(\hat{\boldsymbol{\beta}}_{OLS}) = E\left(\left(\mathbf{X}\mathbf{X}^T\right)^{-1}\mathbf{X}\mathbf{y}\right) = \left(\mathbf{X}\mathbf{X}^T\right)^{-1}\mathbf{X}\mathbf{X}^T\boldsymbol{\beta} = \boldsymbol{\beta}$.

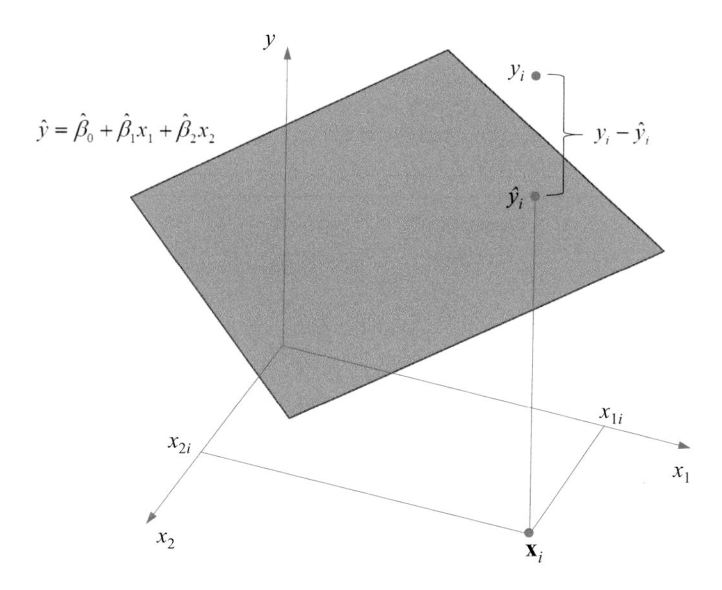

Figure 6.1 Optimal hyperplane for multiple regression example with two independent variables; in this case the relationship is represented by a plane $\hat{y} = \hat{\beta}_0 + \hat{\beta}_1 x_1 + \hat{\beta}_2 x_2$ (two-dimensional hyperplane) in a three-dimensional space of the problem. $\hat{y}_i$ is the point on the hyperplane that represents the predicted value of the response variable for training observation $\mathbf{x}_i = [x_{1i}, x_{2i}]^T$, y_i is the known true value of y for this observation, and $y_i - \hat{y}_i$ is the error made by this prediction.

- $SST = \sum_{i=1}^{N} (y_i - \bar{y})^2$ is the total variation in the dependent variable[9] (or the total sum of squares)
- N is the number of training data points
- y_i is the observed value of the response variable for training point i, $i = 1, \ldots, N$
- $\hat{y}_i$ is the value predicted by the estimated multiple regression equation – the hyperplane (6.9) – for data point i, $i = 1, \ldots, N$
- $\bar{y}$ is the mean of all y_i values

Observe that, since $SST = SSR + SSE$, minimizing SSE (that is, applying the *least squares* criterion) is equivalent to maximizing SSR. By the way, the results of partitioning the total variation in the dependent variable (SST) into the part of this variation that is explained by the independent variables (SSR), and the remaining, unexplained, part of this variation (SSE) are usually summarized in the form of an analysis of variance (ANOVA) table; see Table 6.1.

For regression problems with many independent variables, the coefficient of determination R^2 tends to overestimate the amount of the explained variation. For example, R^2 increases with every independent variable added to the model, whether this

[9] Please note that variance can be interpreted as the variation divided by its degrees of freedom; thus, the total sample variance of the dependent variable, $s_y^2 = \frac{1}{N-1} \sum_{i=1}^{N} (y_i - \bar{y})^2$, can be seen as $s_y^2 = \frac{SST}{df_T}$, where the total degrees of freedom $df_T = N - 1$.

Table 6.1 ANOVA table – partitioning the total variation in the dependent variable. Sum of squares corresponds to variation, mean square – to variance. Variance is calculated as a variation divided by its corresponding degrees of freedom. The ratio of mean square regression (*MSR*) to mean square error (*MSE*) is used as the test statistics for testing the overall significance of the regression model (see Section 6.1.1).

Source of variation	Degrees of freedom	Sum of squares	Mean square	F
Regression	p	SSR	$MSR = \frac{SSR}{p}$	$F = \frac{MSR}{MSE}$
Error (residual)	$N - p - 1$	SSE	$MSE = \frac{SSE}{N - p - 1}$	
Total	$N - 1$	SST		

additional variable actually improves the model or not.[10] To avoid this overestimation, the *adjusted* coefficient of determination R_a^2 (with the adjustment based on the number of independent variables and the sample size) may be used:

$$R_a^2 = 1 - \left(1 - R^2\right)\frac{N-1}{N-p-1}. \tag{6.11}$$

In addition to the coefficients of determination (R^2 and R_a^2),[11] other measures of how well the optimal hyperplane (6.9) **fits the training data** can be calculated; software implementations of the OLS regression often report the mean squared error or the root mean squared error. The mean squared error (*MSE*), which estimates the variance δ^2 of the error term ε, is calculated as the sum of squared errors *SSE* (the variation of the error term) divided by its degrees of freedom:

$$MSE = \frac{SSE}{N-p-1}. \tag{6.12}$$

The root mean squared error, $RMSE = \sqrt{MSE}$, estimates the standard deviation of the error term (and thus, is measured in the units of the response variable), and is called the standard error of the estimate. A word of caution: the same term, *MSE*, is often used not only to measure the fit to the training data, but also to evaluate the performance of a predictive regression model on independent test data. However, Equation (6.12) applies only to the training data situation. When we evaluate the predictive model's performance on test data, we are not estimating any population parameters and thus, the number of degrees of freedom associated with the sum of squared errors is different. Evaluating predictive models on test data is covered in Chapter 4.

[10] This overestimate increases when the number of independent variables p approaches the sample size N (that is, when the error degrees of freedom, $N - p - 1$, is small), and is minimized when $N \gg p$ (Hair et al. 2014).

[11] Note that the adjusted coefficient of determination R_a^2 may be used only during modeling (on the training data) as it has no meaning when testing on any other data ($N - p - 1$ would be meaningless).

6.1.1 Some Other Considerations for Multiple Regression

As previously discussed, in predictive modeling for biomarker discovery, we are *not* focusing on statistical significance. Our focus is on evaluating a predictive model's performance on independent test data. A model that is statistically significant may be useless for the purpose of prediction. On the other hand, if the model is not statistically significant, it is unlikely that it would perform better than random chance. Nonetheless, it may be worth mentioning that for multiple regression – especially, when it is done in a statistical context – two kinds of significance tests may be performed. An F-test for the overall significance of the relationship between the dependent variable and all independent variables (taken together), and p t-tests for individual significance of each of the p independent variables. The F-test (see Table 6.1) is a multivariate test for the regression model:

$$\begin{aligned} &H_0 : \beta_1 = \beta_2 = \cdots = \beta_p = 0 \\ &H_a : \text{At least one population slope is not zero.} \end{aligned} \tag{6.13}$$

The model is considered statistically significant if the null hypothesis is rejected, that is, when at least one of the population slopes is significantly different from zero. If that is the case, then the t-tests may be performed to test whether each of the p population slopes is, individually, significantly different from zero. However, while the F-test may be meaningful, the t-tests are univariate tests and may be misinterpreted. For example, it would be a mistake to retain in the model only the variables that are univariately significant – we already know that variables that are univariately insignificant may be very important for prediction when combined with other variables (see Section 2.3). Removing such variables would be equivalent to univariate filtering, which may decrease or even eliminate chances for developing a powerful predictive model. To select an optimal set of variables, proper multivariate feature selection should be performed. It is also worth noting here that, in classical statistical texts, subset selection for multiple regression was usually described as a stepwise feature selection algorithm based on F statistics, in which variables were added or removed from the set based on some p-value thresholds deciding about their significance. Again, feature selection driven by statistical significance needs to be avoided; criteria related to model performance should be used instead (see Chapter 5).

Recall that the assumptions for the regression model were all about its error term ε – that it is a normally distributed random variable with a mean of zero, constant variance, and independent values. Recall also that if these assumptions are met, then the least square estimate $\hat{\boldsymbol{\beta}}_{OLS}$ (6.8) is unique, and is the best linear unbiased estimator of regression parameters. These assumptions are also necessary for the validity of the significance tests and estimates of confidence intervals for prediction results (software implementations of multiple regression provide confidence intervals for the predicted values). The model assumptions are evaluated by observing residuals. Plotting residuals versus predicted values (the results of prediction for the training observations) allows us to evaluate the mean and variance assumptions. If the plot can be perceived as a horizontal band of points approximately centered around zero, then these assumptions

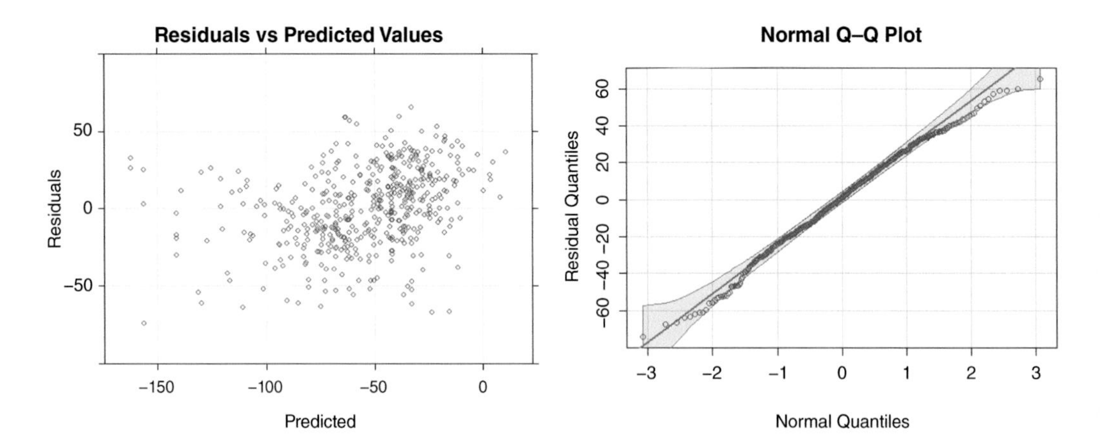

Figure 6.2 Examples of plots for evaluating the regression model assumptions. The left panel shows a plot of residuals versus predicted values. If the cloud of points resembles a horizontal band roughly centered around zero, then the assumptions about zero mean and constant variance of residuals do not seem violated. The right panel shows a quantile–quantile plot for residuals, which also includes a shaded 95% confidence interval area. If all (or almost all) points are within this area, there is no reason to believe that residuals are not distributed normally.

do not seem violated. The normality of residuals may be evaluated via a normal quantile–quantile plot, which plots quantiles of residuals against theoretical quantiles of a normal distribution. Figure 6.2 shows examples of such plots.

6.1.2 Issues with Multiple Regression

One of the issues with the multiple regression method is the *instability* of its results in the case of multicollinearities. Though it is possible (and sometimes quite easy) to remove multicollinearities from data with a small number of variables, it would be very difficult (or practically impossible) to do the same for high-dimensional data. For such data, identifying pairwise correlations between variables would be of little help, since groups of variables can be highly correlated with some other groups of variables, and searching for such multicollinearities would be either impractical or infeasible. Another issue is that multiple regression cannot be directly applied to $p > N$ data, that is, when the number of independent variables p is greater than the number of training observations N. These issues are associated with matrix $\mathbf{XX}^T$. The OLS solution (6.8) requires inversion of this matrix. With singularity, multicollinearity, or when $p > N$, inversion of $\mathbf{XX}^T$ is either impossible, or even if numerically possible, the results would be unstable, and the OLS estimate not unique.

To deal with these issues as well as to properly perform regression-based predictive modeling for high-dimensional data, multivariate feature selection, supervised dimensionality reduction, or regularization (or a combination of these) should be implemented. *Partial Least Squares Regression*, which is a supervised dimensionality reduction method, is one of the approaches that could be considered in such situations.

6.2 Partial Least Squares Regression

The goal of *Partial Least Squares Regression* (PLSR) is to use the partial least squares (PLS) approach to replace a large number of independent variables (possibly with a complicated correlation structure) with a small number of orthogonal (hence uncorrelated) components, and then to perform the least squares regression analysis using those PLS components.[12]

PLSR was introduced by Herman Wold in 1975 (Wold 1975) as an adaptation to the regression problem of his NIPALS[13] algorithm that was originally created for calculating principal components (Wold 1966). Thus, PLSR is somewhat similar to *Principal Component Analysis* (PCA). However, unlike PCA, PLSR considers the response variable; thus, it is a *supervised* method. PLSR identifies the directions (in a cloud of the training data points) that maximize *covariance* between the independent variables and the response variable. In other words, the consecutively identified and orthogonal components (PLS directions) are designed in a way that preserves most of such information in the independent variables that is useful to predict the response variable.

PLS components (directions) are linear functions of the original independent variables. The first of them ($\mathbf{c}_1$) is a vector composed from the univariate slope coefficients calculated by performing simple linear regression individually for each of the p independent variables. This means that independent variables with a stronger correlation to the response variable have more influence on the direction of the first component. After the first PLS component is identified, matrix $\mathbf{X}$ (representing the values of the p independent variables for the N samples) and vector $\mathbf{y}$ (representing the N response values) are deflated (orthogonalized) by removing from them the variability that has already been explained by the first PLS component (this means that $\mathbf{X}$ and $\mathbf{y}$ are replaced by their residuals).

The second PLS component ($\mathbf{c}_2$) is identified in a similar manner to the first one, but only the residual data are used. Hence, the second component is orthogonal to the first one and targets the predictive information that has not been explained by the first component. This process is continued (with each subsequent iteration using residuals from the previous one, and maximizing any remaining covariance) until the desired number of components, m (tuning parameter, $m \leq p$), is identified. The process results in a $p \times m$ matrix $\mathbf{C}$ whose columns represent the m PLS components,

$$\mathbf{C} = \begin{bmatrix} c_{11} & c_{12} & \cdots & c_{1m} \\ c_{21} & c_{22} & \cdots & c_{2m} \\ \vdots & \vdots & \ddots & \vdots \\ c_{p1} & c_{p2} & \cdots & c_{pm} \end{bmatrix}. \tag{6.14}$$

Then, the *Least Squares Criterion* can be used to find an m-dimensional regression hyperplane that best fits the projections of the training data points onto the $(m + 1)$-dimensional

[12] Usually, PLSR algorithms perform these two tasks simultaneously.

[13] NIPALS – Nonlinear Iterative PArtial Least Squares.

space defined by the m PLS directions and the response variable. Although PLS can identify up to p components,[14] we want to use a small number of them; usually, the selection of an optimal m is based either on the proportion of the total variation in the response variable that is explained by a set of components, or through the use of cross-validation to identify the number of components that provides the best predictive ability.[15]

The solutions provided by PLSR can also be described in the context of the bias-variance tradeoff: PLSR can generate a sequence of models with the number of PLS components from $m = 1$ to $m = p$, with the model including only one component being most biased, and the one with p components (same as the OLS solution) being least biased.

Please note that though the dimensionality of the regression problem has been reduced from p to m, each of the m components is a function of all original p independent variables. Thus, we might have addressed the instability issue of the model, but *no feature selection* has been performed, and this means that two problems with the OLS solution have not been solved by PLSR; first, the interpretation of the model still has to be done in the context of all p independent variables, and second, when dealing with high-dimensional data, many of those p independent variables may represent noise (at least from the point of view of the goal of our analysis) and thus, may negatively impact the predictive ability of the PLSR model. This lack of feature selection is perhaps best described by stating that using PLSR may be seen as expressing "*a prior belief that many variables together collectively affect the response with no small subset of them standing out*" (Frank and Friedman 1993).[16]

6.2.1 PLS1 Algorithm

The PLS algorithm is a heuristic approach to multiple regression addressing the instability of the OLS regression caused by multicollinearities. The original version of the algorithm, PLS1, was created for multiple regression with one dependent variable. Later, the method was extended to cover designs with many dependent variables (PLS2). Since in biomarker discovery we are almost exclusively interested in predicting the value of a single response variable, we will focus on the PLS1 regression algorithm.

Assuming that $\mathbf{X}$ is a $p \times N$ matrix representing the values of the p independent variables for the N observations and $\mathbf{y}$ is a vector of N values of the response variable,[17]

[14] If all p components were used, the resulting regression model would be equivalent to the OLS model.

[15] The meta parameter m (the optimal number of PLS components) can be seen as a regularization parameter (Di Ruscio 2000).

[16] The same may be said about Ridge Regression described in the next chapter.

[17] PLS algorithms usually assume (as we do here) that all of the variables have been centered (that is, each of them has a mean of zero). In such a case we have the intercept $\hat{\beta}_0 = 0$. Alternatively, $\hat{\beta}_0$ may be treated in the same way as the slope coefficients, if we add a dummy independent variable x_0 whose values are set to 1 (see Section 6.1).

$$\mathbf{X} = \begin{bmatrix} x_{11} & x_{12} & \cdots & x_{1N} \\ x_{21} & x_{22} & \cdots & x_{2N} \\ \vdots & \vdots & \ddots & \vdots \\ x_{p1} & x_{p2} & \cdots & x_{pN} \end{bmatrix}, \quad \mathbf{y} = \begin{bmatrix} y_1 \\ y_2 \\ \vdots \\ y_N \end{bmatrix}, \tag{6.15}$$

the PLS1 regression algorithm can be described as follows[18] (Helland 1988; Lu et al. 2014):

For each i, $i = 1 \ldots m$, where $m \leq p$:

- Calculate the vector of weights, $\mathbf{w}_i = \dfrac{\mathbf{Xy}}{\|\mathbf{Xy}\|}$
- Calculate the vector of scores, $\mathbf{t}_i = \mathbf{X}^T\mathbf{w}_i$
- Calculate the vector of loadings, $\mathbf{v}_i = \dfrac{\mathbf{Xt}_i}{\mathbf{t}_i^T\mathbf{t}_i}$
- Calculate the regression coefficient, $b_i = \dfrac{\mathbf{t}_i^T\mathbf{y}}{\mathbf{t}_i^T\mathbf{t}_i}$
- Deflate $\mathbf{X}$ and $\mathbf{y}$,[19] $\mathbf{X} := \mathbf{X} - \mathbf{v}_i\mathbf{t}_i^T$, $\mathbf{y} := \mathbf{y} - b_i\mathbf{t}_i$

At each iteration, the three vectors, $\mathbf{w}_i$, $\mathbf{t}_i$, and $\mathbf{v}_i$, define the ith PLS component ($\mathbf{c}_i$). The $p \times 1$ weight vector $\mathbf{w}_i$ defines a direction, and the $N \times 1$ score vector $\mathbf{t}_i$ consists of projections of training data points onto this direction.

The resulting *estimated regression equation* is represented by

$$\hat{\mathbf{y}} = \mathbf{X}^T\mathbf{W}\left(\mathbf{V}^T\mathbf{W}\right)^{-1}\mathbf{b}, \tag{6.16}$$

where $\mathbf{W}$ and $\mathbf{V}$ are $p \times m$ matrices composed of the calculated vectors of weights and loadings respectively, and $\mathbf{b}$ is a $m \times 1$ vector of the regression coefficients. Please note that among the weight vectors only $\mathbf{w}_1$ is directly associated with the original matrix $\mathbf{X}$, vectors $\mathbf{w}_i$, $i = 2, \ldots, m$, are calculated from deflated data. This means that only the first column of $\mathbf{W}$ represents a PLS component ($\mathbf{c}_1$). It can be shown – and we can also infer it from (6.16) – that the $p \times m$ matrix $\mathbf{C}$ of PLS components is represented by

$$\mathbf{C} = \mathbf{W}\left(\mathbf{V}^T\mathbf{W}\right)^{-1}, \tag{6.17}$$

and thus, the estimated regression equation can be re-written as

$$\hat{\mathbf{y}} = \mathbf{X}^T\mathbf{C}\mathbf{b}. \tag{6.18}$$

Since vector $\mathbf{b}$ is composed of the m PLS regression slopes, we may write:

$$\hat{\boldsymbol{\beta}}_{PLS} = \mathbf{b}. \tag{6.19}$$

Therefore, in order to predict the value of the dependent variable y for a new data point $\mathbf{x}_{new}$, we first project the vector $\mathbf{x}_{new}$ onto the m-dimensional space of the PLS directions,[20]

[18] The number of iteration loops of the PLS1 algorithm is the same as the number of the identified components, while PLS2 algorithms iterate also within each loop until convergence (Lu et al. 2014).

[19] It has been shown that, alternatively, only $\mathbf{X}$ needs to be deflated (Rosipal and Krämer 2006).

[20] If one insists, instead of projecting a data point onto the m-dimensional space of the PLS components, one may project the m-dimensional regression coefficient vector $\hat{\boldsymbol{\beta}}_{PLS}$ onto the p-dimensional space of the original independent variables, say $\hat{\boldsymbol{\beta}} = \mathbf{C}\hat{\boldsymbol{\beta}}_{PLS}$, and then predict $\hat{y}(\mathbf{x}_{new})$ using $\mathbf{x}_{new}$ and $\hat{\boldsymbol{\beta}}$.

$$\mathbf{x}_{new} = \begin{bmatrix} x_{1new} \\ x_{2new} \\ \vdots \\ x_{p\,new} \end{bmatrix} \in \Re^p \rightarrow \mathbf{z}_{new} = \begin{bmatrix} z_{1new} \\ z_{2new} \\ \vdots \\ z_{m\,new} \end{bmatrix} \in \Re^m, \quad (6.20)$$

that is,

$$\begin{bmatrix} z_{1new}\ z_{2new} \cdots z_{m\,new} \end{bmatrix} = \begin{bmatrix} x_{1new}\ x_{2new} \cdots x_{p\,new} \end{bmatrix} \begin{bmatrix} c_{11} & c_{12} & \cdots & c_{1m} \\ c_{21} & c_{22} & \cdots & c_{2m} \\ \vdots & \vdots & \ddots & \vdots \\ c_{p1} & c_{p2} & \cdots & c_{pm} \end{bmatrix}, \quad (6.21)$$

and then use this projection $(\mathbf{z}_{new})$ and the $\hat{\boldsymbol{\beta}}_{PLS}$ slopes,

$$\begin{aligned} \hat{y}(\mathbf{x}_{new}) &= \mathbf{x}_{new}^T \mathbf{C}\, \hat{\boldsymbol{\beta}}_{PLS} \\ &= \mathbf{z}_{new}^T \hat{\boldsymbol{\beta}}_{PLS}. \end{aligned} \quad (6.22)$$

There are many versions and descriptions of the PLS1 algorithm. Some (like the one presented here) may assume that both $\mathbf{X}$ and $\mathbf{y}$ are centered (that is, each variable has a zero mean). In such cases we have the intercept $\hat{\beta}_0 = 0$. If only independent variables were centered, we would have $\hat{\beta}_0 = \bar{\mathbf{y}}$. Different scaling is also used by different authors; for example, the vector $\mathbf{w}_i$ is often normalized (that is, calculated as $\mathbf{w}_i = \mathbf{Xy}/\|\mathbf{Xy}\|$); however, it has been shown that this is not necessary and it can be calculated as $\mathbf{w}_i = \mathbf{Xy}$ (Helland 1988). Similarly, only $\mathbf{X}$ needs to be deflated, there is no need to deflate $\mathbf{y}$ in the case of a single response variable.

Since PLS is not scale invariant, we may often encounter a recommendation to standardize each of the independent variables to have a mean of zero and a variance of one. However, like with most blanket recommendations, that is not always advantageous; if all independent variables are measured in the same unit (for example, like gene expression or protein expression levels), such standardization is not only unnecessary, but may be harmful. For more on this, see Chapter 3.

6.2.2 PLS2 Approaches

Since PLS1 is a special case of PLS2, software implementations of PLSR usually cover both situations.[21] Most implementations of PLSR follow either the NIPALS or SIMPLS approach. The general idea of the NIPALS approach is similar to the earlier presented PLS1 algorithm; however, when the number of response variables, R, is greater than one, in addition to the loop for different model cardinalities $(i = 1, \ldots, m)$, the algorithm also loops through the response variables represented in an $R \times N$ matrix $\mathbf{Y}$.[22]

[21] Depending on relations among the response variables as well as on the goals and assumptions of a study, instead of using the PLS2 approach (thus, considering all response variables at once) we may run the PLS1 algorithm separately for each of the response variables.

[22] In this notation, observations correspond to the columns of matrix $\mathbf{Y}$ (the same is true for matrix $\mathbf{X}$).

Furthermore, calculations within the loops are repeated until convergence, and both matrices (**X** and **Y**) have to be deflated (Lu et al. 2014). Repeated calculations as well as deflations of **X** and **Y** increase both the time and storage costs of the algorithm when the number of variables p or the number of observations N, or both become large. The SIMPLS approach (De Jong 1993) improves this by deflating only the $p \times R$ covariance matrix.[23] However, for high-dimensional data, and especially when $p \gg N$, this may still be inefficient. In such situations, kernel algorithms that iteratively use an $N \times N$ kernel matrix may be used.[24] Rännar et al. presented a PLS2 regression algorithm that calculates regression coefficients using three $N \times N$ matrices: the kernel matrix $\mathbf{X}^T\mathbf{X}\mathbf{Y}^T\mathbf{Y}$ and two association matrices, $\mathbf{X}^T\mathbf{X}$ and $\mathbf{Y}^T\mathbf{Y}$ (Rännar et al. 1994).[25]

6.2.3 A Note on Principal Component Regression

Some texts addressing problems with multiple regression present *Principal Component Regression* (PCR) before, or simultaneously with, PLSR. PCR uses Principal Component Analysis (PCA) to reduce dimensionality – by identifying principal components (PCs) that are linear functions of the original independent variables – and then performs multiple regression in a space defined by those principal components. However, to find those PC directions, PCA is using only the independent variables, hence it is blind to the response variable. PCA is thus an *unsupervised* procedure. Using an unsupervised approach to decrease dimensionality in a *supervised* problem (like regression) is an unfortunate, but still common, misconception. The main goal of regression analysis is to maximize the amount of *variation in the response variable* that is explained by the independent variables. Thus, the most important information in predictive modeling are the values of the response variable associated with the training data points. Yet, being totally blind to the response variable, PCA pursues a very different goal – to find directions of most *data variation* in a p-dimensional space of the p independent variables, the space that does not include the response variable. In the result, PCA directions are random from the point of view of the goals of our supervised analysis.[26] For more on this misconception, see Chapter 2.

[23] SIMPLS and NIPALS approaches provide the same results only in the case of a single response variable, otherwise their results are different (Rännar et al. 1994).

[24] It may be worth mentioning that in an opposite situation, that is, when the number of observations is greater than the number of variables, and especially when $N \gg p$, kernel algorithms that use a $p \times p$ kernel matrix may be used (Lindgren et al. 1993). A more general remark: Although kernel algorithms are used here to increase the efficiency of calculations, the main idea behind the kernel approach is associated with nonlinear problems, for which mapping the original data to some feature space makes a linear solution possible. Kernel functions allow for solving such problems by using only the original data without mapping them to the feature space.

[25] The kernel approach may not be advantageous when we have only one response variable. In such a case, the classical PLS1 algorithm does not iterate-until-convergence and thus, is faster.

[26] Although virtually every modern data science text warns about this misconception (or at least about the possibility of inferior results when it is followed), some still include descriptions of methods implementing this approach.

7 Regularized Regression Methods

7.1 Introduction

The *Partial Least Squares Regression* method, presented in Chapter 6, provided improvement over *Multiple Regression* by applying supervised dimensionality reduction. This addressed problems associated with multicollinearities and also allowed us to find a solution in $p > N$ situations. However, there are other – and often better – ways of dealing with these issues. In this chapter, we will discuss regression methods that can efficiently deal with the *curse of dimensionality* by performing regularization, or feature selection, or both.

Generally, regularization can be considered in terms of the *bias-variance tradeoff* (discussed in Chapter 3). Recall that if the complexity of a model increases, then bias decreases but variance increases. In an extreme situation, when the model perfectly fits all training observations but poorly predicts new ones, it will have zero bias and high variance – a case of overfitting. On the other hand, when a model more or less ignores – that is, underfits – the training data, it would have low variance but high bias; such a model would have poor predictive abilities for the training data, as well as for new data. Therefore, we want to find a proper tradeoff between bias and variance – a model that does not overfit, but is complex enough to be generalizable (that is, provides highly accurate predictions for new data). Regularization approaches try to achieve this goal by penalizing models that are too complex.

In this chapter, we will look at regularization techniques for the least squares estimator. The idea behind such techniques is to improve the performance of a predictive model by decreasing its variance in a tradeoff for a slight increase in bias (when compared to the unbiased OLS solution). Recall that the least squares estimator is optimal only among linear and unbiased estimators. If we, however, extend our considerations to also include biased estimators, then it is possible to find estimators that provide smaller *MSE* than the OLS solution.

7.2 Ridge Regression

Ridge Regression (Hoerl and Kennard 1970) is one of the methods implementing a *regularization* approach to deal with instability and overfitting. Such methods provide biased regression models by adding a penalty to the sum of squared errors; hence, they put additional constraints on the solution in order to make it more generalizable.

Recall from Chapter 6 that the OLS regression method fits the prediction model by minimizing the sum of squared errors (*SSE*),

$$SSE = \sum_{i=1}^{N}\left[y_i - \left(\beta_0 + \sum_{k=1}^{p}\beta_k x_{ki}\right)\right]^2 \tag{7.1}$$

where, again:

- N is the number of training data points
- p is the number of independent variables
- y_i is the observed value of the response variable for training point i, $i = 1, \ldots, N$
- β_0 is the y-intercept
- β_k, $k = 1, \ldots, p$, is the slope parameter associated with independent variable x_k

Ridge regression finds a solution to the regression problem by implementing a *penalized* least squares approach, that is, by minimizing the sum of *SSE* and the penalty term,

$$SSE + \lambda \sum_{k=1}^{p} \beta_k^2, \tag{7.2}$$

where $\lambda \geq 0$ is the penalty coefficient (a tuning parameter).[1]

This kind of regularization is called L_2 regularization – the penalty term uses the sum of *squares* of the model coefficients. Models with large values of their coefficients are penalized more, and this regularization has a tendency of shrinking the estimates of slope coefficients toward zero and toward each other (hence, the penalty term can be called a *shrinkage penalty*). The coefficients are only allowed to be large when it leads to reduction in the total error metric, which we may now denote as SSE_{L_2} (sum of squared errors with L_2 penalty),

$$\begin{aligned} SSE_{L_2} &= SSE + \lambda \sum_{k=1}^{p} \beta_k^2 \\ &= \sum_{i=1}^{N}\left[y_i - \left(\beta_0 + \sum_{k=1}^{p}\beta_k x_{ki}\right)\right]^2 + \lambda \sum_{k=1}^{p} \beta_k^2. \end{aligned} \tag{7.3}$$

If we use the same notation and simplification as presented in (6.5), that is, we use a $(p+1) \times N$ matrix $\mathbf{X}$ of training data that includes a dummy independent variable x_0 associated with the intercept, the L_2-penalized sum of squared errors will now be

$$SSE_{L_2}(\boldsymbol{\beta}) = \left(\mathbf{y} - \mathbf{X}^T\boldsymbol{\beta}\right)^T\left(\mathbf{y} - \mathbf{X}^T\boldsymbol{\beta}\right) + \lambda\boldsymbol{\beta}^T\boldsymbol{\beta}, \tag{7.4}$$

which will lead to the following penalized least squares problem:

$$\underset{\boldsymbol{\beta}}{\text{minimize}} \left(\mathbf{y} - \mathbf{X}^T\boldsymbol{\beta}\right)^T\left(\mathbf{y} - \mathbf{X}^T\boldsymbol{\beta}\right) + \lambda\boldsymbol{\beta}^T\boldsymbol{\beta}. \tag{7.5}$$

[1] With $\lambda = 0$, ridge regression reduces to the OLS regression.

To solve (7.5), we will find the partial derivative of SSE_{L_2} with respect to $\boldsymbol{\beta}$, set it to zero, and solve for $\boldsymbol{\beta}$,

$$\frac{\delta SSE_{L_2}}{\delta \boldsymbol{\beta}} = -2\mathbf{X}\left(\mathbf{y} - \mathbf{X}^T\boldsymbol{\beta}\right) + 2\lambda\boldsymbol{\beta} = 0, \tag{7.6}$$

and the regularized estimate of the β parameters would be given by

$$\begin{bmatrix} \hat{\beta}_0 \\ \hat{\beta}_1 \\ \hat{\beta}_2 \\ \vdots \\ \hat{\beta}_p \end{bmatrix} = \left(\mathbf{X}\mathbf{X}^T + \lambda\mathbf{I}\right)^{-1}\mathbf{X}\mathbf{y}. \tag{7.7}$$

However, this solution would penalize all regression coefficients, including the intercept β_0, which is not what was set up by (7.3). To correct this and exclude the intercept from the penalized coefficients, we may replace the $(p+1) \times (p+1)$ identity matrix $\mathbf{I}$ with a $(p+1) \times (p+1)$ diagonal penalty matrix $\mathbf{K}$,

$$\mathbf{K} = \begin{bmatrix} 0 & 0 & \cdots & 0 \\ 0 & 1 & \cdots & 0 \\ \vdots & \vdots & \ddots & \vdots \\ 0 & 0 & \cdots & 1 \end{bmatrix}, \tag{7.8}$$

which acts as the identity matrix for all of the slope coefficients but prevents penalization of the intercept. Thus, the ridge-regularized least squares estimate of the β parameters is given by

$$\hat{\boldsymbol{\beta}}_{Ridge} = \begin{bmatrix} \hat{\beta}_0 \\ \hat{\beta}_1 \\ \hat{\beta}_2 \\ \vdots \\ \hat{\beta}_p \end{bmatrix} = \left(\mathbf{X}\mathbf{X}^T + \lambda\mathbf{K}\right)^{-1}\mathbf{X}\mathbf{y}. \tag{7.9}$$

The only difference between the ridge solution (7.9) and the OLS solution (6.8) is the penalty-associated term $\lambda\mathbf{K}$. By adding this term, we will now be using the regularized matrix $\left(\mathbf{X}\mathbf{X}^T + \lambda\mathbf{K}\right)$, which for sufficiently large values of λ will be invertible even in the case of severe multicollinearity or singularity of $\mathbf{X}\mathbf{X}^T$ (Fahrmeir et al. 2013). The ridge estimator (7.9) is biased,[2] but often a small increase in bias may provide solutions

[2] That is, $E\left(\hat{\boldsymbol{\beta}}_{Ridge}\right) \neq \boldsymbol{\beta}$, since $E\left(\hat{\boldsymbol{\beta}}_{Ridge}\right) = E\left(\left(\mathbf{X}\mathbf{X}^T + \lambda\mathbf{K}\right)^{-1}\mathbf{X}\mathbf{y}\right) = \left(\mathbf{X}\mathbf{X}^T + \lambda\mathbf{K}\right)^{-1}\mathbf{X}\mathbf{X}^T\boldsymbol{\beta}$, and the terms $\left(\mathbf{X}\mathbf{X}^T + \lambda\mathbf{K}\right)^{-1}$ and $\mathbf{X}\mathbf{X}^T$do not cancel themselves out when $\lambda > 0$.

with significantly smaller variance when compared to the OLS solution. This may result in predictions with lower *MSE* than those provided by unbiased models.

Unlike the *Least Squares Criterion* implemented in multiple regression, which only had one solution to the minimization problem, *ridge regression* may have many solutions – one for each value of the penalty coefficient λ. Hence, selecting an optimal value of λ (that is, an optimal tradeoff between bias and variance, which neither overfits nor underfits the training data) is crucial for ridge regression, and this is usually done using cross-validation.

Observe that the overall ridge regression error metrics, SSE_{L_2}, can be seen as the ordinary least squares criterion with an additional constraint:

$$\sum_{k=1}^{p} \beta_k^2 \leq s, \tag{7.10}$$

and thus, the alternative way of presenting the ridge regularization problem (with s as the tuning parameter) is:

$$\begin{aligned} &\underset{\boldsymbol{\beta}}{\text{minimize}} \ \left(\mathbf{y} - \mathbf{X}^T\boldsymbol{\beta}\right)^T \left(\mathbf{y} - \mathbf{X}^T\boldsymbol{\beta}\right) \\ &\text{subject to} \ \sum_{k=1}^{p} \beta_k^2 \leq s. \end{aligned} \tag{7.11}$$

Constraint (7.10) represents an origin-centered hypersphere in a p-dimensional space of p slope parameters β_k, $k = 1, \ldots, p$ (a circular area $\beta_1^2 + \beta_2^2 \leq s$ when we only have two independent variables; see Figure 7.1). Since each of the infinite number of points on

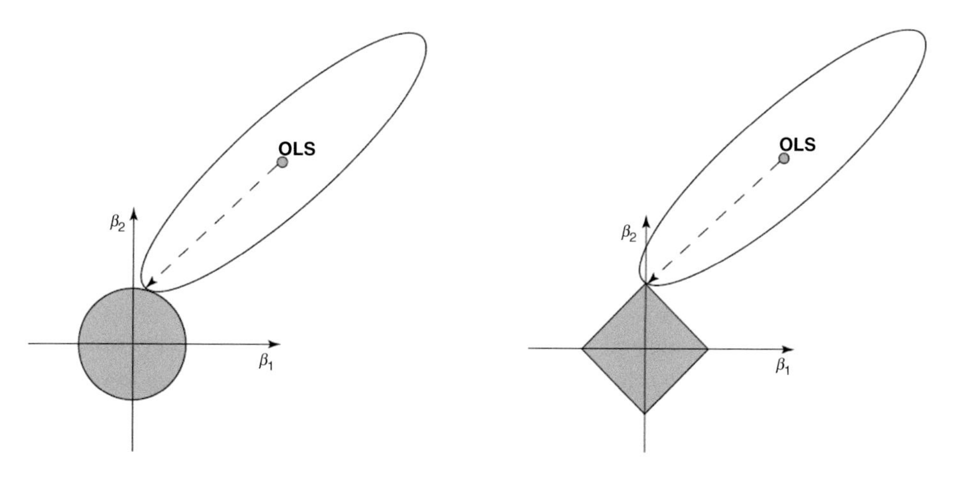

Figure 7.1 Geometry of *ridge regression* and the *lasso regression*. In this toy example (only two independent variables), the circular area $\beta_1^2 + \beta_2^2 \leq s$ represents the ridge regression constraint, and the rhombus area $|\beta_1| + |\beta_2| \leq s$ represents the lasso constraint. The Ordinary Least Squares solution (obtained by applying the regular least squares criterion only) is marked as the OLS point. The expanding elliptical areas (centered at the OLS) represent simplified paths of searching for a solution to each of these two regularized regression problems. Each of the two paths starts at OLS and ends when the elliptical contour meets the constraint area.

the continuous surface of a hypersphere has the same probability of being selected, it is very unlikely that any point corresponding to the zero value of a slope parameter will be selected. Thus, we expect that models generated by the ridge regression method will have to be interpreted in the context of all independent variables.

To sum up, ridge regression allows us to address instability and overfitting problems by penalizing complex models and shrinking model coefficients toward zero. However, none of the coefficients is actually set to zero; hence, ridge regression does not perform feature selection.

7.3 The Lasso (Least Absolute Shrinkage and Selection Operator)

Similar to ridge regression, the *lasso* (Tibshirani 1996) implements a *regularization* approach to deal with overfitting by penalizing models that are too complex. However, unlike ridge regression (which shrinks slope parameter values toward zero without setting any of them to zero), lasso sets some of the slope parameters to be equal to zero, and thus simultaneously performs parameter estimation and feature selection. Like ridge regression, the lasso implements a *penalized* least squares approach, but instead of L_2 regularization, the lasso uses L_1 regularization and minimizes the sum of SSE and the L_1 penalty term,

$$SSE + \lambda \sum_{k=1}^{p} |\beta_k|, \tag{7.12}$$

where $\lambda \geq 0$ is the penalty coefficient (a tuning parameter).

In L_1 regularization, the penalty term uses the sum of absolute values of the model coefficients. As in ridge regression, the coefficients are allowed to be large only when it leads to a reduction in the total error metric, SSE_{L_1} (sum of squared errors with L_1 penalty):

$$\begin{aligned} SSE_{L_1} &= SSE + \lambda \sum_{k=1}^{p} |\beta_k| \\ &= \sum_{i=1}^{N} \left[y_i - \left(\beta_0 + \sum_{k=1}^{p} \beta_k x_{ki} \right) \right]^2 + \lambda \sum_{k=1}^{p} |\beta_k|. \end{aligned} \tag{7.13}$$

However, while the ridge quadratic penalty has a very strong impact on large coefficients but a weak impact on small ones, the lasso absolute value penalty strongly affects small coefficients by shrinking them toward zero, but has a relatively lower impact on larger ones.

If we again assume that a dummy independent variable x_0 associated with the intercept is included in a $(p+1) \times N$ matrix $\mathbf{X}$ representing training data, as well as assume that the intercept is not penalized, then the L_1-penalized sum of squared errors will be represented by

$$SSE_{L_1}(\boldsymbol{\beta}) = \left(\mathbf{y} - \mathbf{X}^T\boldsymbol{\beta}\right)^T \left(\mathbf{y} - \mathbf{X}^T\boldsymbol{\beta}\right) + \lambda \sum_{k=1}^{p} |\beta_k|. \tag{7.14}$$

Since SSE_{L_1} includes absolute values of the regression coefficients, it is not differentiable. Consequently, the lasso regularized least squares optimization problem,

$$\underset{\boldsymbol{\beta}}{\text{minimize}} \left(\mathbf{y} - \mathbf{X}^T\boldsymbol{\beta}\right)^T \left(\mathbf{y} - \mathbf{X}^T\boldsymbol{\beta}\right) + \lambda \sum_{k=1}^{p} |\beta_k|, \tag{7.15}$$

has no exact (closed-form) solution, and regression parameters are estimated using numerical procedures. One of them is least angle regression (LARS), which can very efficiently determine the entire path of lasso solutions, from which the value of λ for an optimal solution can be determined via cross-validation. The LARS algorithm for the lasso is described in Section 7.3.1.

Like ridge regression, the lasso can be seen as the ordinary least squares criterion with an additional constraint; for the lasso, this constraint is

$$\sum_{k=1}^{p} |\beta_k| \le s, \tag{7.16}$$

and the lasso optimization problem can be presented as:

$$\begin{aligned} &\underset{\boldsymbol{\beta}}{\text{minimize}} \left(\mathbf{y} - \mathbf{X}^T\boldsymbol{\beta}\right)^T \left(\mathbf{y} - \mathbf{X}^T\boldsymbol{\beta}\right) \\ &\text{subject to } \sum_{k=1}^{p} |\beta_k| \le s. \end{aligned} \tag{7.17}$$

The constraint for the lasso represents a hyper rhombus (a multidimensional rhombus). When we only have two independent variables (see Figure 7.1), it would be a rhombus $|\beta_1| + |\beta_2| \le s$ with its corners on the axes β_1 and β_2 (i.e., with a zero value of one or the other of the slope parameters). When we have many independent variables, this constraint will represent a hyper-rhombus with many corners. Since the corners (singularities, or the "sticking out" points of a hyper-rhombus) are more likely to be selected by the minimizing criterion than other hyper-rhombus' points, the lasso will set some slope parameters to exactly zero. Thus, by removing the corresponding independent variables from the model, the lasso does perform feature selection.

Of course, the penalty coefficient λ and the size of the constraint shape, s, are related – each value of λ has its corresponding value of s. By selecting a sufficiently large λ (or a sufficiently small s), we may force the lasso model to include as few independent variables as we want.[3]

7.3.1 Least Angle Regression for the Lasso

The Least Angle Regression (LARS) (Efron et al. 2004) is a version of the stepwise variable selection algorithm that is very efficient in determining the entire lasso path (that is, the sequence of all lasso solutions for different values of the regularization

[3] Which also means that there is a value of λ for which (and for all values greater that this one) all slope coefficients are set to zero.

parameter λ), especially for high-dimensional $p \gg N$ data. It starts with all slope coefficient estimates equal to zero (that is, an empty set of variables), and identifies the variable that has the highest absolute correlation (the least angle) with the response variable. However, instead of setting the slope estimate of this variable to its least squares value, it starts moving it away from zero and toward this least squares value. This decreases the absolute correlation between the variable and the continuously changing residual (that is, the currently unexplained portion of the variation in the response variable). The move in this direction continues until another variable has the same absolute correlation with the current residual as the first variable. This second variable is then added to the model and the slope estimates of both variables are moved in their joint least squares direction (that is, equiangularly between the two variables), until some other variable has a correlation with the residual that is the same as the variables in the model. This process of adding variables to the model and moving in the equiangular direction of a set of variables (that is, along the least angle direction) continues; however, whenever a slope coefficient reaches zero, its variable is removed from the model and the current joint least squares direction is recalculated. The process ends when either the set includes a predetermined number of variables, or when it includes $\min(p, N)$ variables, whichever comes first.[4] The LARS algorithm is very efficient due to the fact that the path of lasso solutions is piecewise linear; thus, at each step (that is, whenever a variable is added or removed) the algorithm needs to calculate only the new direction and the length of the move along this direction (Rish and Grabarnik 2015).

7.4 The Elastic Net

The lasso method described in the previous section seems to have a clear advantage over ridge regression due to its ability to provide sparse solutions – models that include only a small number of independent variables. In predictive modeling, we almost always strive for this kind of sparsity, that is, to identify parsimonious models that include as few independent variables as necessary for efficient generalization, and none of the other variables that would increase noise or lead to overfitting. Thus, the theoretical limitation of the lasso that allows it to select at most N of the p independent variables (when $p > N$) should not be practically prohibitive – we should not seriously consider data sets that have thousands of variables and only a dozen or so observations (we may assume that *any* method applied to such a small sample would lead to overfitting). Nonetheless, it is also worth considering the ways the lasso and ridge regression treat groups of highly correlated variables. The lasso tends to select only one of such variables and ignores the rest of the group. Although ridge regression does not perform any feature selection, we may want to utilize its tendency to shrink together the slopes of highly correlated variables. Thus, in some situations (for example, when data

[4] This means that for $p > N$, the lasso can select at most N variables.

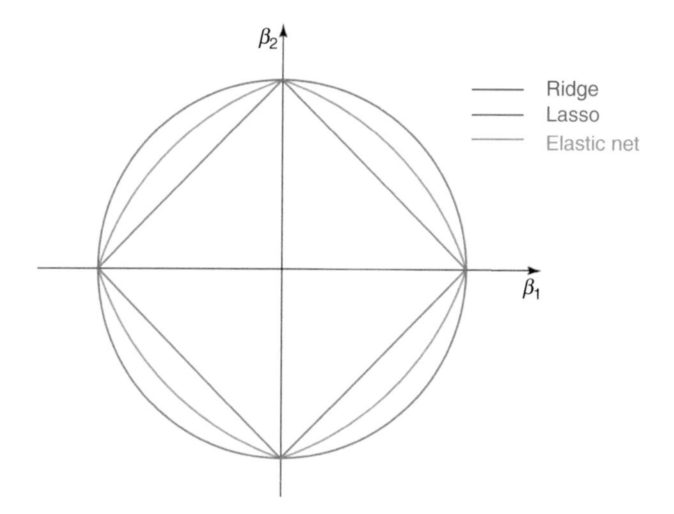

Figure 7.2 Contours of constant value for *ridge regression* (red), *lasso* (blue), and *elastic net* (green). In this two-dimensional example, both the lasso and the elastic net have four corners, which are graphical representations of sparsity. All three shapes associated with these contours are convex; however, ridge and elastic net are strictly convex. (For a convex shape, any two points of the shape can be connected by a straight-line segment that belongs to the shape. For the strict convexity, any such straight-line segment – with the exception of its endpoints – must be inside the shape.)

include groups of highly correlated variables and – for whatever reason – we want to preserve them), we may want to combine the feature selection property of the lasso with the way ridge regression treats groups of correlated variables.

The *elastic net* (Zou and Hastie 2005) combines the best of the two approaches by applying both L_1 and L_2 penalties,

$$\begin{aligned} SSE_{enet} &= SSE + \lambda_1 \sum_{k=1}^{p} \beta_k^2 + \lambda_2 \sum_{k=1}^{p} |\beta_k| \\ &= \sum_{i=1}^{N} \left[y_i - \left(\beta_0 + \sum_{k=1}^{p} \beta_k x_{ki} \right) \right]^2 + \lambda_1 \sum_{k=1}^{p} \beta_k^2 + \lambda_2 \sum_{k=1}^{p} |\beta_k|. \end{aligned} \tag{7.18}$$

where $\lambda_1 \geq 0$ is the ridge penalty coefficient, and $\lambda_2 \geq 0$ is the lasso penalty coefficient.

The geometry of the elastic net constraint, in comparison to lasso and ridge regression, is presented in Figure 7.2. When $\lambda_1 = 0$, the elastic net is equivalent to the lasso, and when $\lambda_2 = 0$ it is ridge regression. Since SSE_{enet} is not differentiable, its minimization (and finding an optimal combination of the two penalty coefficients) is done via numerical procedures. Zou and Hastie (2005) introduced the LARS-EN algorithm, which is a stage-wise algorithm based on LARS (see Section 7.3.1), and can efficiently identify the entire path of elastic net solutions for each fixed value of the ridge penalty coefficient λ_1. Then, an optimal combination of (λ_1, λ_2) can be identified via cross-validation.

Combining L_1 and L_2 penalties allows the elastic net to perform as well as the lasso in situations when the lasso provides the best solution, but – when it is necessary or advantageous – it also allows selecting more than N variables (that is, to have more than N non-zero slope coefficients), or to include in the solution a group of strongly correlated variables.[5]

Consider, for example, a search for a multivariate gene expression biomarker for predicting the probability of relapse in patients with a particular cancer disease. It is conceivable that the lasso would provide a parsimonious biomarker consisting of a few genes whose combined expression pattern can accurately predict this probability.[6] However, since gene expression levels are not independent from each other and are usually correlated (or highly correlated) within some groups of genes, each gene selected into the biomarker can be seen as representing a group of genes with similar expression patterns. This means that our biomarker – even if it happens to have the best predictive abilities – is not unique; we could easily imagine the existence of other parsimonious (and possibly quite efficient) biomarkers consisting of different sets of genes representing the same groups[7] of genes with similar patterns. This also means that a small change in the parameters of the predictive modeling process may lead to a different multivariate biomarker. In the early 'omic' era, this phenomenon was sometimes labeled as an indicator of the intrinsic instability of the results. However, the apparently different results may still represent a stable solution if they point to the same set of biological processes underlying the prediction.

Hence, we are again looking at the two important aspects of biomarker discovery – predictive power and facilitating biological interpretation. If we were satisfied with the high predictive power of the biomarker (when it properly predicts the probability of relapse for new patients), even if we do not know why it works, then our lasso predictor could be all that we need. Yet, no biomarker discovery project is fully finished if we cannot facilitate a plausible interpretation of the biomarker. However, providing biological interpretation of a parsimonious multivariate biomarker may be a very challenging task.[8] We may try to make this task easier by using the elastic net to combine (or compromise between) the sparsity of its lasso component and the group treatment of its ridge component. We would still take the advantage of feature selection, but our new biomarker would be less sparse as it would include groups of highly correlated variables.[9]

[5] Zou and Hastie liken the elastic net to "*a stretchable fishing net that retains 'all the big fish*'" (Zou and Hastie 2005).

[6] Assuming that such a prediction is possible in the context of available data. However, for high-dimensional data, we would rather perform many parallel *lasso*-based feature selection experiments, and aggregate their results (see Chapter 5).

[7] A different combination of different groups can also be imagined and possible.

[8] Recall that providing univariate biological interpretations of the variables included in the biomarker is not a proper interpretation of the multivariate biomarker.

[9] In Part IV, however, a method that allows for identification of a parsimonious biomarker while simultaneously facilitating its biological interpretation by linking the biomarker with essential patterns is described.

Finally, let's take another look at the elastic net penalty term from (7.18),

$$\lambda_1 \sum_{k=1}^{p} \beta_k^2 + \lambda_2 \sum_{k=1}^{p} |\beta_k|. \tag{7.19}$$

Instead of using λ_1 and λ_2, the separate ridge and lasso penalty coefficients, we may set

$$\lambda = \lambda_1 + \lambda_2, \quad \text{and} \quad \alpha = \frac{\lambda_1}{\lambda_1 + \lambda_2}, \tag{7.20}$$

and use the following form of the elastic net penalty:

$$\lambda \sum_{k=1}^{p} \left(\alpha \beta_k^2 + (1 - \alpha) |\beta_k| \right), \tag{7.21}$$

where λ can be interpreted as the coefficient controlling the strength of the penalty, and α as the one controlling the balance between the ridge and the lasso components of the elastic net. Since α may only assume values from the range [0, 1], it makes it convenient for setting a grid of values to consider when selecting an optimal balance between the ridge and lasso components via cross-validation.

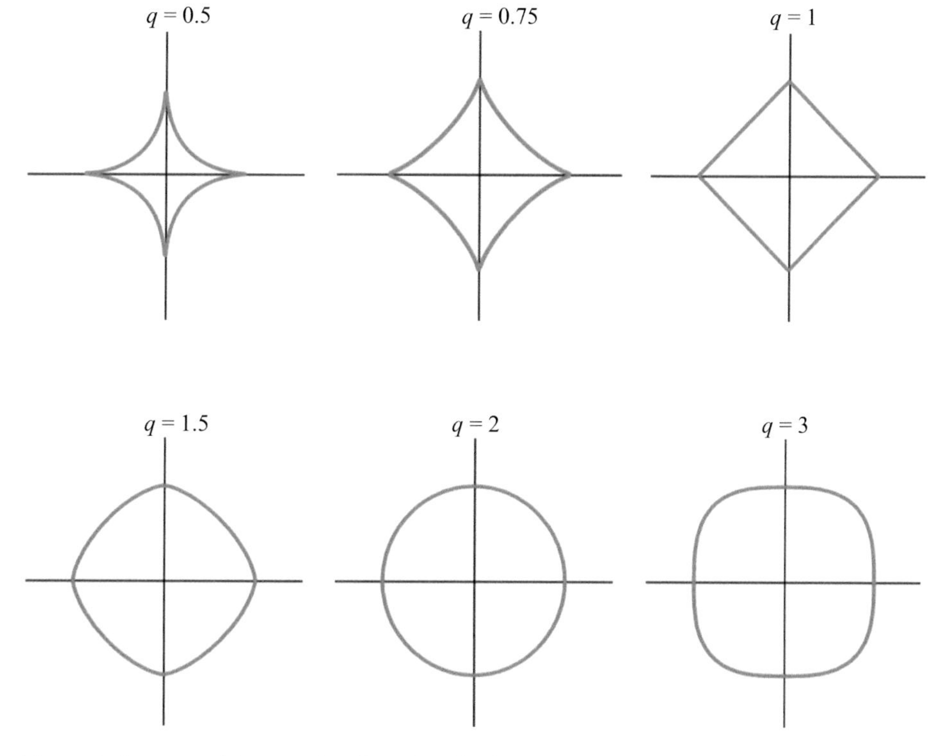

Figure 7.3 Contours of L_q constraints for selected values of q for two-variable problems; for $p = 2$, the contours are represented by $\sum_{k=1}^{p} |\beta_k|^q = |\beta_1|^q + |\beta_2|^q = s$. The constraints with $q \geq 1$ are convex, and those with $q \leq 1$ have sharp corners and thus, support sparse solutions. Hence, the only single L_q constraint that is convex and allows for feature selection is the lasso with $q = 1$.

7.5 Notes on L_q-Penalized Least Squares Estimates

Ridge regression and the *lasso* are examples of a more general class of L_q-regularized least squares optimization problems:

$$\begin{aligned} &\underset{\boldsymbol{\beta}}{\text{minimize}}\ \left(\mathbf{y} - \mathbf{X}^T\boldsymbol{\beta}\right)^T\left(\mathbf{y} - \mathbf{X}^T\boldsymbol{\beta}\right) \\ &\text{subject to}\ \sum_{k=1}^{p} |\beta_k|^q \leq s, \end{aligned} \tag{7.22}$$

where $q \geq 0$.

For $q = 0$, we have L_0 penalty, which performs feature selection by restricting only the number of nonzero coefficients. With $q = 1$ corresponding to the lasso and $q = 2$ to ridge regression, penalties with $1 < q < 2$ provide a compromise between the lasso and ridge regression. Observe, however, that here we are referring to the compromise provided by a single L_q penalty, whereas the elastic net described in the previous section represents a different kind of compromise – between the lasso and ridge regression – that is achieved by combining their two penalties.

In the same manner as for the lasso and ridge regression, any L_q penalty can be geometrically represented as an origin-centered p-dimensional hypershape (sometimes called an L_q ball), which is convex for $q \geq 1$, and concave for $0 < q < 1$ (see

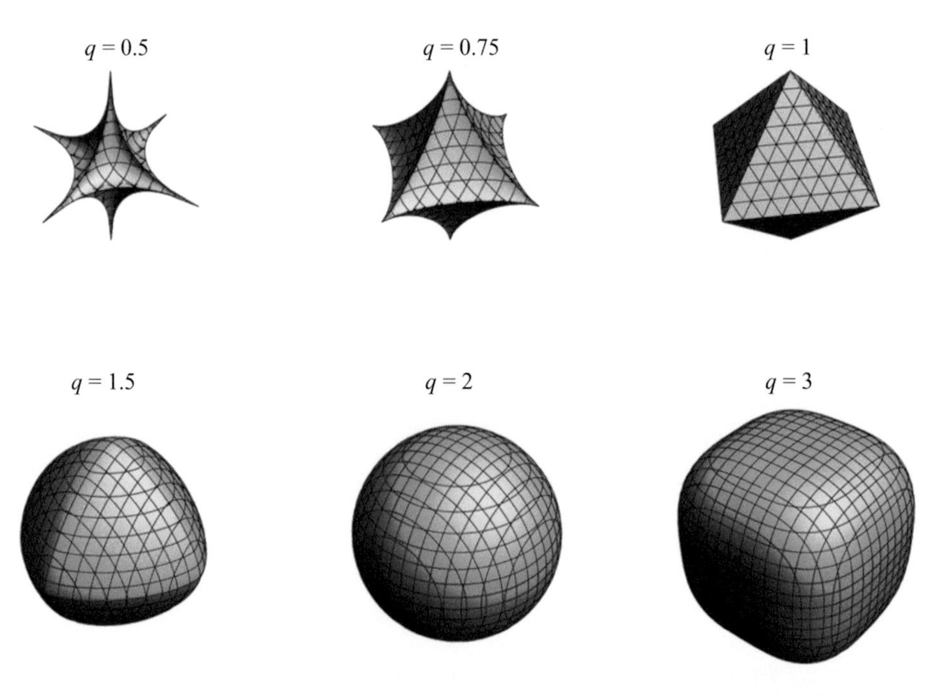

Figure 7.4 Contours of L_q constraints for selected values of q for three-variable problems; for $p = 3$, the contours are represented by $\sum_{k=1}^{p} |\beta_k|^q = |\beta_1|^q + |\beta_2|^q + |\beta_3|^q = s$. L_1 constraint corresponds to the lasso, and L_2 to ridge regression.

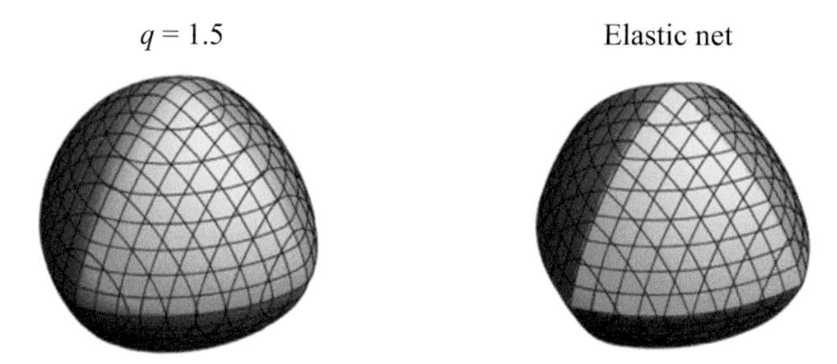

Figure 7.5 Contours for L_q constraint with $q = 1.5$ and for the elastic net (for three-variable problems; $p = 3$). Although they look very similar, the elastic net performs feature selection (sharp nondifferentiable corners of the contour), while L_q constraints with $q > 1$ do not.

Figures 7.3 and 7.4 for two- and three-dimensional examples). Furthermore, only the L_q functions with $q \leq 1$ provide sparse solutions. This means that the only L_q function that is both convex and performs feature selection is the L_1 penalty of the lasso. That is important since solving optimization problems with concave (nonconvex) constraints is more difficult.

Sparsity of a solution corresponds to sharp non-differentiable corners of penalty hypershapes (see Figures 7.3 and 7.4), which are associated with the feature selection capability. Since such corners are only associated with L_q functions with $q \leq 1$, we may now look again at the difference between the *elastic net* and the L_q penalty with $1 < q < 2$. Such a penalty is convex and does not support sparsity. Thus, even if the contour of this penalty seems to be very similar to the elastic net contour (see Figure 7.5), the latter one (being a combination of the lasso and ridge regression) has sharp corners and allows feature selection.

8 Regression with Random Forests

8.1 Introduction

The regression methods discussed in Chapter 7 could be seen as extensions or modifications to *Multiple Regression*. Although they were implementing feature selection or regularization via penalization, or a combination thereof, they were still based on the *Least-Squares* method. However, other supervised learning algorithms can also be used for predictive regression modeling. Some of them, especially relatively newer ones (like *Support Vector Machines*, *Neural Networks*, and *Random Forests*), may – at least in some situations – outperform the least-squares-based methods.

In this chapter, we will look at the *Random Forests* learning algorithm that represents ensemble-based supervised data science methods. This algorithm generates an ensemble of decision tree models, while employing two levels of randomization.

The ***decision tree*** algorithm is a conceptually simple nonparametric method that creates a hierarchical heuristic structure consisting of a sequence of binary decisions. This approach can be used either for classification or regression (Breiman et al. 1984). Here, we will look at the regression tree algorithm. As any decision tree, it starts with the *root node* that includes all training data, and splits the data into two subsets – associated with two child nodes – in a way that maximizes the homogeneity of the child nodes. Each variable is tested for any potential split value. The variable and its split value that provide the lowest sum of squared errors when combined over the two child nodes are selected to partition the data. Then, the data at each child node is split in exactly the same manner; and this recursive binary partitioning process continues until no further decrease in the overall sum of squared errors is possible, or until each of the terminal nodes (which are also called *leaf nodes*) includes fewer observations than some predetermined threshold. Since such a fully-grown tree with many node levels is prone to overfit the training data, it is usually pruned, that is, some of the latest splits are reversed and the number of terminal nodes is decreased. This pruning process can be seen as regularization that penalizes a model that is too complex.

The decision tree algorithm has been popular for low-dimensional data – decisions based on a small tree with a few variables that can be easily traced and interpreted. However, the high variance of decision trees (due to their hierarchical structure) makes a single regression tree a weak and unstable predictive model,[1] which is practically

[1] A model is unstable if a slight change in training data may result in a very different tree.

useless for predictions based on high-dimensional data. Nevertheless, predictive models based on the decision tree approach have some advantages – they are simple and easy to build, and they can handle both quantitative and categorical variables. Therefore, instead of using a single tree, we may grow many trees based on randomized versions of the training data and thus, build an ensemble model that should have a smaller variance than any of its trees.

The ***bootstrap*** resampling approach can be used to generate randomized versions of the training data. The most popular bootstrap algorithm is Efron's nonparametric bootstrap that randomly selects observations from the training data with replacement, and creates bootstrap samples of the same size as the size of the original training data. Since observations are selected with replacement, some of them may be selected more than once, and some not at all. Observations that are not selected into a particular bootstrap sample are called its *out-of-bag* sample.[2] Once many bootstrap samples are created, each of them is used to grow a large tree – usually without pruning – and thus, together, they build an ensemble of tree models – a ***bagged tree model***. This method is called ***bagging***, that is, *bootstrap aggregating* (Breiman 1996a).[3] By aggregating predictions of many individual tree models built on different bootstrap samples of the training data, bagging produces an ensemble model that may have a smaller variance than any of the trees of the ensemble. However, even if each tree is grown from a different bootstrap sample of the training data, it is possible (and even quite likely) that the structure of the relationships between variables and response may result in trees that are similar to each other and thus, are correlated.[4] This will undermine – at least to some degree – the effect of reducing variance of the ensemble model. The random forests algorithm improves on bagged trees by introducing a second level of randomization that allows for the decorrelation of the trees.

8.2 Random Forests Algorithm for Regression

The *Random Forests* algorithm (Breiman 2001) creates an ensemble of decision tree models. Like in the bagged tree method, each tree is unpruned and based on a different bootstrap sample of the training data, and each bootstrap sample is also generated via Efron's nonparametric bootstrap method (that is, each bootstrap sample is of the same size as the training data, and the observations are randomly selected by repeated sampling with replacement). However, unlike bagged trees, the random forests approach does not consider all variables at the tree nodes. Instead, it introduces a second level of randomization by considering at each tree node a smaller number m

[2] For more information on bootstrapping, see Chapter 4.

[3] Bagging is especially useful for reducing variance of noisy (i.e. high-variance) and low-bias models, like decision trees.

[4] For example, many trees may start by splitting their root node on the same variable, or their first few splits may be done by using only a small subset of variables. Hence, since early splits may have high influence on the hierarchical architecture of the trees, even if the trees are different, they still may be correlated, or even highly correlated.

of variables, which are independently randomly selected for that node. For regression, the default value of m is $m = \lfloor p/3 \rfloor$, where p is the number of independent variables in the training data.[5] By considering a different subset of variables for splitting each node, correlations among trees are significantly decreased, and the resulting ensemble model has a significantly lower variance than any of its trees. This also decreases the cost of building each tree as only a subset of variables needs to be considered at each node.

To split a node of a tree, m variables are randomly selected, specifically for this node. For each of these m variables, all possible splits are evaluated and the best one is identified (that is, the one that provides the smallest sum of squared errors, combined over the child nodes), and then the best of those m best splits is selected and used to split the node.

Each random forest's tree t_b, $b = 1, \ldots, B$ is grown from its own bootstrap sample b that includes N non-unique observations[6] and p variables. Each observation i, $i = 1, \ldots, N$, is represented by a $p \times 1$ vector $\mathbf{x}_i = [x_{1i}, \ldots, x_{pi}]^T$ and is associated with the observed value of the continuous response variable, y_i. Let's consider splitting node $\mathcal{P}$ (a parent node) into two child nodes C_L and C_R (left and right). Assume that the parent node includes $N_\mathcal{P} \leq N$ observations (where $N_\mathcal{P} = N$ only for the root node), which we may denote by $\mathbf{x}_i \in \mathcal{P}$, $i = 1, \ldots, N_\mathcal{P}$. We want to split these $N_\mathcal{P}$ observations into two disjoint subsets, $\mathbf{x}_i \in C_L$, $i = 1, \ldots, N_L$, and $\mathbf{x}_i \in C_R$, $i = 1, \ldots, N_R$, where $N_L + N_R = N_\mathcal{P}$, in a way that minimizes the sum of the sums of squared errors of the child nodes.

8.2.1 Splitting a Node

The sum of squared errors of the parent node is

$$SSE_\mathcal{P} = \sum_{\mathbf{x}_i \in \mathcal{P}} (y_i - \bar{y}_\mathcal{P})^2, \tag{8.1}$$

where $\bar{y}_\mathcal{P}$ is the average of response values associated with the observations included in node $\mathcal{P}$,[7]

$$\bar{y}_\mathcal{P} = \frac{1}{N_\mathcal{P}} \sum_{\mathbf{x}_i \in \mathcal{P}} y_i. \tag{8.2}$$

To split the node, we will first randomly select m out of p variables, and then – for each of them – evaluate all potential splits.[8] Hence, for each of the selected variables

[5] Floor brackets $\lfloor x \rfloor$ denote the largest integer that is less than or equal to x.

[6] Recall that each bootstrap sample is randomly selected from the original training data (of N unique observations) *with replacement.*

[7] Note that if node $\mathcal{P}$ was one of the terminal nodes of a tree, then $\bar{y}_\mathcal{P}$ would be the predicted value of the response variable that would be assigned – for this tree – to any observation which after traversing the tree would land in this particular terminal node.

[8] For example, all midpoints between sorted values of the variable.

x_k, $k = 1, \ldots, m$, and each potential split value s on this variable, the observations from the parent node would be assigned as follows:

$$\begin{aligned} \mathbf{x}_i &\in C_L \quad \text{if } x_{ki} \leq s, \\ \mathbf{x}_i &\in C_R \quad \text{if } x_{ki} > s. \end{aligned} \tag{8.3}$$

And the sums of squared errors associated with split (k, s) would be

$$\begin{aligned} SSE_L(k,s) &= \sum_{\mathbf{x}_i \in C_L(k,s)} \left(y_i - \overline{y}_{C_L(k,s)}\right)^2, \\ SSE_R(k,s) &= \sum_{\mathbf{x}_i \in C_R(k,s)} \left(y_i - \overline{y}_{C_R(k,s)}\right)^2, \end{aligned} \tag{8.4}$$

where:

$$\begin{aligned} \overline{y}_{C_L(k,s)} &= \frac{1}{N_L(k,s)} \sum_{\mathbf{x}_i \in C_L(k,s)} y_i, \\ \overline{y}_{C_R(k,s)} &= \frac{1}{N_R(k,s)} \sum_{\mathbf{x}_i \in C_R(k,s)} y_i. \end{aligned} \tag{8.5}$$

The parent node will be split on variable k and its split value s that will

$$\underset{k,s}{\text{minimize}}\ SSE_L(k,s) + SSE_R(k,s). \tag{8.6}$$

Alternatively, we may calculate the decrease in the overall sum of squares provided by each considered split (k, s),

$$\Delta(k,s) = SSE_{\mathcal{P}} - [SSE_L(k,s) + SSE_R(k,s)], \tag{8.7}$$

and split the parent node on variable k and its split value s that will

$$\underset{k,s}{\text{maximize}}\ \Delta(k,s). \tag{8.8}$$

Both approaches, (8.6) and (8.8), provide the same solution; however, the latter may also provide a suggestion to declare node $\mathcal{P}$ a terminal node if the decrease in SSE is negligible.

8.2.2 Modeling and Evaluation

Like in bagged trees, each bootstrap sample in random forests is associated with its own *out-of-bag* sample (OOB sample). On average, when resampling with replacement, the bootstrap sample includes about 63.2% of the unique training data observations, with the remaining observations constituting the OOB sample. While the bootstrap sample is used to grow a tree, the OOB sample can be used to estimate the performance of the tree on the observations in its OOB sample, as none of those observations were used to train this particular tree model (Breiman 1996b). By averaging such performance estimates over all of the ensemble trees, we compute the OOB estimate of the prediction error for the random forest model (see Figure 8.1).

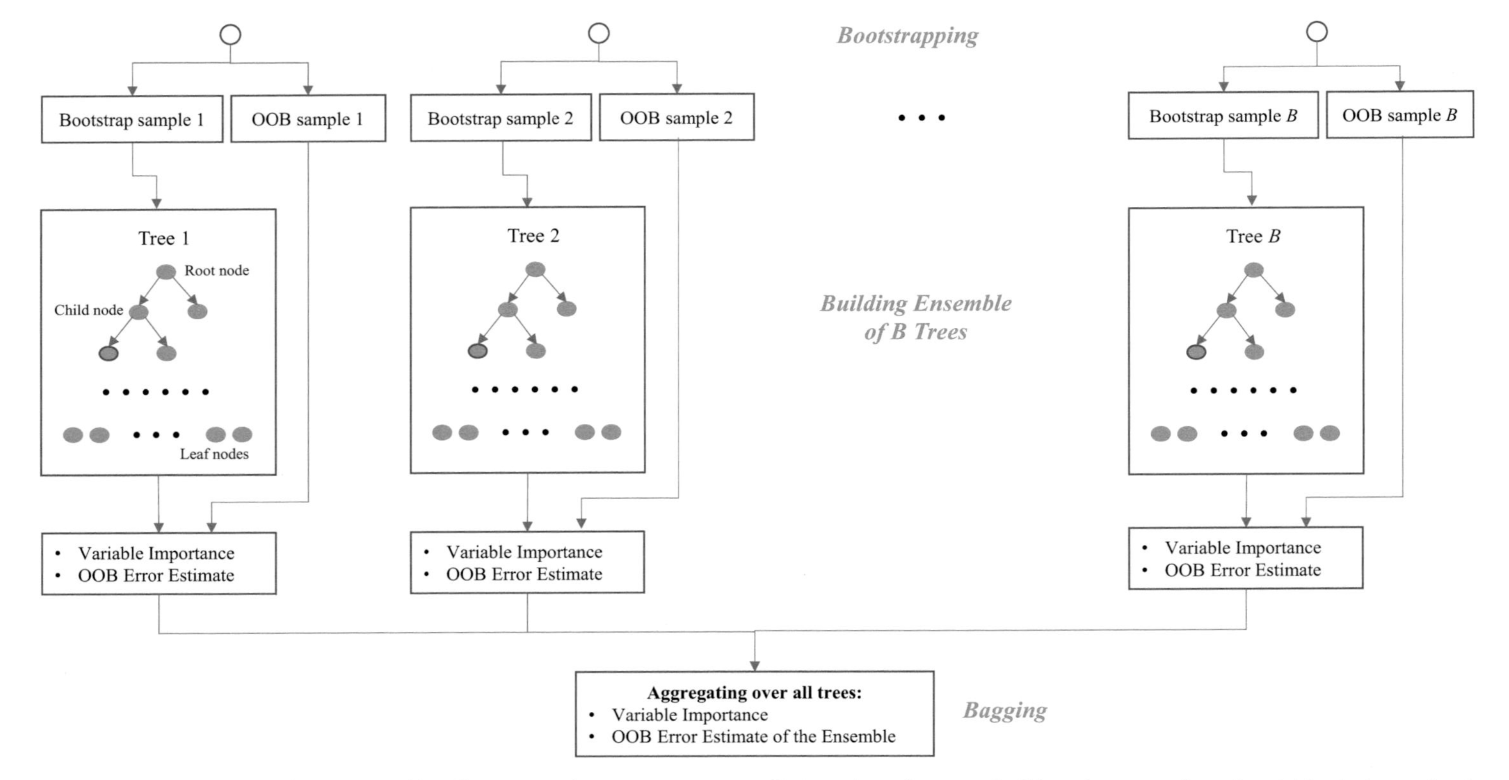

Figure 8.1 Building a random forest ensemble. The process has two parameters: B – number of trees to build, and m – number of variables to be randomly selected at each node, to be considered for splitting this node. The bootstrapping step generates B bootstrap samples (and their own OOB samples). The trees are built without pruning. Each tree is based on its own bootstrap sample. Variable importance and OOB prediction error estimate are calculated for each tree, and then aggregated over all trees (the bagging step). Please note that this diagram represents only the building stage; after the predictive model is built (and eventually tuned), it should be tested on an independent test data set.

To find the OOB prediction for each training observation i, $i = 1, \ldots, N$, we average predictions for this observation over the trees in which this observation was not in the bootstrap sample,

$$\widehat{y}(\mathbf{x}_i) = \frac{1}{B_i} \sum_{b:\mathbf{x}_i \in OOB_b} \widehat{y}(\mathbf{x}_i, t_b), \tag{8.9}$$

where B_i is the number of trees where observation i was in the OOB sample,[9] and $\widehat{y}(\mathbf{x}_i, t_b)$ is the prediction for this observation from tree t_b. Then, we can calculate the OOB prediction error estimate for the random forest as the mean squared error:

$$error_{OOB} = \frac{1}{N} \sum_{i=1}^{N} \left(y_i - \widehat{y}(\mathbf{x}_i)\right)^2. \tag{8.10}$$

Predictions for new observations would be calculated as the average of the predicted values of the response variable over all of the ensemble's trees:

$$\widehat{y}(\mathbf{x}_{new}) = \frac{1}{B} \sum_{b=1}^{B} \widehat{y}(\mathbf{x}_{new}, t_b). \tag{8.11}$$

Usually, a random forest consists of hundreds or thousands of trees that are built without pruning. Although increasing the number of trees in random forests does not result in overfitting (Breiman 2001), generating significantly more trees than necessary increases both the time of building the model, as well as the time for making predictions when the model is implemented. When building the model, increasing the number of trees in the ensemble initially results in a significant improvement in the estimated prediction accuracy, which usually stabilizes somewhere between 500 and 1,000 trees, though it depends on a data set – sometimes it may stabilize at fewer than 500 trees (see Figure 8.2), and sometimes more than 1,000 trees are required. To decide on an optimal number of trees (B) for a particular data set, we may first run the random forests algorithm with 1,000 or 2,000 trees, and then – if the OOB estimate of the mean squared error does not stabilize yet – try a larger number of trees to find out at what range it does stabilize.

In addition to the number of trees, random forests have only one more parameter – m, the already described number of variables considered at a node that are randomly selected at each node of each tree. Although we may tune this parameter – by finding its optimal value corresponding to the lowest OOB estimate of the prediction error – its default value usually works well.

For low-dimensional data, especially when $p \ll N$, it may be quite reasonable to build just a single random forest predictive model. Nevertheless, for biomarker discovery (or any project based on high-dimensional data), limiting the project to building a single predictive model (even an ensemble-based one) on the training data including thousands of variables is not recommended. **Multivariate feature selection should be**

[9] On average, each observation is in an OOB sample in about one-third of the forest trees.

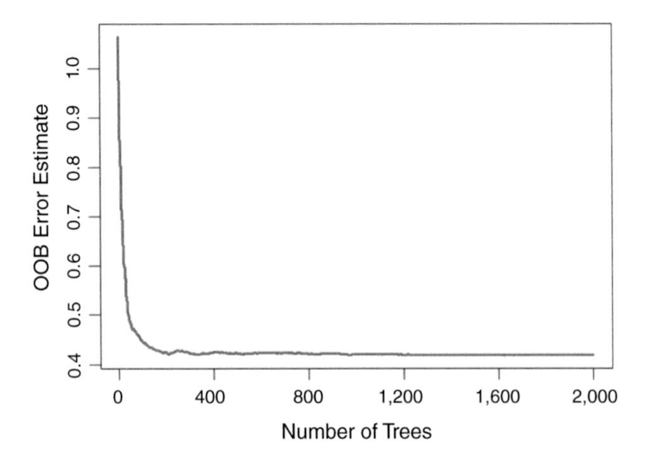

Figure 8.2 Out-of-bag error estimate plotted against the number of trees in a random forest's ensemble.

performed first.[10] One of the problems with building a random forest model on high-dimensional data without performing feature selection first is that most of the variables in such data may be irrelevant for the problem (and thus, represent noise). Models built in such a way, especially when m is small, are likely to have poor performance. Another serious problem is that such an approach would not fulfill the main goals of biomarker discovery – no parsimonious multivariate biomarker would be identified, and there would be no way to facilitate a reasonable biological interpretation of the predictive model.

If we build a random forest predictive model *after* the feature selection step, the model would be using a small number of variables constituting our optimal parsimonious multivariate biomarker. However, it is very important to realize that such a situation is substantially different from building a single model for low-dimensional data, with no feature selection performed first. For the latter, using OOB samples to estimate prediction error is considered as good as testing the predictive model on independent test data. However, in our biomarker discovery situation, if the model is built *after* feature selection, using the OOB samples for evaluating its performance would amount to useless *internal* validation – all training observations were already used during the feature selection process (to identify the optimal biomarker) and thus, none of those observations can be seen as independent (see Chapter 4). In such situations, the random forest model has to be tested on an independent test data set (or on a holdout set that has not been used in any of the previous modeling steps), see Figure 8.3.

[10] We are mentioning it here again, because one may argue that by selecting random subsets of variables to consider at nodes, and by not necessarily including all variables in the ensemble model, random forests perform a *kind* of feature selection. However, this is at most a side effect of the algorithm rather than one of its goals, and it is extremely unlikely that with hundreds or thousands of trees, the set of variables used in the ensemble model can be considered a reasonable result of feature selection, let alone a parsimonious multivariate biomarker.

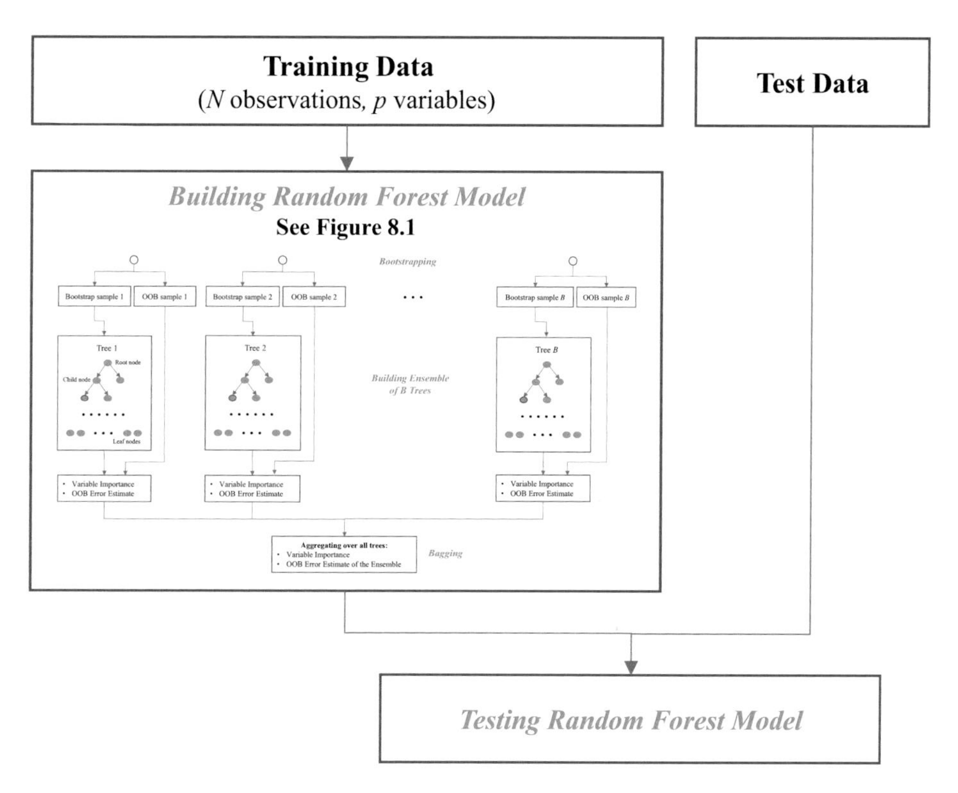

Figure 8.3 After a random forest model is built on a multivariate biomarker identified during the feature selection step, it has to be tested on a test data set that was unseen by any previous step of the modeling process. Using OOB samples now to estimate the model's performance would be a mistake as it would constitute internal validation (that is, testing the model on the same training observations that were already used during the feature selection process).

Since building even thousands of trees is simple and relatively quick, and since the random forests algorithm provides its internal measures of variable importance, it is a very good candidate for a learning algorithm employed within such feature selection schemas like *Recursive Feature Elimination* (see Chapter 5). Let us then look at random forests' variable importance measures.

8.3 Variable Importance Measures

Measures of variable importance offered by the random forests learning algorithm are based either on permutation experiments or, alternatively, on the decrease in node impurity when a node is split on a variable. For example, the increase in the OOB estimate of the prediction error after the values of variable k, $k = 1, \ldots, p$, are randomly permuted in OOB samples may be used as a variable importance measure. Larger increases will indicate more important variables. For each regression tree t_b, $b = 1, \ldots, B$ and for each of its OOB observations $\mathbf{x}_i \in \mathrm{OOB}_b$, $i = 1, \ldots, N(\mathrm{OOB}_b)$, we can find the prediction error for the observation,

$$e_i(t_b) = y_i - \widehat{y}(\mathbf{x}_i, t_b), \tag{8.12}$$

where $\widehat{y}(\mathbf{x}_i, t_b)$ is again the prediction for this observation from tree t_b, and $N(\mathrm{OOB}_b)$ is the number of observations in the OOB sample associated with tree t_b. Thus, the OOB error estimate for tree t_b can be calculated as:

$$error_{OOB}(t_b) = \frac{1}{N(\mathrm{OOB}_b)} \sum_{i=1}^{N(\mathrm{OOB}_b)} e_i(t_b)^2. \tag{8.13}$$

Then, we will randomly permute values of variable k in all of the OOB samples, run the altered OOB observations again down each tree, and calculate a new OOB error estimate for each tree t_b,

$$error_{OOB}(t_b, k) = \frac{1}{N(\mathrm{OOB}_b)} \sum_{i=1}^{N(\mathrm{OOB}_b)} e_i(t_b, k)^2. \tag{8.14}$$

The importance of variable k in tree t_b can be defined as the difference between these two error estimates,

$$\mathcal{I}_{OOB}(k, t_b) = error_{OOB}(t_b, k) - error_{OOB}(t_b), \tag{8.15}$$

and the importance of this variable for the random forest as:

$$\mathcal{I}_{OOB}(k) = \frac{1}{B} \sum_{b=1}^{B} \mathcal{I}_{OOB}(k, t_b). \tag{8.16}$$

Another importance measure may be calculated as the average decrease in impurities (which can be represented by the decrease in the overall sum of squared errors) due to splits on a particular variable. Such importance of variable k in tree t_b can be calculated as

$$\mathcal{I}_{impurity}(k, t_b) = \sum_{nodes(t_b, k)} \Delta(k, *), \tag{8.17}$$

where $\Delta(k, *)$ represents the decrease in the overall sum of squares provided by a split on variable k regardless of the split value s used, see Equation (8.7), and the summation is over all nodes of tree t_b that were split on this variable.

By averaging over all B trees of the forest, we can calculate the importance of variable k for the random forest,

$$\mathcal{I}_{impurity}(k) = \frac{1}{B} \sum_{b=1}^{B} \mathcal{I}_{impurity}(k, t_b). \tag{8.18}$$

Although implementations of variable importance measures may differ in details of their calculations, the general ideas are the same – either quantifying the increase in prediction error estimate after the variable is perturbed or aggregating the increase in homogeneity (purity) when nodes are split on the variable. For example, some implementations may have an option to additionally scale the variable importance by dividing it by its standard error, and some may assign a score of 100 to the most important

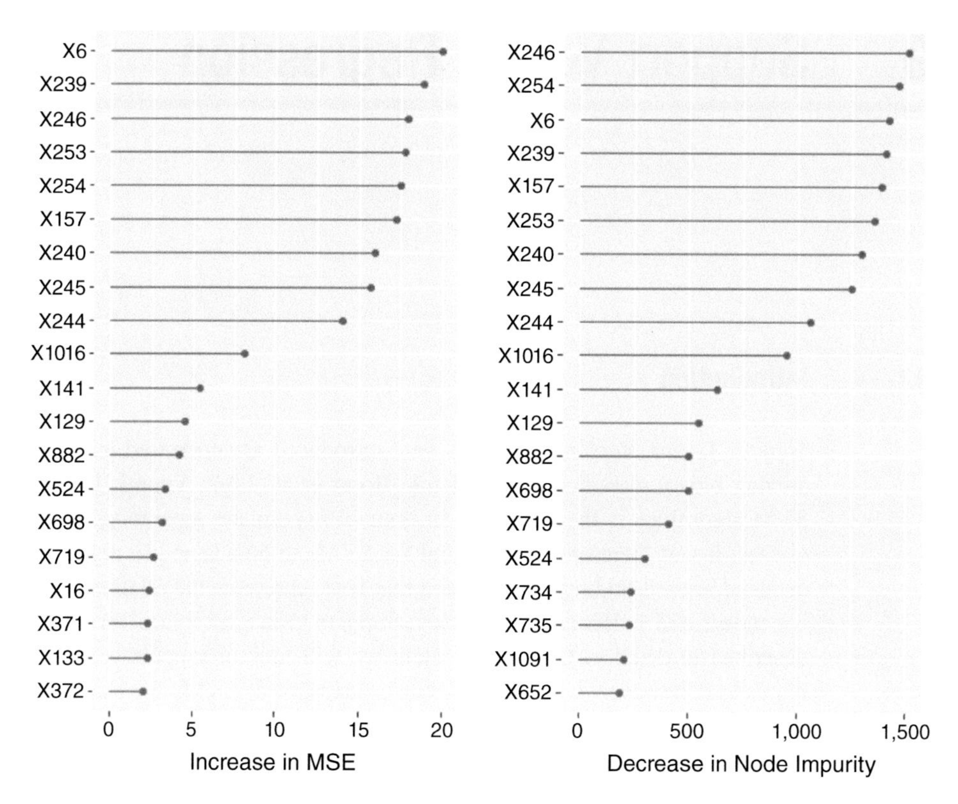

Figure 8.4 An example of variable importance plots. The left panel shows variables ordered by their importance measured by the increase in the OOB estimate of the prediction error after the values of the variable are randomly permuted in the OOB samples. In the right panel, variable importance is measured by the average decrease in node impurity when a node is split on the variable. Although the order of variables is different, the sets of 16 most important variables for both importance measures are the same (the data included more than a thousand variables).

variable and express the other scores as a percent of this maximal score. Usually the results are presented graphically (see example in Figure 8.4).

Observe, however, that importance calculations are performed independently for each variable, and that variables have the same chance of being selected for consideration at a node. Hence, correlations between variables may heavily influence their importance scores. For example, a group of highly correlated variables may have their individual importance decreased as random forests tend to "distribute" their importance across all of them.

9 Support Vector Regression

9.1 Introduction

Support Vector Machines (SVMs) were introduced as supervised learning algorithms solving binary classification problems (Boser et al. 1992; Vapnik 1998).[1] Central to SVM algorithms is the notion of an optimal hyperplane separating the differentiated classes. It was therefore quite natural for SVMs to be adapted to regression problems centered around finding a hyperplane that best represents the relationship between the dependent and independent variables. Support vector machines for classification are described in detail in Chapter 11. Here we will look at support vector machines for regression; however, we will start with a snapshot of SVMs for classification, which will allow for drawing analogies between SVMs for regression and classification. We will also describe here a few alternative error functions for regression, including *ε-insensitive* error functions that are used by *Support Vector Regression* (SVR).

9.1.1 A Snapshot of Support Vector Machines for Classification

SVMs designed for linearly-separable training data are called *hard-margin support vector machines*. Let's assume that training data include N data points and p variables, and that each data point $i, i=1, \ldots, N$, is represented by a $p \times 1$ vector $\mathbf{x}_i = \left[x_{1i}, \ldots, x_{pi}\right]^T$ and a class label $y_i \in \{+1, -1\}$. Then, the solution for hard-margin SVMs is represented by an optimal separating hyperplane, $\beta_0 + \mathbf{x}^T\boldsymbol{\beta} = 0$ and two support hyperplanes, $\beta_0 + \mathbf{x}^T\boldsymbol{\beta} = +1$ and $\beta_0 + \mathbf{x}^T\boldsymbol{\beta} = -1$, which are identified by solving the following convex optimization problem:

$$\begin{aligned} &\underset{\boldsymbol{\beta}, \beta_0}{\text{minimize}} \ \frac{1}{2}\|\boldsymbol{\beta}\|^2 \\ &\text{subject to} \ \ y_i\left(\beta_0 + \mathbf{x}_i^T\boldsymbol{\beta}\right) \geq 1, \quad i=1, \ldots, N, \end{aligned} \tag{9.1}$$

where vector $\boldsymbol{\beta}$ determines orientation of the separating hyperplane, and β_0 is the offset of the hyperplane from the origin. For the hard-margin SVMs, no margin violations are allowed, that is, no training data points are allowed in the area between the two support hyperplanes.

[1] There are extensions of SVMs to multiclass problems.

SVM classifiers that allow for margin violations (including misclassification of some of the training data points if they are on the wrong side of the separating hyperplane) are called *soft-margin support vector machines*. The optimal solution for them is based on a tradeoff between maximizing the margin (the distance between the support hyperplanes) and minimizing the cost of its violations. This solution can be found by solving the following optimization problem:

$$\begin{aligned} &\underset{\boldsymbol{\beta},\beta_0}{\text{minimize}} \ \frac{1}{2}\|\boldsymbol{\beta}\|^2 + C\sum_{i=1}^{N}\xi_i \\ &\text{subject to} \ \ y_i\left(\beta_0 + \mathbf{x}_i^T\boldsymbol{\beta}\right) \geq 1 - \xi_i, \\ &\qquad \xi_i \geq 0, \quad \text{for } i = 1, \ldots, N, \end{aligned} \tag{9.2}$$

where ξ_i, $i = 1, \ldots, N$, are the slack variables quantifying margin violations of training data points, and C is the tradeoff parameter regulating the extent of overlap between the classes.

Problems (9.1) and (9.2) are finding optimal solutions with regard to $p + 1$ variables – the p-dimensional orientation vector $\boldsymbol{\beta}$ and the scalar offset β_0. These optimization problems may be converted into their *dual representations*, and then solved with regard to N variables (see Chapter 11); this would be much more efficient when we deal with high-dimensional data, especially when $p \gg N$.

Neither of the two solutions would, however, work if the boundary between the classes is inherently nonlinear. In such situations, we may map the p-dimensional space defined by the p independent variables into such a higher-dimensional *feature space*, in which the classes are linearly separable. Furthermore, the *kernel trick* allows, in such situations, for finding an optimal solution without explicit mapping of the data into the feature space.

9.1.2 ε-Insensitive Error Functions

Recall that – in regression – the error (or residual) is defined for each of the N training observations $\mathbf{x}_i$, $i = 1, \ldots, N$ as the difference between the observed and the predicted values of the response variable, $y_i - \hat{y}_i$. Recall also (from Chapter 6) that multiple regression minimizes the sum of squared errors, $SSE = \sum_{i=1}^{N}(y_i - \hat{y}_i)^2$, in order to find an optimal regression hyperplane via the least squares criterion. This means that the training data points with large errors (especially outliers or influential observations) may have a very significant impact on the solution. To make the solution more robust (that is, less sensitive to large residuals), the Huber error (or loss) function has been proposed (Huber 1964) – it uses squared errors when they are not greater than a specified threshold δ, and the absolute values of the errors when they are above the threshold,

$$loss_{Huber} = \begin{cases} \frac{1}{2}(y_i - \hat{y}_i)^2 & \text{if } |y_i - \hat{y}_i| \leq \delta \\ \delta|y_i - \hat{y}_i| - \frac{1}{2}\delta^2 & \text{otherwise.} \end{cases} \tag{9.3}$$

The Huber loss function is represented by the blue curve in Figure 9.1.

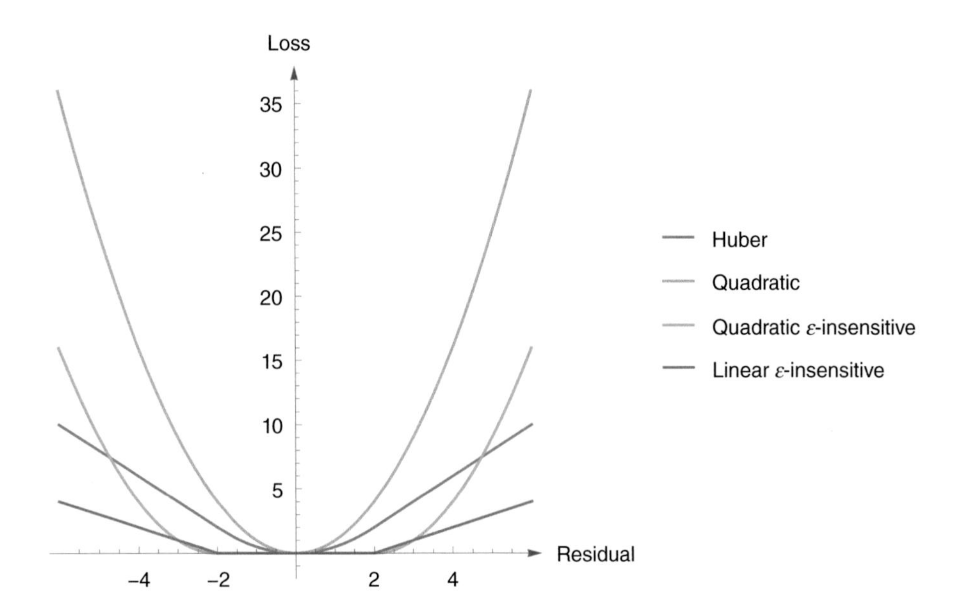

Figure 9.1 Loss functions: quadratic (orange curve), Huber (blue curve), linear ε-insensitive (red curve), and quadratic ε-insensitive (green curve). In this example, both threshold δ for the Huber loss function and ε are equal to 2.

There are other robust loss functions that can be used in regression. Support vector regression usually uses the *linear ε-insensitive loss function* (Vapnik 2000), which disregards the errors that are not greater than ε (a parameter) and uses the absolute values of the errors that are greater than the ε threshold,[2]

$$loss_{\varepsilon,\, linear} = \begin{cases} 0 & \text{if } |y_i - \hat{y}_i| \leq \varepsilon \\ |y_i - \hat{y}_i| - \varepsilon & \text{otherwise.} \end{cases} \tag{9.4}$$

Some versions of support vector regression may, instead, use the *quadratic ε-insensitive loss function* (Vapnik 2000), in which squared errors are used for the errors above the ε threshold,

$$loss_{\varepsilon,\, quadratic} = \begin{cases} 0 & \text{if } |y_i - \hat{y}_i| \leq \varepsilon \\ (|y_i - \hat{y}_i| - \varepsilon)^2 & \text{otherwise.} \end{cases} \tag{9.5}$$

The two ε-insensitive loss functions are also graphed in Figure 9.1 – the linear is represented by the red curve, and the quadratic by the green one.

9.2 Linear Support Vector Regression

The solution for a regression problem is represented by a hyperplane that best approximates the relationship between a continuous response variable y and p independent variables x_1,

[2] Alternatively, loss function (9.4) can be written in the form $loss_{\varepsilon,\, linear} = \max\{0, |y_i - \hat{y}_i| - \varepsilon\}$.

$x_2, \ldots, x_p$. Since the SVM solution for a classification problem is also a hyperplane (though in this case such a hyperplane that best separates the classes), it seems quite natural to adapt SVMs to regression problems. Although we use the same $\boldsymbol{\beta}$ and β_0 notation for both the support vector classification and support vector regression, these parameters have different interpretations (for classification, $\boldsymbol{\beta}$ represents a vector of weights of the p independent variables, this vector is normal to the separating hyperplane and thus, defines its orientation, and β_0 is the offset of the hyperplane from the origin of the p-dimensional space).

Observe also that for classification SVMs, the solution is a $(p - 1)$-dimensional hyperplane in a p-dimensional space of the p independent variables; the dependent variable, the class label, is a meta-attribute of each training observation, rather than an additional geometrical dimension. For regression, however, the solution is a p-dimensional hyperplane in a $(p + 1)$-dimensional space defined by one dependent and p independent variables.

Let's start with the linear regression hyperplane,

$$y = \beta_0 + \mathbf{x}^T\boldsymbol{\beta}, \tag{9.6}$$

and then imagine a *hypertube* with a radius of ε surrounding this hyperplane in such a way that all training data points are within the hypertube. Let's call this hypertube the *ε-tube*. If we assume that each training data point is represented by a p-dimensional vector $\mathbf{x}_i = \left[x_{1i}, \ldots, x_{pi}\right]^T$ and a value of a continuous response variable y_i, where $i = 1, \ldots, N$, then it means that the residual for each training point (which is calculated here as the distance between the observed value of the response variable for training point i and the value predicted by the hyperplane), is bounded by ε,

$$|\, y_i - \left(\beta_0 + \mathbf{x}_i^T\boldsymbol{\beta}\right) | \leq \varepsilon. \tag{9.7}$$

Clearly, there is an infinite number of such hyperplane/hypertube pairs, and the optimal solution would have the training observations as close to the surface of the ε-tube as possible. This could be seen as maximizing the margin $2\varepsilon/\|\boldsymbol{\beta}\|$ (Schölkopf et al. 2000). Since this margin corresponds to the width of the ε-tube in the direction orthogonal to the y axis,[3] maximizing this margin for fixed $\varepsilon > 0$ can be geometrically interpreted as searching for a hyperplane $y = \mathbf{x}^T\boldsymbol{\beta} + \beta_0$ that is as flat as possible (which corresponds to seeking a small norm of $\boldsymbol{\beta}$), under the condition that all training data points are within the ε-tube (that is, not farther form this central hyperplane than ε). This can be presented as a convex optimization problem:[4]

$$\begin{aligned} &\underset{\boldsymbol{\beta}, \beta_0}{\text{minimize}} \ \frac{1}{2}\|\boldsymbol{\beta}\|^2 \\ &\text{subject to } \left|y_i - \left(\beta_0 + \mathbf{x}_i^T\boldsymbol{\beta}\right)\right| \leq \varepsilon \quad \text{for } i = 1, \ldots, N. \end{aligned} \tag{9.8}$$

[3] That is, when the ε-tube defined in the $(p + 1)$-dimensional space spanned by y and the p independent variables is sliced by a p-dimensional hyperplane orthogonal to the y axis.

[4] Note that the factor ½ is sometimes included in the literature on this topic, and sometimes not; whether it is included or not, the solution remains the same (Boser et al. 1992). One may find, however, that it is convenient to include it; for example, when calculating the derivative (9.11).

Since the constraint of problem (9.8) means that none of the training data points can be outside of the ε-tube, this version of support vector regression is analogous to the hard-margin SVMs problem (9.1) for classification. Satisfying this condition is often infeasible. Therefore, analogous to the soft-margin classification SVMs, we will allow for some training data points to lie outside the ε-tube. To quantify their distance from the tube (as well as the cost of the error), we will introduce nonnegative slack variables associated with training observations $\mathbf{x}_i$, $i=1, \ldots, N$. However, we will now need two slack variables, $\xi_i \geq 0$ and $\xi_i^* \geq 0$, $i=1, \ldots, N$, to account for both positive and negative residuals. The training points that are outside of the ε-tube will have one of their slack variables positive, and for the training points inside the ε-tube (including on its surface) both slack variables will have a value of zero.

9.2.1 The Primal Optimization Problem for Linear SVR

The *primal* optimization problem will now be analogous to (9.2), which for SVR means that the optimal regression hyperplane will be a "tradeoff" between maximizing the flatness of the ε-tube and minimizing the penalty based on the slack variables (Smola and Schölkopf 2004),

$$\begin{aligned} \underset{\boldsymbol{\beta}, \beta_0}{\text{minimize}} \; & \frac{1}{2}\|\boldsymbol{\beta}\|^2 + C\sum_{i=1}^{N}(\xi_i + \xi_i^*) \\ \text{subject to} \; & y_i - (\beta_0 + \mathbf{x}_i^T\boldsymbol{\beta}) \leq \varepsilon + \xi_i, \\ & (\beta_0 + \mathbf{x}_i^T\boldsymbol{\beta}) - y_i \leq \varepsilon + \xi_i^*, \\ & \xi_i \geq 0, \xi_i^* \geq 0 \quad \text{for } i=1, \ldots, N, \end{aligned} \tag{9.9}$$

where $C > 0$ is a regularization (or tradeoff) parameter. Since the penalty term is a linear function of the slack variables and since these slack variables quantify the excess of residuals beyond ε, this optimization problem implements the *linear ε-insensitive loss function* (Vapnik 2000).

An important (and somewhat counterintuitive) aspect of support vector regression is that the resulting predictive model does not depend on the training data points that are inside the ε-tube (since they all have $\xi_i = \xi_i^* = 0$). Depending on the value of ε, this may mean a majority or even almost all training data points; thus, in extreme situations the model design could be influenced mostly (or only) by outliers.

9.2.2 The Dual Optimization Problem for Linear SVR

The primal optimization problem (9.9) is a convex objective function with linear constraints. An efficient way of solving such problems is to use a Lagrange function with a dual set of variables. We need four N-dimensional vectors of nonnegative Lagrange multipliers, $\boldsymbol{\alpha}$, $\boldsymbol{\alpha}^*$, $\boldsymbol{\eta}$, $\boldsymbol{\eta}^*$, whose elements, α_i, α_i^*, η_i, η_i^*, are associated with training data points $\mathbf{x}_i$, $i = 1, \ldots, N$. They will be used in the following *primal* Lagrange function:[5]

[5] A general form of a Lagrangian includes the objective function and the terms in the form of (Lagrange multiplier $*$ constraint); since problem (9.9) has four constraints, we need four Lagrange multipliers for each training data point.

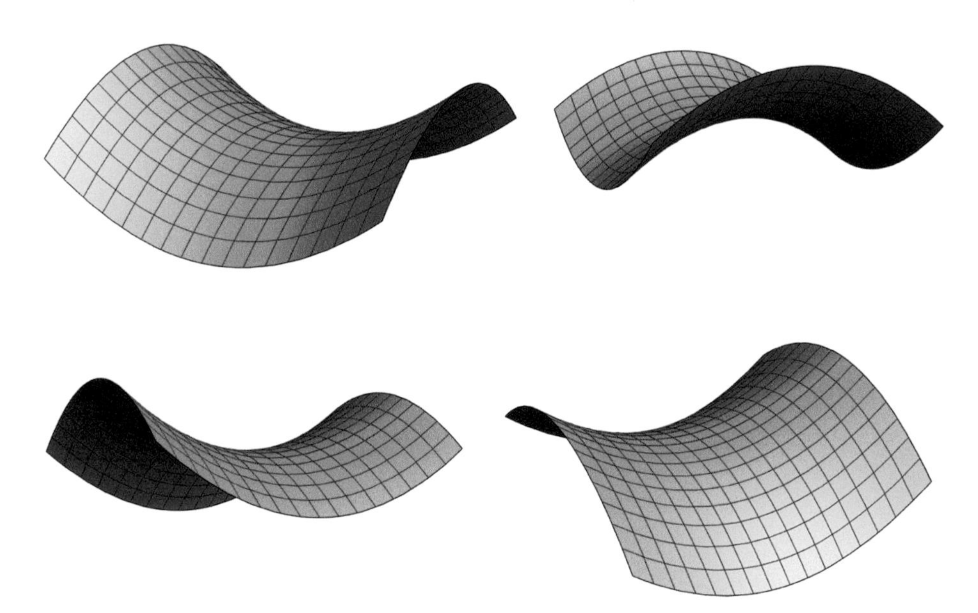

Figure 9.2 An example of a saddle point for a quadratic optimization problem represented by a simple function of only two variables; the surface of this function is shown from different angles.

$$
\begin{aligned}
L_P(\boldsymbol{\beta}, \beta_0, \xi_i, \xi_i^*) = \frac{1}{2}\|\boldsymbol{\beta}\|^2 + C\sum_{i=1}^{N}(\xi_i + \xi_i^*) - \sum_{i=1}^{N}(\eta_i\xi_i + \eta_i^*\xi_i^*) \\
- \sum_{i=1}^{N}\alpha_i(\varepsilon + \xi_i - y_i + \beta_0 + \mathbf{x}_i^T\boldsymbol{\beta}) \\
- \sum_{i=1}^{N}\alpha_i^*(\varepsilon + \xi_i^* + y_i - \beta_0 - \mathbf{x}_i^T\boldsymbol{\beta}).
\end{aligned}
\tag{9.10}
$$

This Lagrangian has a saddle point (for an example of a saddle point see Figure 9.2); in order to identify it, we will start with finding its minimum with respect to the primal variables $\boldsymbol{\beta}$, β_0, ξ_i, and ξ_i^* by setting the partial derivatives to zero,

$$\frac{\delta L_P}{\delta \boldsymbol{\beta}} = \boldsymbol{\beta} - \sum_{i=1}^{N}(\alpha_i - \alpha_i^*)\mathbf{x}_i^T = 0, \quad \text{hence } \boldsymbol{\beta} = \sum_{i=1}^{N}(\alpha_i - \alpha_i^*)\mathbf{x}_i^T \tag{9.11}$$

$$\frac{\delta L_P}{\delta \beta_0} = \sum_{i=1}^{N}(\alpha_i^* - \alpha_i) = 0 \tag{9.12}$$

$$\frac{\delta L_P}{\delta \xi_i} = C - \eta_i - \alpha_i = 0, \quad \text{hence } \eta_i = C - \alpha_i, \quad i = 1, \ldots, N, \tag{9.13}$$

$$\frac{\delta L_P}{\delta \xi_i^*} = C - \eta_i^* - \alpha_i^* = 0, \quad \text{hence } \eta_i^* = C - \alpha_i^*, \quad i = 1, \ldots, N. \tag{9.14}$$

By plugging the results of (9.11)–(9.14) into (9.10), we will obtain the *dual* Lagrange function, which depends only on the values of vectors $\boldsymbol{\alpha}$ and $\boldsymbol{\alpha}^*$,

$$L_D(\boldsymbol{\alpha}, \boldsymbol{\alpha}^*) = \left[-\frac{1}{2} \sum_{i,j=1}^{N} (\alpha_i - \alpha_i^*)(\alpha_j - \alpha_j^*) \mathbf{x}_i^T \mathbf{x}_j \right. \\ \left. -\varepsilon \sum_{i=1}^{N} (\alpha_i + \alpha_i^*) + \sum_{i=1}^{N} y_i (\alpha_i - \alpha_i^*) \right]. \tag{9.15}$$

This leads to the *dual representation* of the optimization problem (Smola and Schölkopf 2004):

$$\underset{\boldsymbol{\alpha}, \boldsymbol{\alpha}^*}{\text{maximize}} \left[-\frac{1}{2} \sum_{i,j=1}^{N} (\alpha_i - \alpha_i^*)(\alpha_j - \alpha_j^*) \mathbf{x}_i^T \mathbf{x}_j \right. \\ \left. -\varepsilon \sum_{i=1}^{N} (\alpha_i + \alpha_i^*) + \sum_{i=1}^{N} y_i (\alpha_i - \alpha_i^*) \right] \tag{9.16} \\ \text{subject to} \quad \sum_{i=1}^{N} (\alpha_i - \alpha_i^*) = 0 \quad \text{and} \quad \alpha_i, \alpha_i^* \in [0, C].$$

For high-dimensional $p \gg N$ data, the dual representation of the optimization problem is not only easier to solve, but is also computationally much more efficient since it is solved with respect to only N variables. Using the dual representation also allows for easy extension of SVR to nonlinear cases.

A solution to (9.16) has to satisfy the following Karush–Kuhn–Tucker (KKT) complementarity conditions:

$$\alpha_i \left(\varepsilon + \xi_i - y_i + \beta_0 + \mathbf{x}_i^T \boldsymbol{\beta} \right) = 0 \tag{9.17}$$

$$\alpha_i^* \left(\varepsilon + \xi_i^* + y_i - \beta_0 - \mathbf{x}_i^T \boldsymbol{\beta} \right) = 0 \tag{9.18}$$

$$(C - \alpha_i)\xi_i = 0 \tag{9.19}$$

$$\left(C - \alpha_i^* \right) \xi_i^* = 0. \tag{9.20}$$

These KKT conditions specify that each product of (Lagrange multiplier $*$ constraint) – these products were added to the objective function in Lagrangian (9.10) – must equal zero (that is, has to vanish) at the solution to the dual optimization problem (9.16).

From (9.17) we have that α_i can be nonzero only when $\varepsilon + \xi_i - y_i + \beta_0 + \mathbf{x}_i^T \boldsymbol{\beta} = 0$, that is, when the residual of training data point i, $y_i - \left(\beta_0 + \mathbf{x}_i^T \boldsymbol{\beta} \right)$, is either equal to ε and $\xi_i = 0$, or when this residual is equal to $\varepsilon + \xi_i$ and $\xi_i > 0$. In the first case, the data point lies on the upper surface of the ε-tube, and in the second – above the ε-tube (see Figure 9.3). Similarly, constraint (9.18) ensures that α_i^* can be nonzero only when the training data point lies on

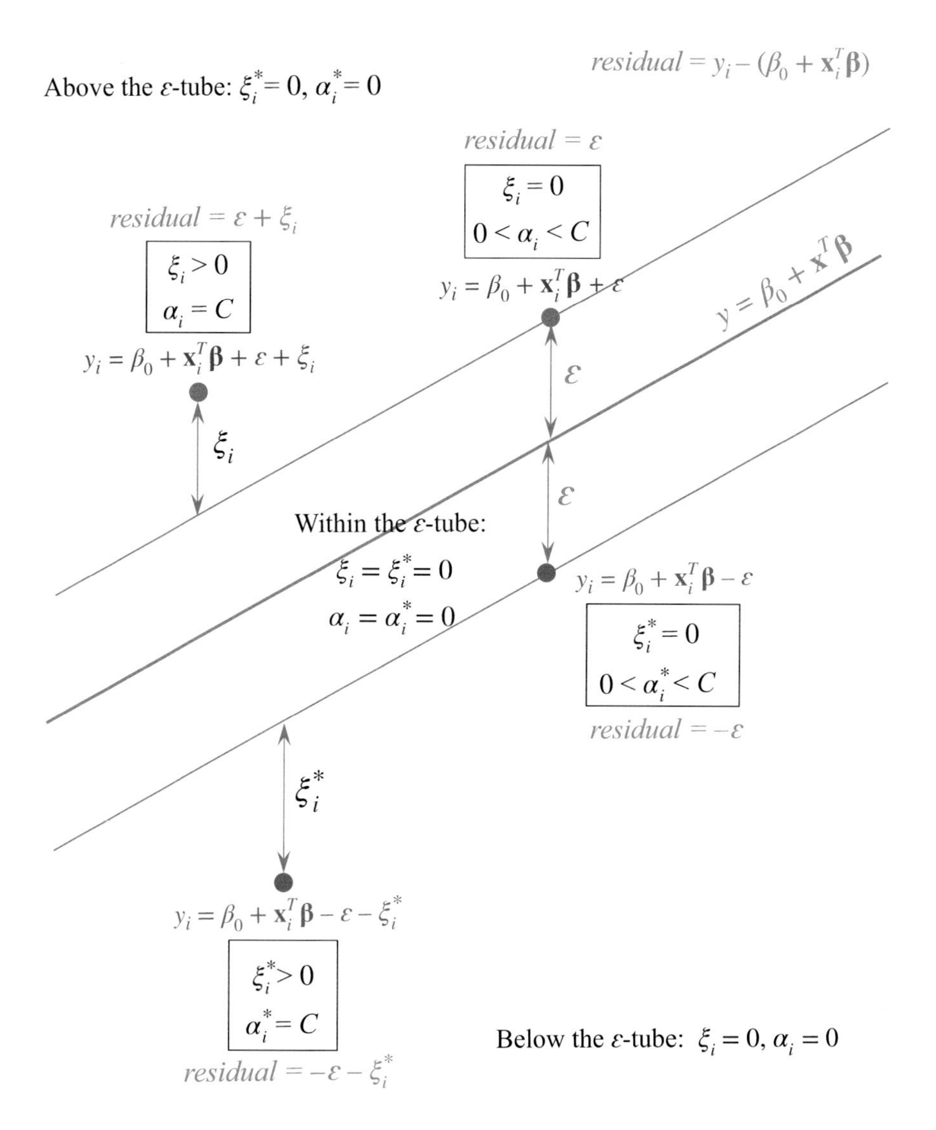

Figure 9.3 ε-tube support vector regression. Training data points that are within the ε-tube have $\alpha_i = \alpha_i^* = 0$ and hence, do not contribute to the solution. The points that are either above the ε-tube $(\alpha_i = C)$ or below it $(\alpha_i^* = C)$ and the points on the surface of the ε-tube ($0 < \alpha_i < C$ or $0 < \alpha_i^* < C$) are the support vectors (that is, they contribute to prediction).

the lower surface of the ε-tube or below it.[6] This means that all data points within the ε-tube must have $\alpha_i = \alpha_i^* = 0$. Furthermore, from (9.19) and (9.20) we have that all points outside the ε-tube (for which either $\xi_i > 0$ or $\xi_i^* > 0$) must have $\alpha_i = C$ or $\alpha_i^* = C$, respectively,

[6] "above" or "below" refer here to positions measured along the y axis, that is, along the single direction – in the $(p + 1)$-dimensional space – corresponding to the values of the response variable.

and the points on the surface of the ε-tube have either $0 < \alpha_i < C$ (upper surface) or $0 < \alpha_i^* < C$ (lower surface).

9.2.3 Linear Support Vector Regression Machine

Solving dual optimization problem (9.16) provides the estimate of $\boldsymbol{\beta}$,

$$\hat{\boldsymbol{\beta}} = \sum_{i=1}^{N} (\alpha_i - \alpha_i^*) \mathbf{x}_i, \tag{9.21}$$

as a linear combination of the vectors representing the training data points that are either outside or on the surface of the ε-tube. For this reason, Equation (9.21) is called *support vector expansion.*

An estimate of β_0 can be calculated from (9.17) for any training data point on the upper surface of the ε-tube; for such points $0 < \alpha_i < C$ and $\xi_i = 0$. Alternatively, it can be calculated from (9.18) for any training point on the lower surface of the ε-tube; for such points $0 < \alpha_i^* < C$ and $\xi_i^* = 0$,[7]

$$\hat{\beta}_0 = \begin{cases} y_i - \varepsilon - \mathbf{x}_i^T \hat{\boldsymbol{\beta}} & \text{for } 0 < \alpha_i < C \\ y_i + \varepsilon - \mathbf{x}_i^T \hat{\boldsymbol{\beta}} & \text{for } 0 < \alpha_i^* < C. \end{cases} \tag{9.22}$$

Therefore, the *estimated regression equation* $\hat{y}(\mathbf{x}) = \hat{\beta}_0 + \mathbf{x}^T \hat{\boldsymbol{\beta}}$ can now be presented as:

$$\hat{y}(\mathbf{x}) = \hat{\beta}_0 + \sum_{i=1}^{N} (\alpha_i - \alpha_i^*) \mathbf{x}^T \mathbf{x}_i. \tag{9.23}$$

Since only the training data points that have either $\alpha_i > 0$ or $\alpha_i^* > 0$ (that is, the points that are either outside of the ε-tube or on its surface) are used for predictions performed via (9.23), they are the *support vectors*, and the predictive model itself can be called a *support vector regression machine.*

Furthermore, from (9.12), which can be presented as $\sum_{i=1}^{N} \alpha_i = \sum_{i=1}^{N} \alpha_i^*$, we may infer that the optimal SVR solution is balancing the influence of the support vectors on both sides of the ε-tube.

Linear support vector regression has two hyperparameters that require tuning: ε – the radius of the hypertube, and C – the penalty (or regularization) parameter. The values of an optimal combination of ε and C may be found by a grid search driven by some estimate of the prediction accuracy. This is usually done via either the cross-validation or bootstrap-based approach.

Alternatively, instead of controlling the size of the tube directly via ε (such a support vector regression algorithm is called *ε-SVR*), we may modify the algorithm to use the proportion of training data points, ν, $0 \leq \nu \leq 1$, that are allowed to be outside the tube. This *ν-SVR* algorithm (which also uses the ε-insensitive loss function) automatically determines ε by controlling the hyperparameter ν (Schölkopf and Smola 2002).

[7] In practice, however, it is better to calculate the estimate of β_0 for every support vector lying on the upper or lower surface of the ε-tube, and set $\hat{\beta}_0$ to the average of those estimates (Bishop 2006).

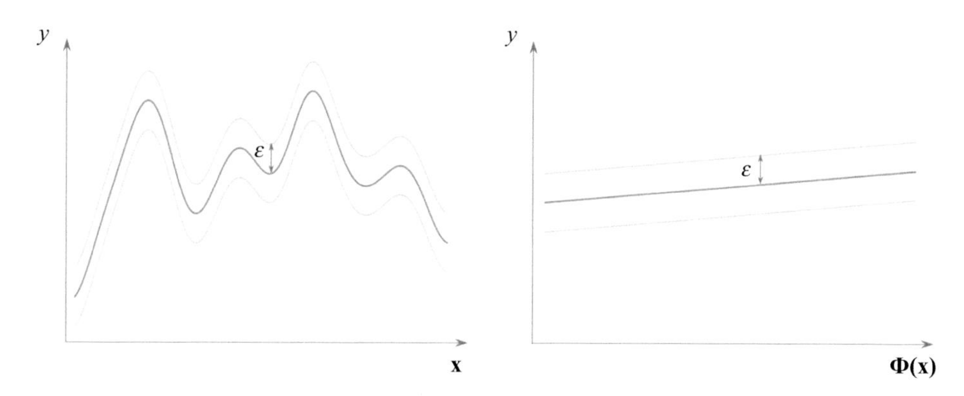

Figure 9.4 Nonlinear support vector regression. Left panel shows ε-tube for nonlinear support vector regression in the original space of the regression problem. The right panel shows the results of mapping the original data space (and the ε-tube) into a feature space in which the relationship is linear. Using a kernel function allows for solving such nonlinear regression problems without explicitly performing the mapping.

9.3 Nonlinear Support Vector Regression

If the relationship between the dependent variable and the p independent variables is inherently nonlinear (and thus, cannot be well approximated by a hyperplane), we may map the original $(p + 1)$-dimensional data space into a *feature space* (usually a higher-dimensional one) in which the relationship is linear. Theoretically, instead of using each training data point $\mathbf{x}_i$ we could use its mapping $\Phi(\mathbf{x}_i)$ and find the SVR solution explicitly in the higher-dimensional feature space (see Figure 9.4). However, this approach would often prove to be computationally infeasible. An efficient approach is based on the observation that dual optimization problem (9.16) depends only on the inner products[8] $\mathbf{x}_i^T\mathbf{x}_j$, $i,j = 1, \ldots, N$ between the training observations. Therefore, instead of explicitly using the inner products of their mappings $\Phi(\mathbf{x}_i)^T\Phi(\mathbf{x}_j)$, it is enough to know and use some nonlinear kernel function $K(\mathbf{x}_i, \mathbf{x}_j) = \Phi(\mathbf{x}_i)^T\Phi(\mathbf{x}_j)$ whose parameters are vectors representing the original training observations. Applying this *kernel trick* to (9.16) will result in the following dual optimization problem:

$$\underset{\boldsymbol{\alpha},\,\boldsymbol{\alpha}^*}{\text{maximize}} \left[-\frac{1}{2}\sum_{i,j=1}^{N}\left(\alpha_i - \alpha_i^*\right)\left(\alpha_j - \alpha_j^*\right)K\left(\mathbf{x}_i, \mathbf{x}_j\right) - \varepsilon\sum_{i=1}^{N}\left(\alpha_i + \alpha_i^*\right) + \sum_{i=1}^{N} y_i\left(\alpha_i - \alpha_i^*\right) \right] \tag{9.24}$$

$$\text{subject to } \sum_{i=1}^{N}\left(\alpha_i - \alpha_i^*\right) = 0 \text{ and } \alpha_i, \alpha_i^* \in [0, C].$$

[8] The inner product of two p-dimensional vectors $\mathbf{x}_i$ and $\mathbf{x}_j$ is calculated as $\mathbf{x}_i^T\mathbf{x}_j = \sum_{k=1}^{p} x_{ik}x_{jk}$; the result is a scalar.

By finding a linear solution in the feature space, this optimization problem provides a solution to a nonlinear relationship in the original data space without explicitly performing the mapping of the training data points. Hence, for nonlinear cases, predictions can be made by using the following estimated regression equation:

$$\hat{y}(\mathbf{x}) = \hat{\beta}_0 + \sum_{i=1}^{N} (\alpha_i - \alpha_i^*) K(\mathbf{x}_i, \mathbf{x}). \tag{9.25}$$

Among popular kernel functions are: the radial basis function (RBF) kernel, the polynomial kernel, and the hyperbolic tangent (sigmoid) kernel. The RBF kernel,

$$K(\mathbf{x}_i, \mathbf{x}_j) = e^{-\gamma \|\mathbf{x}_i - \mathbf{x}_j\|^2}, \tag{9.26}$$

is an example of kernels that have an infinite-dimensional feature space. This means that the explicit mapping of the training data points into the feature space would be impossible. The RBF kernel is quite universal since linear combinations of this kernel can approximate any continuous function. For the polynomial kernel,

$$K(\mathbf{x}_i, \mathbf{x}_j) = (c + \gamma \mathbf{x}_i^T \mathbf{x}_j)^d, \tag{9.27}$$

the linear relationship in the feature space will correspond to a polynomial curve of degree d in the original data space. The sigmoid function,

$$K(\mathbf{x}_i, \mathbf{x}_j) = \tanh(c + \gamma \mathbf{x}_i^T \mathbf{x}_j), \tag{9.28}$$

is a kernel only for some values of c and γ, but is quite popular in some applications.

Parameters in kernel functions (9.26)–(9.28) are $\gamma > 0$, $c \geq 0$, and d is a positive integer. They also require tuning (in addition to the two SVR hyperparameters described earlier). Choosing a kernel for a particular nonlinear modeling situation is generally not a trivial task. Often, we start with trying the RBF kernel or a low-degree polynomial kernel (Izenman 2008).

Part III

Classification Methods

10 Classification with Random Forests

10.1 Introduction

Random Forests is a supervised learning algorithm employing the ensemble approach. It randomly resamples the training data to generate many (hundreds or thousands) bootstrap samples by selecting them with replacement. Although each of the bootstrap samples is of the same size as the original training set, they include, on average, only about 63.2 percent of the unique training observations. Each bootstrap sample is treated as a second-order training set that is used to build a *classification tree*. Once the tree is built, its classification performance may be tested by running the training observations that were not included in the bootstrap sample (the remaining about 36.8 percent of the observations from the original training set) down the tree; those observations are called the *out-of-bag* (OOB) sample. After all the trees are built, without pruning, they constitute an ensemble classifier.

Recall from Chapter 8 that a decision tree (for either classification or regression) is a nonparametric method generating a hierarchical structure of binary decisions. A tree starts with its *root node*, in which all observations from its training set are split into two child nodes in a way that maximizes the homogeneity of the child nodes. For classification, this homogeneity may be generally described in terms of proportions of observations from each of the differentiated classes assigned to the node (a node would be fully homogeneous if it includes observations belonging to only one class). The split is based on the values of only one of the $m < p$ independent variables (which are randomly selected for each node) – the one that provides the best homogeneity for the child nodes. Each of the child nodes is then split in the same way, and this binary partitioning continues until either the terminal nodes (*leaf nodes*) are fully homogeneous, or some other stopping criterion is met (for example, the minimum size of terminal nodes). No pruning of a tree is performed. Similar to a regression tree, a single classification tree is a weak classifier with high variance. Although it may be used with low-dimensional data (mainly due to its easy interpretation for such data), a single classification tree is useless for high-dimensional data. However, growing many trees based on random resamples of the training data and then aggregating their individual predictions may lead to an ensemble classifier with a smaller variance than any of its trees. Nevertheless, a single level of randomization – provided by growing trees from bootstrap samples – is not enough. Such bagged trees are likely to be correlated, which causes the reduction in variance to be smaller than it would be for uncorrelated trees. Hence, the random forests algorithm implements an additional level of randomization, in the node splitting approach, which decorrelates the trees and further decreases the variance of the ensemble classifier.

10.2 Random Forests Algorithm for Classification

The *Random Forests* algorithm for *regression* has been described in Chapter 8. Although the general idea of building a forest of trees from random subsamples of the training data is the same for both classification and regression approaches (see Figure 8.1), there are, of course, significant differences in the details. As with regression, the random forests algorithm for classification (Breiman 2001) performs resampling of the training data using Efron's nonparametric bootstrap approach. Each bootstrap sample is the result of selecting with replacement, is of the same size as the original training data, and is associated with its own *out-of-bag* (OOB) sample. The second level of randomization is also implemented – at each node, a small number of variables, m, is randomly selected, and only those m variables are considered for splitting this node. The best split from among the best splits on each of the m variables is selected to split the node. For classification, the default value of m is $m = \lfloor \sqrt{p} \rfloor$, where p is the number of independent variables in the training set. The advantage of considering small and different subsets of variables at each node is twofold: (i) it decorrelates the trees and thus further reduces the variance of the ensemble classifier (in addition to the reduction resulting from using the bootstrap samples), and (ii) it significantly speeds up the process of building a random forest classifier.[1]

As it was for regression, the random forests algorithm for classification also has only two (and the same) parameters – the number of trees to build and the number of variables m to be randomly selected at each node. Though it depends on the data, the performance (measured, for example, by accuracy, or AUC) usually stabilizes in the range between 500 and 1,000 trees; thus, 1,000 trees should be enough in most situations. However, it is a good idea to check it for new data by building a forest with a larger number of trees (say 2,000), plotting performance measures versus number of trees, and evaluating where it stabilizes (see Figure 10.1). Similarly to regression, the default value of m also usually works well, but of course it may be tuned if one chooses to do so.

Assume that the original training data (that include N observations and p independent variables) were resampled B times, and that bootstrap sample b is used to grow tree $t_b, b = 1, \ldots, B$. Of course, the size of each bootstrap sample is N, the same as that of the original training data; however, unlike the original data, a bootstrap sample consists of N *nonunique* observations. Each observation included in the bootstrap sample is represented by vector $\mathbf{x}_i = [x_{1i}, \cdots, x_{pi}]^T, i = 1, \ldots, N$, and by the label of the class to which the observation belongs, that is, the value y_i of the categorical response variable (with J possible values corresponding to J differentiated classes). When growing tree t_b, each parent node $\mathcal{P}$ is split into two child nodes; let us call them left child node (C_L) and right

[1] When compared, for example, to building an ensemble using the *bagged tree* method, where all p variables are considered at each tree node.

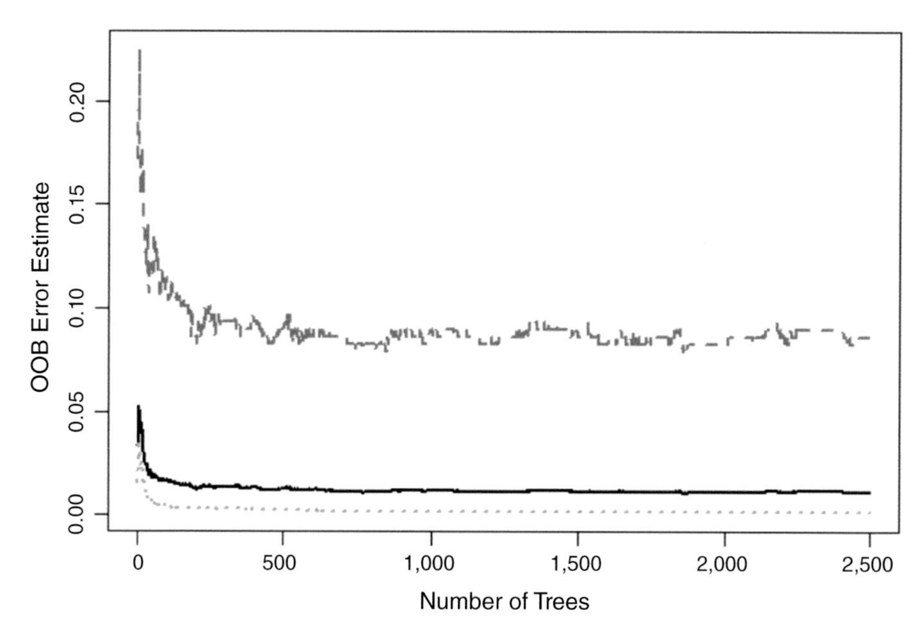

Figure 10.1 An example of a plot showing how random forest classifier's OOB error estimates change with the number of trees, for a binary classification. The black line shows the overall OOB error rate of the classifier (this error rate is associated with accuracy), and the green and red lines correspond to the OOB error rates for each of the two differentiated classes (associated with sensitivity and specificity).

child node (C_R). If the parent node includes $N_{\mathcal{P}}$ observations, $\mathbf{x}_i \in \mathcal{P}, i = 1, \ldots, N_{\mathcal{P}}$ (where $N_{\mathcal{P}} < N$ for all parent nodes except the root one), we want to partition them into two groups (one assigned to the left child node, $\mathbf{x}_i \in C_L, i = 1, \ldots, N_L$, and the other to the right one, $\mathbf{x}_i \in C_R, i = 1, \ldots, N_R, N_L + N_R = N_{\mathcal{P}}$) in such a way that maximizes homogeneity of the child nodes.

10.2.1 Node-Splitting Criteria

A popular criterion used to decide about the best node split is the *Gini criterion*. It is based on the Gini impurity index that is actually a measure of node heterogeneity. This means that a node that is fully homogenous (that is, includes only observations from one class) would have a zero value of the Gini index, and that this index would be maximized if a node includes an equal number of observations from each class. The general form of the Gini index for node $*$ is

$$\begin{aligned} Gini(*) &= \sum_{j \neq h} p_j(*) p_h(*) \\ &= \sum_{j=1}^{J} p_j(*)\big(1 - p_j(*)\big) \\ &= 1 - \sum_{j=1}^{J} \big[p_j(*)\big]^2, \end{aligned} \tag{10.1}$$

where $p_j(*)$ is the probability that an observation assigned to node $*$ belongs to class j, $j, h = 1, \ldots, J$. This probability is estimated by the proportion of class j observations in the node,

$$p_j(*) = \frac{N_*(j)}{N_*}. \tag{10.2}$$

To split any parent node $\mathcal{P}$, we randomly select m out of p variables, and then for each of the selected variables $x_k, k = 1, \ldots, m$, and for each potential split s on this variable, we consider the decrease in heterogeneity associated with split (k, s), measuring this decrease via the *Gini criterion*,[2]

$$\Delta_{Gini}(\mathcal{P}, k, s) = Gini(\mathcal{P}) - \frac{N_L}{N_{\mathcal{P}}} Gini(C_L) - \frac{N_R}{N_{\mathcal{P}}} Gini(C_R). \tag{10.3}$$

Consequently, parent node $\mathcal{P}$ will be split into child nodes C_L and C_R using the split (k, s) that will

$$\underset{k,s}{\text{maximize}} \ \Delta_{Gini}(\mathcal{P}, k, s). \tag{10.4}$$

Another splitting criterion is based on the *entropy impurity* (*EI*), which is calculated for node $*$ as

$$EI(*) = -\sum_{j=1}^{J} p_j(*) \log_2 p_j(*). \tag{10.5}$$

Since entropy may be interpreted as the degree of disorder, the entropy impurity of zero corresponds to a node with all observations belonging to the same class.[3] On the other hand, if the probability for each class is the same, the entropy impurity has its largest value. Following the same reasoning as for the Gini index, the decrease in the entropy impurity for potential split (k, s) is measured by

$$\Delta_{EI}(\mathcal{P}, k, s) = EI(\mathcal{P}) - \frac{N_L}{N_{\mathcal{P}}} EI(C_L) - \frac{N_R}{N_{\mathcal{P}}} EI(C_R), \tag{10.6}$$

and to split the parent node, we will use split (k, s) that will

$$\underset{k,s}{\text{maximize}} \ \Delta_{EI}(\mathcal{P}, k, s). \tag{10.7}$$

The third popular splitting criterion uses the *misclassification impurity* (*MI*),

$$MI(*) = 1 - \max_j p_j(*). \tag{10.8}$$

Like the other two impurity metrics, $MI(*)$ is zero for a homogeneous node and largest for evenly distributed classes. The decrease in the misclassification impurity for potential split (k, s) is

[2] To simplify notation, explicit references to split (k, s) have been dropped from the right side of the equation.

[3] Equation (10.5) is used under the assumption that $0 \cdot \log_2 0$ is replaced with zero.

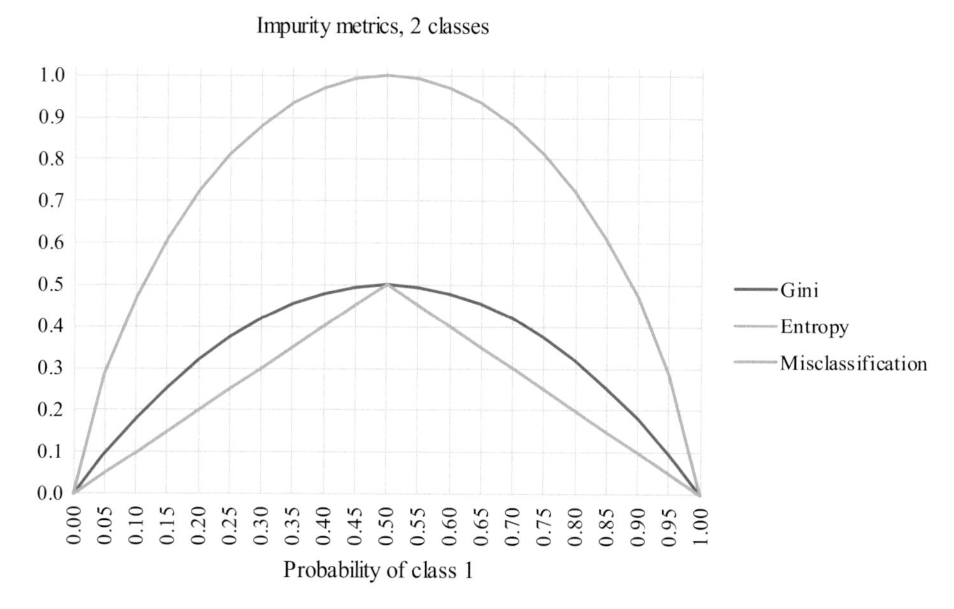

Figure 10.2 Impurity metrics for binary classification ($J = 2$). The horizontal axis represents the probability of class 1, $p_1(*)$, and thus $p_2(*)=1-p_1(*)$. Each of the three impurity metrics is equal to zero when a node includes only observations from one of the classes, and they reach their maximal values when the class probabilities are the same. Observe that the Gini and entropy impurity metrics are differentiable, while the misclassification one is not.

$$\Delta_{MI}(\mathcal{P},k,s)=MI(\mathcal{P})-\frac{N_L}{N_{\mathcal{P}}}MI(C_L)-\frac{N_R}{N_{\mathcal{P}}}MI(C_R), \tag{10.9}$$

and the best split will

$$\underset{k,s}{\text{maximize}}\ \Delta_{MI}(\mathcal{P},k,s). \tag{10.10}$$

The values of the three impurity metrics for binary classification are illustrated in Figure 10.2. In this case, they are symmetric across the range of class probabilities – reaching their maximal values when $p_1(*)=p_2(*)=0.5$, and are equal to zero when the node includes observations from only one of the classes. As an example of classification with more than two classes, Figure 10.3 shows the values of the impurity metrics for classification among four classes.

10.2.2 Classification

Once a tree is built, each of its terminal nodes (leaves) is assigned to one of the J classes, so it may be called a class j terminal node, $j=1,\ldots,J$. A node is deemed terminal either if it is homogeneous or if it was not split because no significant decrease in impurity was possible (Breiman et al. 1984).[4] In the latter case, the node is assigned to

[4] Assuming that no other stopping criterion was set.

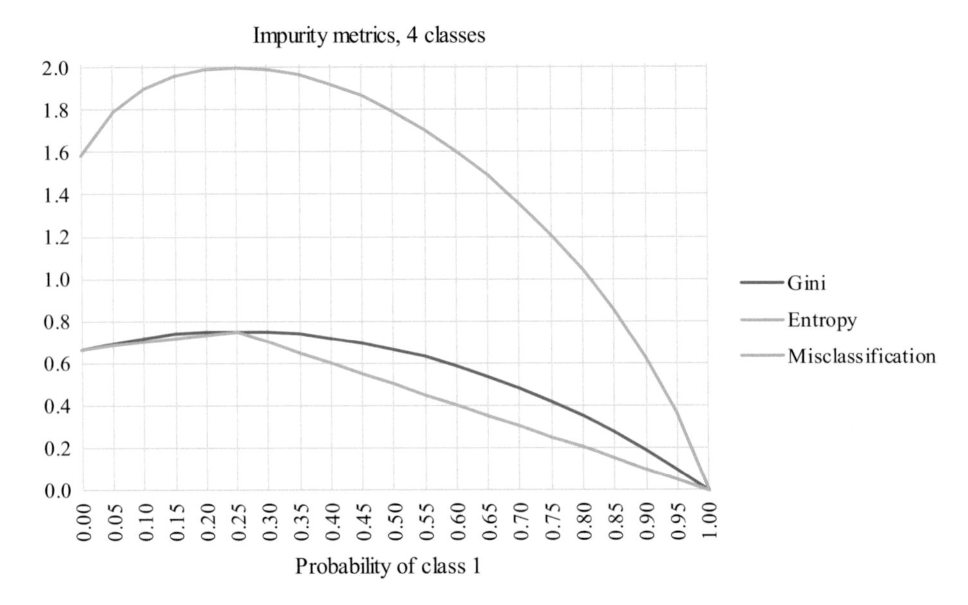

Figure 10.3 An example of impurity metrics for multiclass classification, with $J = 4$. The horizontal axis represents the probability of class 1, $p_1(*)$, and the probabilities of the three other classes are kept the same, $p_2(*) = p_3(*) = p_4(*) = (1 - p_1(*))/3$. In this example, a node is homogeneous only when $p_1(*) = 1$; and only then do all three metrics reach zero. Each of them reaches its maximal value when all of the four class probabilities are the same and equal to 0.25.

the class with the highest representation in the node (the plurality rule). If an observation represented by vector $\mathbf{x}$ is run down tree t_b, it will end up in one of its leaf nodes and will be classified into the class assigned to this node,

$$\widehat{y}(\mathbf{x}, t_b) = Class(\mathbf{x}, t_b). \tag{10.11}$$

To find the OOB estimate of the classification error of a random forest classifier, each training observation $\mathbf{x}_i, i = 1, \ldots, N$ will be classified by only those trees for which the observation was not in the bootstrap sample used to train the tree (that is, $b : \mathbf{x}_i \in OOB_b$). Each of those trees will *vote* for a class, and the observation will be assigned to the class with the largest number of votes (the plurality vote),

$$\widehat{y}(\mathbf{x}_i) = \text{plurality vote } \left\{\widehat{y}(\mathbf{x}_i, t_b)\right\}_{b:\mathbf{x}_i \in OOB_b}. \tag{10.12}$$

Then, the results for all training observations will be aggregated, and the OOB estimate of the classification error will be calculated as the proportion of wrong classifications (when the predicted class $\widehat{y}(\mathbf{x}_i)$ is different from the true class y_i),

$$error_{OOB} = \frac{1}{N} \sum_{i=1}^{N} \mathbf{1}_{y_i \neq \hat{y}(\mathbf{x}_i)}, \tag{10.13}$$

where $\mathbf{1}_{y_i \neq \hat{y}(\mathbf{x}_i)}$ represents the function equal to 1 if the condition is true (i.e., the predicted class of the observation is different from its true class) and otherwise is equal to zero.

To classify a new observation, it will be passed down each of the B trees of the forest and then assigned to the class with the most votes,

$$\begin{aligned} \hat{y}(\mathbf{x}_{new}) &= \text{plurality vote } \{\hat{y}(\mathbf{x}_{new}, t_b)\}_{b=1}^{B} \\ &= \arg\max_j \sum_{b=1}^{B} \mathbf{1}_{\hat{y}(\mathbf{x}_{new}, t_b) = j}. \end{aligned} \tag{10.14}$$

10.3 Variable Importance

Two types of variable importance measures are commonly used – one is based on permuting values of the variable and measuring the decrease in forest classification accuracy (or increase in its OOB misclassification error), and the other on the increase in homogeneity (or decrease in impurity) when a node is split on the variable.

10.3.1 Permutation-Based Variable Importance

Let us denote the set of out-of-bag observations in each tree $t_b, b = 1, \ldots, B$, by OOB_b, and their number by $N(\text{OOB}_b)$. When each of these observations $\mathbf{x}_i \in \text{OOB}_b$ is run down the tree, it is assigned either to its true class y_i, or to any other class and thus the accuracy of classifying observation $\mathbf{x}_i$ by tree t_b is

$$\text{Accuracy}_i(t_b) = \begin{cases} 1 & \text{if } \; y_i = \hat{y}(\mathbf{x}_i, t_b) \\ 0 & \text{if } \; y_i \neq \hat{y}(\mathbf{x}_i, t_b) \end{cases}. \tag{10.15}$$

Hence, the accuracy of classifying all OOB_b observations of tree t_b (OOB accuracy for the tree) is

$$\text{Accuracy}_{OOB}(t_b) = \frac{1}{N(\text{OOB}_b)} \sum_{i=1}^{N(\text{OOB}_b)} \text{Accuracy}_i(t_b). \tag{10.16}$$

This is the "original" OOB accuracy for tree t_b. Now we will randomly permute values of variable k in all OOB_b observations, run those altered observations down tree t_b, save the new accuracy measure for each of them, $\text{Accuracy}_i(t_b, k)$, and calculate, for the tree, the OOB accuracy with variable k permuted,

$$\text{Accuracy}_{OOB}(t_b, k) = \frac{1}{N(\text{OOB}_b)} \sum_{i=1}^{N(\text{OOB}_b)} \text{Accuracy}_i(t_b, k). \tag{10.17}$$

The importance of variable k in tree t_b is then calculated as the difference between the two accuracies (that is, the decrease in accuracy after variable k is permuted),

$$\mathcal{I}_{OOB}(k, t_b) = \text{Accuracy}_{OOB}(t_b) - \text{Accuracy}_{OOB}(t_b, k). \tag{10.18}$$

Averaging over all B trees of the forest provides the overall measure of importance of variable k,

$$\mathcal{I}_{OOB}(k) = \frac{1}{B}\sum_{b=1}^{B}\mathcal{I}_{OOB}(k, t_b), \quad k = 1, \ldots, p. \tag{10.19}$$

10.3.2 Impurity-Based Variable Importance

Since the trees of the forest are grown in such a way that any of the three described impurity indices is greater for a parent node than for its two child nodes, the decrease in impurity averaged over all nodes split by a variable may be used as the measure of variable importance. A commonly used importance measure of this type is based on the Gini impurity index. If we denote the decrease in the Gini index at parent node $\mathcal{P}$ that has been split on variable $k, k = 1, \ldots, p$, as $\Delta_{Gini}(\mathcal{P}, k)$, then from (10.3) and (10.4) we have

$$\Delta_{Gini}(\mathcal{P}, k) = \max \Delta_{Gini}(\mathcal{P}, k, s). \tag{10.20}$$

Since a variable may be used in more than one split in a tree, the importance of variable k for tree $t_b, b = 1, \ldots, B$, will be calculated as

$$\mathcal{I}_{Gini}(k, t_b) = \sum_{\mathcal{P} \in t_b} \Delta_{Gini}(\mathcal{P}, k). \tag{10.21}$$

By averaging over all B trees, we will have the Gini-based importance of variable k for the forest,

$$\mathcal{I}_{Gini}(k) = \frac{1}{B}\sum_{b=1}^{B}\mathcal{I}_{Gini}(k, t_b). \tag{10.22}$$

Figure 10.4 shows a typical visualization of the results of variable importance. It should be noted, however, that – similar to random forests for regression (see Chapter 8) – correlations among variables may have a high impact on their importance measures. For example, consider variable k with high values of its important measures. If we add to the training data a new variable that is redundant to variable k (or highly correlated with it), the importance measures of variable k will most likely decrease to about half of their previous values.

10.4 Notes on Feature Selection

Another important remark about random forests is related to feature selection. The random forests algorithm is sometimes described as a method for feature selection, but that is not the case. Although some of the original variables are not used in the ensemble, some of those used may be irrelevant (especially when the number of variables m to be randomly selected for a node is very small). Even more importantly, the random forests algorithm tends to utilize a large number of available variables, as

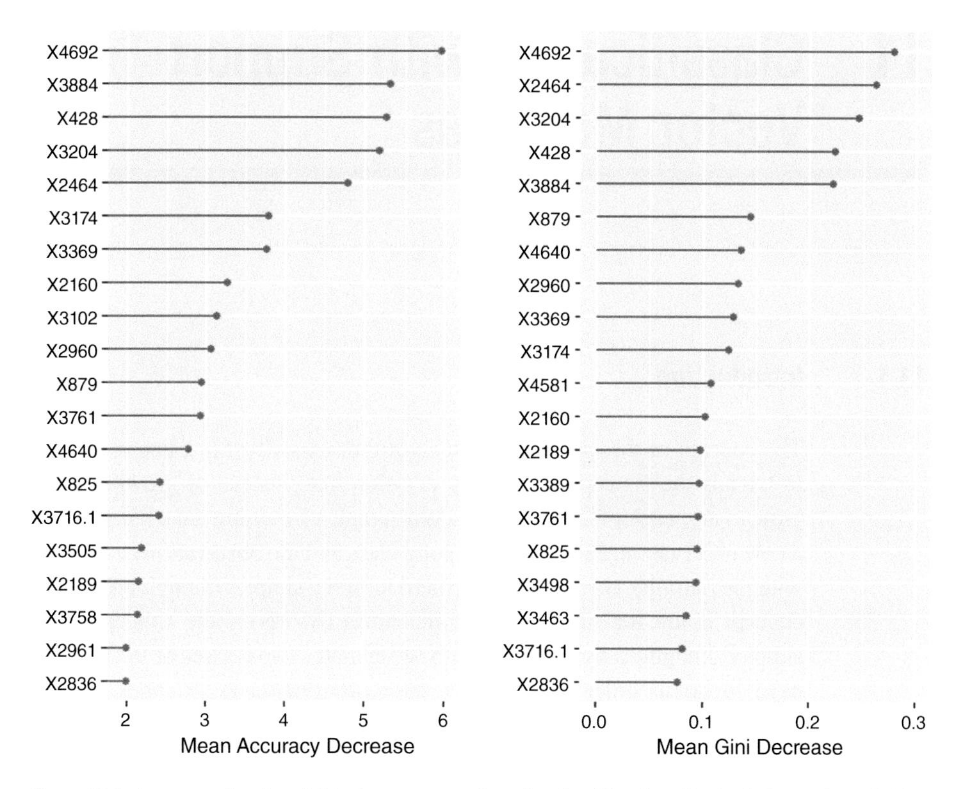

Figure 10.4 An example of variable importance plots for classification. In the left panel, variables are ordered by the average decrease in the classification accuracy after the values of the variable are randomly permuted in the OOB samples (the permutation-based accuracy measures were scaled by their standard errors). The right panel shows variable importance represented by the decrease of the Gini impurity index when the variable is used to split nodes (averaged over all trees in the forest).

feature selection is not one of its goals. This again leads to the conclusion that a single random forest predictive model (whether for classification or for regression) should not be used for high-dimensional data without first performing a proper multivariate feature selection. Thus, the best place for random forests in biomarker discovery studies is as a learning algorithm used within parallel feature selection experiments. Averaging random forests' performance as well as variable importance measures over hundreds of such parallel experiments (on top of aggregations within each of the random forest models), when used in a properly designed feature selection schema (see Chapter 5), is one of the efficient ways of finding a parsimonious multivariate biomarker.

11 Classification with Support Vector Machines

11.1 Introduction

The concept of the random forests algorithm (described in Chapter 10) is very simple. It extends the concept of a binary decision tree: instead of growing a single tree, we grow many, each of them on a random sample of training data. Then, when we classify a new observation, each tree votes for a class and the observation is assigned to the class with the plurality vote. It would be hard to find a simpler (though brilliant) classification concept if not for support vector machines (SVMs) – the concept of SVM is even simpler. Imagine a hyperplane that best separates two classes of the training data; all we need to do in order to classify a new observation is to check which side of the hyperplane it is on. After understanding this, we only need to learn how to identify the hyperplane that optimally separates the classes.

Support vector regression has been described in Chapter 9. However, the *Support Vector Machines* algorithm has been originally introduced for *binary classification* (Boser et al. 1992; Vapnik 1998) and only later adapted to regression problems. Even if the concept of a separating hyperplane may suggest linearity, we can build SVM classifiers for linear as well as nonlinear boundaries between classes. For the latter (nonlinear classifiers), SVM would map the original variable space into a sufficiently higher-dimensional feature space in which the classes can be separated by a hyperplane. Moreover, the kernel trick allows nonlinear solutions to be found without explicitly performing the mapping. Although SVMs may be extended to multiclass classification, in this chapter we will focus on their most common application – binary classification.

11.2 Linear Support Vector Classification

Support vector machines are supervised learning algorithms for binary classification. Their key concept is the optimal separating hyperplane. Assume that our training data include N observations – each of them assigned to one of the two differentiated classes (corresponding to the two categories of the response variable) – and p variables. Thus, each of the training observations is represented by a $p \times 1$ vector $\mathbf{x}_i = [x_{1i}, \cdots, x_{pi}]^T$ and a class label y_i, $i = 1, \ldots, N$. Assume also that the class labels are denoted by $+1$ and –1 (we will show later that this arbitrary notation is quite useful for SVMs). Let us first consider the simplest situation when the two classes can be completely separated by a hyperplane

$$\beta_0 + \mathbf{x}^T\boldsymbol{\beta} = 0. \tag{11.1}$$

Here we will use notation similar to that for support vector regression; however, the parameters of the hyperplane now have different interpretations. Recall from Chapter 9 that the SVM solution to a regression problem was a p-dimensional hyperplane in a $(p + 1)$-dimensional space, the space whose dimensions represented p independent variables plus the response variable. Thus, for regression, the parameters of the hyperplane $y = \beta_0 + \mathbf{x}^T\boldsymbol{\beta}$ were directly related to the y dimension (with β_0 representing the y-intercept and $\boldsymbol{\beta}$ being a vector of the slope coefficients, each of them associated with response y and one of the independent variables). Now, however, an SVM solution to a binary classification problem is a $(p - 1)$-dimensional hyperplane in a p-dimensional space defined only by p independent variables, since now – unlike in regression – the response variable does not represent a geometrical dimension; rather, it is a metadata defining the class of each of the training observations. Consequently, vector $\boldsymbol{\beta}$ now represents the weights of p independent variables and thus defines the orientation of the hyperplane, and β_0 is the offset of the hyperplane from the origin of the p-dimensional space.

If the classes are linearly separable, then there is an infinite number of hyperplanes separating them. The one that maximizes the distance between itself and the nearest observations from both classes (on either side of the hyperplane) is an *optimal separating hyperplane*. Thus, the closest observations from either class are equidistant from the optimal separating hyperplane. To find this hyperplane (that is, its β_0 and $\boldsymbol{\beta}$), we assume that for each training observation belonging to the positive class, $(\mathbf{x}_i, y_i = +1)$, we will have $\beta_0 + \mathbf{x}_i^T\boldsymbol{\beta} \geq 1$, and for each observation from the negative class, $(\mathbf{x}_i, y_i = -1)$, we will have $\beta_0 + \mathbf{x}_i^T\boldsymbol{\beta} \leq -1$. Multiplying those equations by y_i (positive for the former and negative for the latter) gives us the constraint

$$y_i\left(\beta_0 + \mathbf{x}_i^T\boldsymbol{\beta}\right) \geq 1. \tag{11.2}$$

This means that the training observations closest to the optimal hyperplane will lie on one of the two hyperplanes,

$$\begin{aligned} \beta_0 + \mathbf{x}^T\boldsymbol{\beta} &= +1, \text{or} \\ \beta_0 + \mathbf{x}^T\boldsymbol{\beta} &= -1. \end{aligned} \tag{11.3}$$

These hyperplanes are called *support hyperplanes*, and the Euclidean distance between them, M, is called *the margin*, which should be maximized in order to identify the optimal separating hyperplane (see Figure 11.1).

To find the margin, denote the vectors representing those training observations that lie on one of the support hyperplanes as $\mathbf{x}_+$ and $\mathbf{x}_-$, respectively; observe also from constraint (11.2) that they satisfy $y_i(\beta_0 + \mathbf{x}^T\boldsymbol{\beta}) = 1$, and thus $\mathbf{x}_+^T\boldsymbol{\beta} = 1 - \beta_0$, and $\mathbf{x}_-^T\boldsymbol{\beta} = -1 - \beta_0$. Consequently, the margin measured in the direction of the unit vector $\boldsymbol{\beta}/\|\boldsymbol{\beta}\|$, normal to the separating hyperplane, is

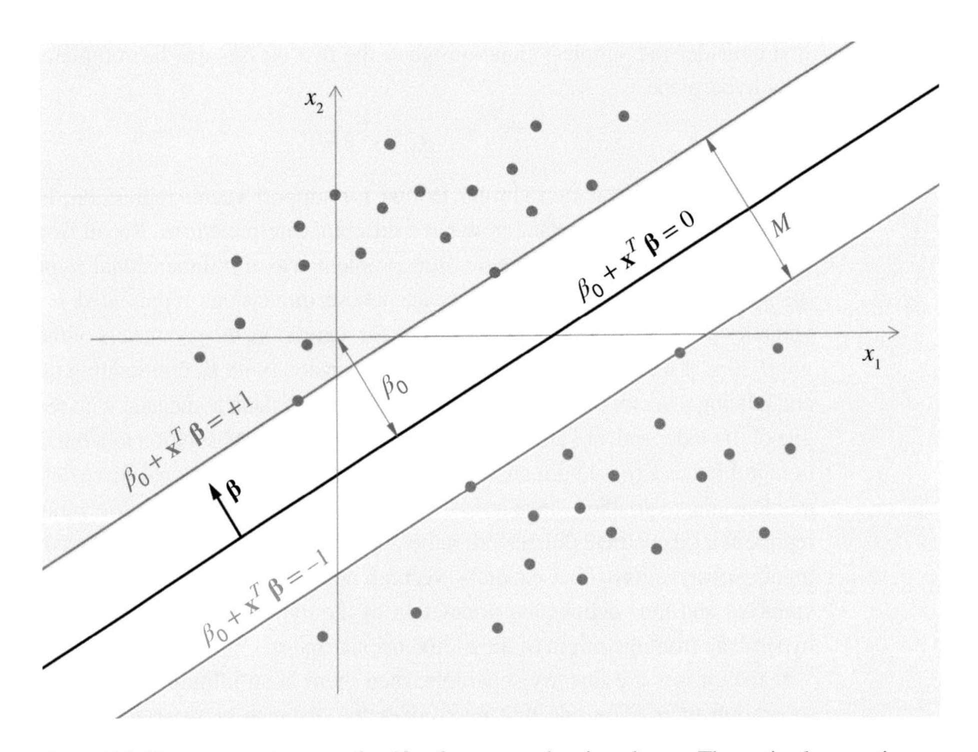

Figure 11.1 Linear support vector classifier for nonoverlapping classes. The optimal separating hyperplane $\beta_0 + \mathbf{x}^T\boldsymbol{\beta} = 0$ is the one that maximizes the margin (*M*), that is, the distances between itself and two equidistant support hyperplanes $\beta_0 + \mathbf{x}^T\boldsymbol{\beta} = \pm 1$, which are defined by those training observations from both classes that are closest to the separating hyperplane. No training observation is allowed to violate the margin of its class, that is, to lie outside of its class area (marked here with the class-assigned background color). For this reason, such classifiers are called hard-margin SVMs. β_0 is the offset of the separating hyperplane from the origin of the p-dimensional space defined by the p independent variables of the classification problem (here $p = 2$). Vector $\boldsymbol{\beta}$ is normal to the separating hyperplane and thus defines its orientation.

$$\begin{aligned} M &= \left(\mathbf{x}_+^T - \mathbf{x}_-^T\right)\frac{\boldsymbol{\beta}}{\|\boldsymbol{\beta}\|} \\ &= (1 - \beta_0 + 1 + \beta_0)\frac{1}{\|\boldsymbol{\beta}\|} \\ &= \frac{2}{\|\boldsymbol{\beta}\|}, \end{aligned} \tag{11.4}$$

where $\|\boldsymbol{\beta}\|$ is the Euclidean norm, that is, the length of vector $\boldsymbol{\beta}$ in the p-dimensional space of p independent variables.

To find the optimal separating hyperplane, we may maximize the margin $M = 2/\|\boldsymbol{\beta}\|$ or, just for mathematical convenience, minimize $\frac{1}{2}\|\boldsymbol{\beta}\|^2$. Including the constraint (11.2) these considerations were based on, the optimization problem can be stated as

$$\begin{aligned} &\underset{\boldsymbol{\beta},\beta_0}{\text{minimize}}\ \frac{1}{2}\|\boldsymbol{\beta}\|^2 \\ &\text{subject to } y_i\left(\beta_0 + \mathbf{x}_i^T\boldsymbol{\beta}\right) \geq 1, \quad i = 1, \ldots, N. \end{aligned} \tag{11.5}$$

By solving this optimization problem, we would find a *hard-margin* support vector classifier, which can also be called the maximum margin classifier. Recall, however, that all our considerations so far were made under the assumption that classes are linearly separable, without error, which means that no training observations were allowed in the area between the two support hyperplanes. However, with the exception of very simple (or trivial) data, satisfying this condition is rarely feasible. Therefore, we will only use these considerations as a starting point for a more realistic situation – where the classes overlap and, thus, are not perfectly linearly separable. This means that we will allow for some of the training observations to lie in the area between the support hyperplanes (thus violating the margin of their class), including situations where they may be on the wrong side of the separating hyperplane (even beyond the area between the support hyperplanes), and thus would be misclassified.

The idea of such *soft-margin SVMs* is to find an optimal balance between maximizing the margin and minimizing the effects of its violations. This will be done by adding a regularization term to the optimization problem (11.5) that will control the penalty for margin violations. To quantify the extent of such violations, we will introduce nonnegative slack variables $\xi_i, i = 1, \ldots, N$, which will be associated with the training observations $\mathbf{x}_i, i = 1, \ldots, N$. Training observations that do not violate the margin of their class will have $\xi_i = 0$; those that violate their margin, but are on the correct side of the separating hyperplane will have $0 < \xi_i < 1$; and those that are on the wrong side of the separating hyperplane (and thus are misclassified) will have $\xi_i > 1$. Finally, if it happens that a training observation lies exactly on the separating hyperplane (and thus cannot be classified), it will have $\xi_i = 1$.

11.2.1 The Primal Optimization Problem for Linear SVM

The *primal* optimization problem for linear SVM classifiers can now be formulated as

$$\begin{aligned} &\underset{\boldsymbol{\beta}, \beta_0}{\text{minimize}} \ \frac{1}{2}\|\boldsymbol{\beta}\|^2 + C\sum_{i=1}^{N}\xi_i \\ &\text{subject to} \ y_i\left(\beta_0 + \mathbf{x}_i^T\boldsymbol{\beta}\right) \geq 1 - \xi_i, \\ &\qquad \xi_i \geq 0, \quad \text{for } i = 1, \ldots, N, \end{aligned} \tag{11.6}$$

where $C > 0$ is a tunable cost (or regularization) parameter that controls the magnitude of the slack variables and thus allows for balancing the two contradictory aspects of the solution – maximizing the margin and minimizing its violations. If the value of C – the cost of margin violations – increases, then both the margin and the number of violations decrease.

Problem (11.6) is a quadratic optimization problem with linear constraints. Similar to support vector regression (Chapter 9), we will solve this convex problem using Lagrange multipliers – by including in the resulting Lagrange function the terms combining the multipliers and the original constraints.

11.2.2 The Dual Optimization Problem for Linear SVM

Since problem (11.6) has two constraints, we will need two $N \times 1$ vectors of nonnegative Lagrange multipliers, $\boldsymbol{\alpha}$ and $\boldsymbol{\eta}$, whose elements α_i and η_i will be associated with the training observation $\mathbf{x}_i, i = 1, \ldots, N$. Hence, the *primal* Lagrange function constructed from problem (11.6) will be

$$L_P(\boldsymbol{\beta}, \beta_0, \boldsymbol{\xi}) = \frac{1}{2}\|\boldsymbol{\beta}\|^2 + C\sum_{i=1}^{N}\xi_i - \sum_{i=1}^{N}\eta_i\xi_i - \sum_{i=1}^{N}\alpha_i\left[y_i\left(\beta_0 + \mathbf{x}_i^T\boldsymbol{\beta}\right) - (1-\xi_i)\right]. \tag{11.7}$$

To find the saddle point of this Lagrangian, we will first find its partial derivatives with respect to the primal variables, $\boldsymbol{\beta}, \beta_0$, and ξ_i, and set them to zero,

$$\frac{\delta L_P}{\delta \boldsymbol{\beta}} = \boldsymbol{\beta} - \sum_{i=1}^{N}\alpha_i y_i \mathbf{x}_i = 0, \text{ hence } \boldsymbol{\beta} = \sum_{i=1}^{N}\alpha_i y_i \mathbf{x}_i, \tag{11.8}$$

$$\frac{\delta L_P}{\delta \beta_0} = -\sum_{i=1}^{N}\alpha_i y_i = 0, \text{ hence } \sum_{i=1}^{N}\alpha_i y_i = 0, \tag{11.9}$$

$$\frac{\delta L_P}{\delta \xi_i} = C - \eta_i - \alpha_i = 0, \text{ hence } \eta_i = C - \alpha_i, i = 1, \ldots, N. \tag{11.10}$$

Then, by substituting the results of (11.8)–(11.10) into (11.7), we will have the *dual representation* of the Lagrange function, which depends only on the values of $\boldsymbol{\alpha}$,

$$L_D(\boldsymbol{\alpha}) = \sum_{i=1}^{N}\alpha_i - \frac{1}{2}\sum_{i=1}^{N}\sum_{j=1}^{N}\alpha_i\alpha_j y_i y_j \mathbf{x}_i^T\mathbf{x}_j. \tag{11.11}$$

Furthermore, from (11.10) we have $C - \alpha_i = \eta_i$ and thus (since η_i is nonnegative) $C - \alpha_i \geq 0$, that is, $\alpha_i \leq C$. However, since α_i is also nonnegative, the lower and upper bounds for α_i values are zero and C,

$$0 \leq \alpha_i \leq C. \tag{11.12}$$

The dual representation of the optimization problem for linear SVM classifiers can be now written as

$$\begin{aligned} &\underset{\boldsymbol{\alpha}}{\text{maximize}} \sum_{i=1}^{N}\alpha_i - \frac{1}{2}\sum_{i=1}^{N}\sum_{j=1}^{N}\alpha_i\alpha_j y_i y_j \mathbf{x}_i^T\mathbf{x}_j \\ &\text{subject to } \sum_{i=1}^{N}\alpha_i y_i = 0 \text{ and } 0 \leq \alpha_i \leq C, i = 1, \ldots, N. \end{aligned} \tag{11.13}$$

In addition, the following Karush-Kuhn-Tucker (KKT) complementarity conditions have to be satisfied by a solution to (11.13):

$$\alpha_i\left[y_i\left(\beta_0 + \mathbf{x}_i^T\boldsymbol{\beta}\right) - (1-\xi_i)\right] = 0, \tag{11.14}$$

$$\xi_i(\alpha_i - C) = 0, \tag{11.15}$$

$$y_i(\beta_0 + \mathbf{x}_i^T \boldsymbol{\beta}) - (1 - \xi_i) \geq 0. \tag{11.16}$$

From the KKT condition (11.14) we find that α_i can be nonzero (positive) only when $y_i(\beta_0 + \mathbf{x}_i^T \boldsymbol{\beta}) = (1 - \xi_i)$. This means that the training observations that do not violate their class margin, and thus have $\xi_i = 0$, may have $\alpha_i > 0$ only when they lie on their class support hyperplane. This also means that all training observations that violate their class margin have both $\alpha_i > 0$ and $\xi_i > 0$. Furthermore, from (11.15) we find that if $\xi_i > 0$ (which is true for all observations violating their class margin), we must have $\alpha_i = C$ (see Figure 11.2).

11.2.3 Linear Support Vector Classification Machine

The solution of the dual optimization problem (11.13) gives us vector $\boldsymbol{\alpha}$, which allows calculating the solution for the weight vector $\boldsymbol{\beta}$,

$$\hat{\boldsymbol{\beta}} = \sum_{i=1}^{N} \alpha_i y_i \mathbf{x}_i, \tag{11.17}$$

which defines the orientation of the optimal separating hyperplane.[1] Observe that this orientation only depends on the training observations for which $\alpha_i > 0$. Since only they support the solution, they are called *support vectors*. They include all training observations that violate their class margin (and thus have $\alpha_i = C$) and the observations that lie on their support hyperplane (and have $0 < \alpha_i < C$).

The only other parameter needed to fully define the optimal separating hyperplane is its offset from the origin, β_0. It can be calculated from any support vector that lies on one of the support hyperplanes (for which we have $\beta_0 + \mathbf{x}^T \boldsymbol{\beta} = \pm 1$). Depending on which support hyperplane is used, we will have

$$\hat{\beta}_0 = \begin{cases} 1 - \mathbf{x}_i^T \hat{\boldsymbol{\beta}} & \text{for the "+" hyperplane} \\ -1 - \mathbf{x}_i^T \hat{\boldsymbol{\beta}} & \text{for the "−" hyperplane.} \end{cases} \tag{11.18}$$

However, for numerical stability, we usually calculate it for every support vector lying on either of the support hyperplanes and average the results.

To classify an observation, we only need to find out on which side (positive or negative) of the separating hyperplane, $\hat{\beta}_0 + \mathbf{x}^T \hat{\boldsymbol{\beta}} = 0$, the vector $\mathbf{x}$ representing the observation lies; hence

$$Class(\mathbf{x}) = sign\left(\hat{\beta}_0 + \mathbf{x}^T \hat{\boldsymbol{\beta}}\right). \tag{11.19}$$

[1] Recall that this weight vector is the solution for our particular training data set, and thus it is only an estimate of its true value in the target population. This is why we use the notation $\hat{\boldsymbol{\beta}}$ and $\hat{\beta}_0$ for the parameters of the separating hyperplane that is the solution to the SVM classification problem. However, for simplicity, we do not extend this notation to other variables (such as α_i).

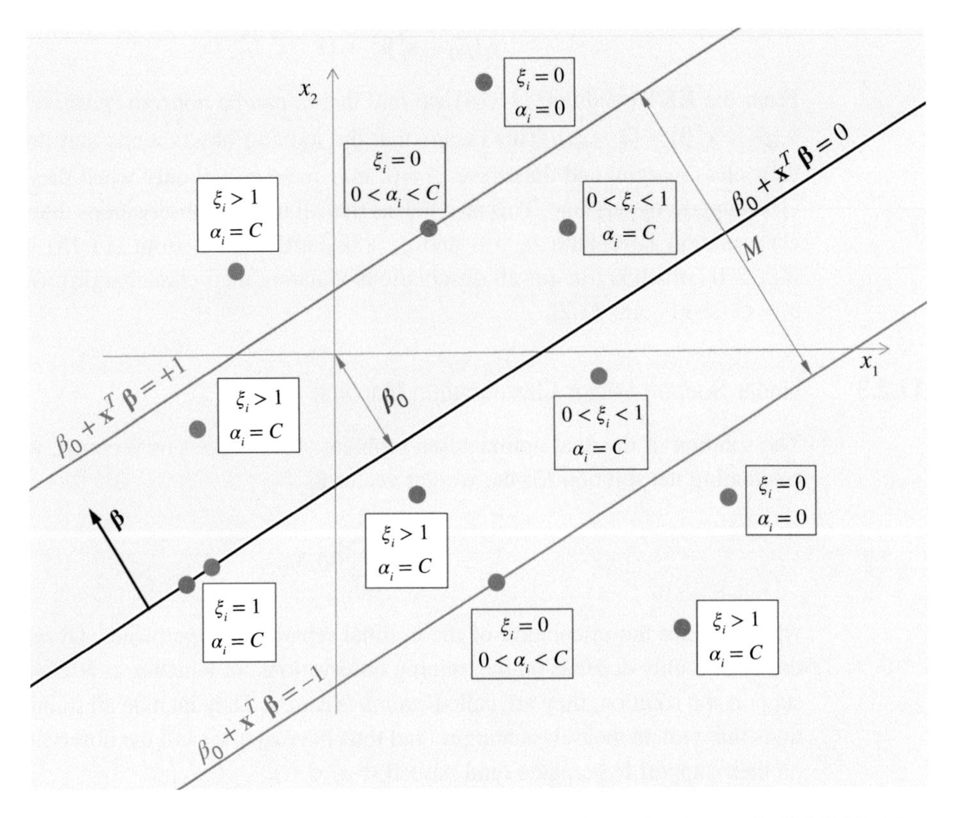

Figure 11.2 Linear support vector classifier for overlapping classes (a soft-margin SVM). Training observations are allowed to violate the margin of their class; if they are on the right (for their class) side of the separating hyperplane, they are correctly classified, otherwise they are misclassified. The optimal separating hyperplane is identified as a tradeoff between maximizing the margin (M) and minimizing the cost of margin violations. The violations are quantified by the slack variables ξ_i associated with training observations. Observations that do not violate the margin have $\xi_i = 0$, if they are in their class area, they have $\alpha_i = 0$; however, if they lie on their support hyperplane, they have $0 < \alpha_i < C$. All training observations that violate their class margin have $\xi_i > 0$ and $\alpha_i = C$. For those that are on the correct side of the separating hyperplane $0 < \xi_i < 1$, for those on the separating hyperplane $\xi_i = 1$, and for those on the wrong side of the separating hyperplane $\xi_i > 1$. Parameters of the SVM classifier depend only on the support vectors, which are all training observations that violate their class margin plus those that lie on their support hyperplane.

Alternatively, by combining Equations (11.19) and (11.17), we may express this classification function in terms of the dual problem,

$$Class(\mathbf{x}) = sign\left(\hat{\beta}_0 + \sum_{i=1}^{N} \alpha_i y_i \mathbf{x}^T \mathbf{x}_i\right), \tag{11.20}$$

from which it is clear that the class assignment of observation $\mathbf{x}$ only depends on the inner products $\mathbf{x}^T\mathbf{x}_i$ associated with the support vectors. Hence, this classifier is called a *support vector classification machine*.

11.3 Nonlinear Support Vector Classification

So far we have described the linear SVM classifier that works well when the differentiated classes can be linearly separated, with or without overlap. However, it would not work if the class boundaries are intrinsically nonlinear and thus cannot be approximated by a hyperplane. In such situations, we may map the original p-dimensional space into a higher-dimensional (even infinite-dimensional) *feature space* in which the classes can be separated by a hyperplane. Let $\Phi(\mathbf{x}_i)$ denote the mapping of each training observation $\mathbf{x}_i, i=1, \ldots, N$ from the original data space into a feature space. To find an optimal separating hyperplane in the feature space, we would adjust (11.13) to now represent the dual optimization problem in the feature space,

$$\begin{aligned} &\underset{\boldsymbol{\alpha}}{\text{maximize}} \sum_{i=1}^{N} \alpha_i - \frac{1}{2} \sum_{i=1}^{N} \sum_{j=1}^{N} \alpha_i \alpha_j y_i y_j \Phi(\mathbf{x}_i)^T \Phi(\mathbf{x}_j) \\ &\text{subject to } \sum_{i=1}^{N} \alpha_i y_i = 0 \text{ and } 0 \leq \alpha_i \leq C, i=1, \ldots, N. \end{aligned} \tag{11.21}$$

The solution to this problem would be

$$Class(\mathbf{x}) = sign\left(\hat{\beta}_0 + \sum_{i=1}^{N} \alpha_i y_i \Phi(\mathbf{x}_i)^T \Phi(\mathbf{x}_j)\right). \tag{11.22}$$

However, performing such computations in a very high- or infinite-dimensional feature space would be either very costly or infeasible. What we can do is to first observe that both the optimization problem (11.21) and its solution (11.22) use mapping Φ only in the form of inner products $\Phi(\mathbf{x}_i)^T \Phi(\mathbf{x}_j)$. Then, instead of the costly calculations of the inner products in the feature space, we may use a nonlinear kernel function $K(\mathbf{x}_i, \mathbf{x}_j) = \Phi(\mathbf{x}_i)^T \Phi(\mathbf{x}_j)$, which performs the calculations in the original p-dimensional space of training observations $\mathbf{x}_i, i=1, \ldots, N$. The resulting optimization problem,

$$\begin{aligned} &\underset{\boldsymbol{\alpha}}{\text{maximize}} \sum_{i=1}^{N} \alpha_i - \frac{1}{2} \sum_{i=1}^{N} \sum_{j=1}^{N} \alpha_i \alpha_j y_i y_j K(\mathbf{x}_i, \mathbf{x}_j) \\ &\text{subject to } \sum_{i=1}^{N} \alpha_i y_i = 0 \text{ and } 0 \leq \alpha_i \leq C, i=1, \ldots, N, \end{aligned} \tag{11.23}$$

is now a *linear* SVM problem, which utilizes a nonlinear kernel function to find an optimal separating hyperplane in the feature space, but it does so *without* performing any mapping into this feature space. The solution,

$$Class(\mathbf{x}) = sign\left(\hat{\beta}_0 + \sum_{i=1}^{N} \alpha_i y_i K(\mathbf{x}_i, \mathbf{x}_j)\right), \tag{11.24}$$

represents nonlinear class boundaries in the original data space. This quite remarkable approach is called *the kernel trick* (Cortes and Vapnik 1995). By using a kernel

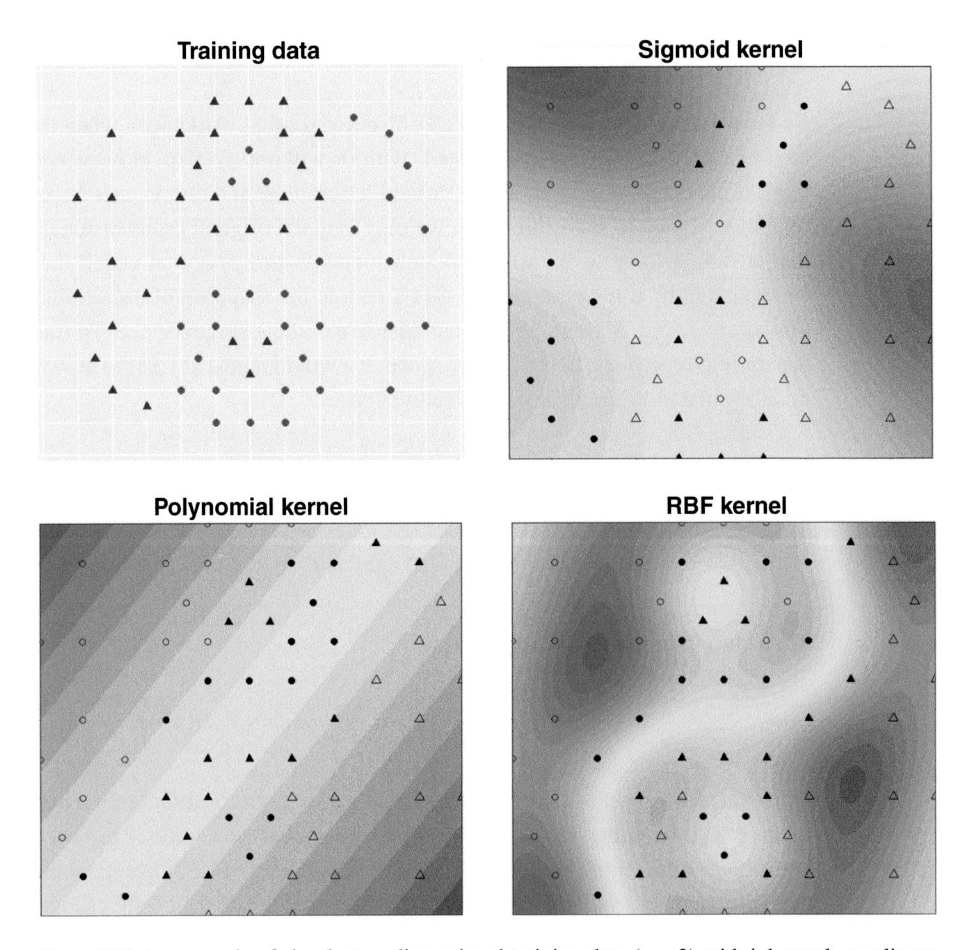

Figure 11.3 An example of simple two-dimensional training data ($p = 2$) with inherently nonlinear boundaries between classes. Each of the two differentiated classes includes an enclave of observations totally surrounded by observations from the other class. Three SVM classifiers using different kernel functions were trained on this data set (without extensive tuning). Only the classifier using the RBF kernel seems to be able to identify the enclaves and the proper boundaries between the classes. On the graphics representing these three SVM classifiers, training observations corresponding to support vectors have their markers filled in. The white areas correspond to a separating hyperplane in an appropriate feature space.

function, we not only avoid explicit mapping into the feature space but also do not even need to know the exact form of mapping Φ.

Kernel functions that are commonly used for nonlinear SVM classifiers are the same ones used for nonlinear support vector regression. The radial basis function (RBF) kernel,

$$K(\mathbf{x}_i, \mathbf{x}_j) = e^{-\gamma \|\mathbf{x}_i - \mathbf{x}_j\|^2}, \quad (11.25)$$

is frequently used due to its versatility – it can quite often be successfully applied to highly nonlinear situations, for which other commonly used kernels may have a difficulty in identifying proper class boundaries (see the example in Figure 11.3). In this kernel, the term $\|\mathbf{x}_i - \mathbf{x}_j\|^2$ represents the Euclidean distance between vectors $\mathbf{x}_i$ and $\mathbf{x}_j$ in the

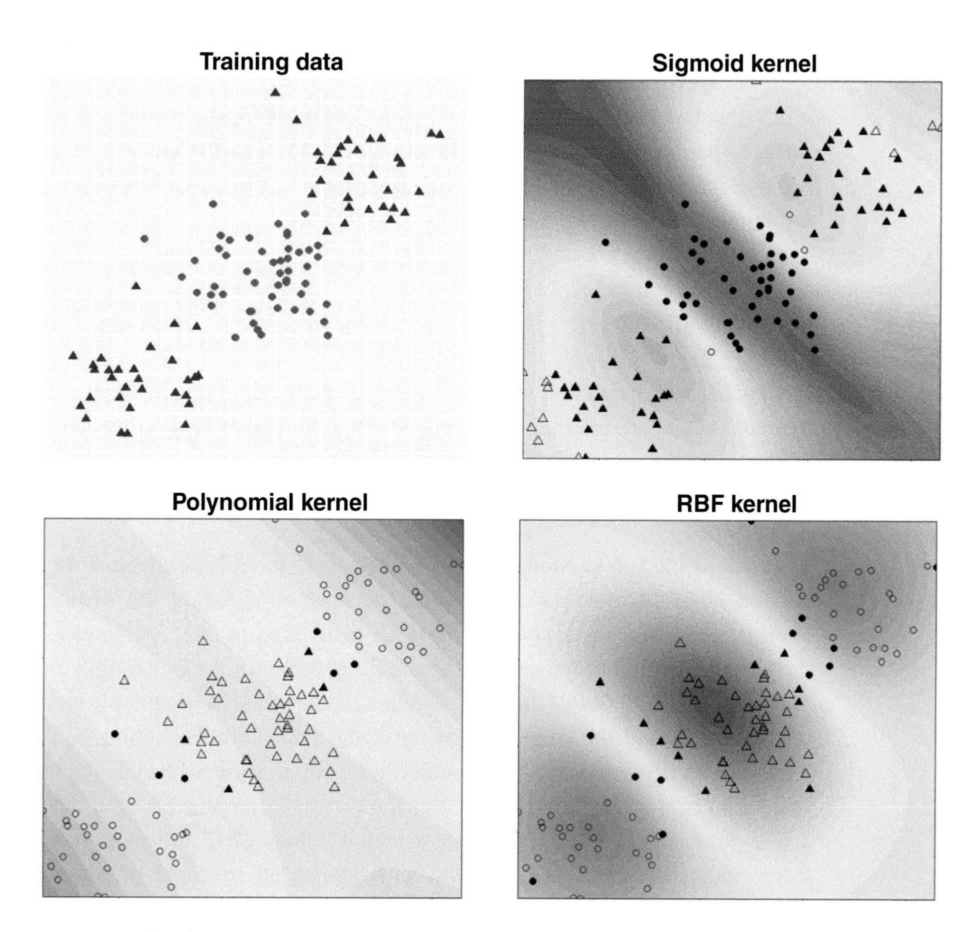

Figure 11.4 Another toy example of two-dimensional training data ($p = 2$). One of the classes has observations in two areas separated by observations from the other class, grouped between them. In this case all three of the common kernels can identify some boundaries that separate the classes reasonably well. Yet, even in this situation, the class boundaries of the SVM classifier using the RBF kernel seem to be most appropriate. The other two kernels extend the red class area toward the northwest and southeast corners of the plot. Although we do not know whether this is correct or not, there are no observations in the training data set that support such extensions. This is why the RBF kernel's solution appears to be better.

original p-dimensional data space, and $\gamma > 0$ is its tunable parameter. The feature space for the RBF kernel may be infinite-dimensional; hence, without the kernel trick, we would not be able to find a solution in such a space. If we use the polynomial kernel of degree d,

$$K(\mathbf{x}_i, \mathbf{x}_j) = \left(c + \gamma \mathbf{x}_i^T \mathbf{x}_j\right)^d, \tag{11.26}$$

then an optimal separating hyperplane in the kernel's feature space will be represented by polynomial curve boundaries (of degree d) in the original data space.[2] The hyperbolic tangent kernel (also called the sigmoid kernel),

[2] Generally, when $c > 0$, it is an inhomogeneous polynomial kernel; it will be homogeneous when $c = 0$. A homogeneous polynomial has all its nonzero terms of the same degree; for example, $x^5 + 3x^2y^3 + 2xy^4$ is a homogeneous polynomial as its terms are all of degree 5.

$$K(\mathbf{x}_i, \mathbf{x}_j) = \tanh(c + \gamma \mathbf{x}_i^T \mathbf{x}_j), \tag{11.27}$$

is also a popular one, though it is only a proper kernel for some values of its parameters. Tunable parameters appearing in functions (11.25)–(11.27) are $\gamma > 0$, $c \geq 0$, and a positive integer d. Figure 11.4 shows examples of SVM classifiers using these kernels for simple data with two independent variables.

11.4 Hyperparameters

In Chapter 3, a distinction was made between parameters and hyperparameters. Recall that parameters are internal characteristics of a predictive model that can be calculated or estimated directly from the training data. Hyperparameters, however, are external to the model, and they either are set manually or require tuning. Hence, they are tunable parameters.

The hard-margin version of linear SVM has no tunable parameters. Its soft-margin version has one hyperparameter – C, the cost of violating the model constraints. Nonlinear SVMs, in addition to cost C, also have tunable parameters defined by the kernel function they use (see the previous section). As an example, let us look at a nonlinear SVM with the RBF kernel. It has two hyperparameters – the cost parameter, C, and the single parameter of the RBF kernel function, γ. To tune more than one hyperparameter, we usually perform a *grid search* using either cross-validation or a validation data set. For the two hyperparameters here, we would have a two-dimensional grid, with some selected values of C and γ covering a possibly large range of their values. Figure 11.5 shows an example of a nonlinear SVM classifier with the RBF kernel for few values of C and γ. In practice, a grid would usually include significantly more values for each of the hyperparameters. It is worth noting that kernel functions may have their idiosyncrasies; for example, using the RBF kernel with a too large γ may lead to severe overfitting of the training data (see Figure 11.6).

Although support vector machines are capable of dealing with high-dimensional data, it is not recommended to build a single SVM classifier for a biomarker discovery project, as it would not result in a parsimonious multivariate biomarker.[3] Feature selection should be performed first. Since variable importance metrics can be calculated for SVMs, this has placed SVM among the learning algorithms that can be used within parallel multivariate feature selection experiments implementing recursive feature elimination. It is worth noting that recursive feature elimination had been originally designed for support vector machines (Guyon et al. 2002; Ambroise and McLachlan 2002).

[3] Also recall that in typical high-dimensional biomedical data, many variables would be irrelevant to the goal of the project, which would negatively impact the model's performance.

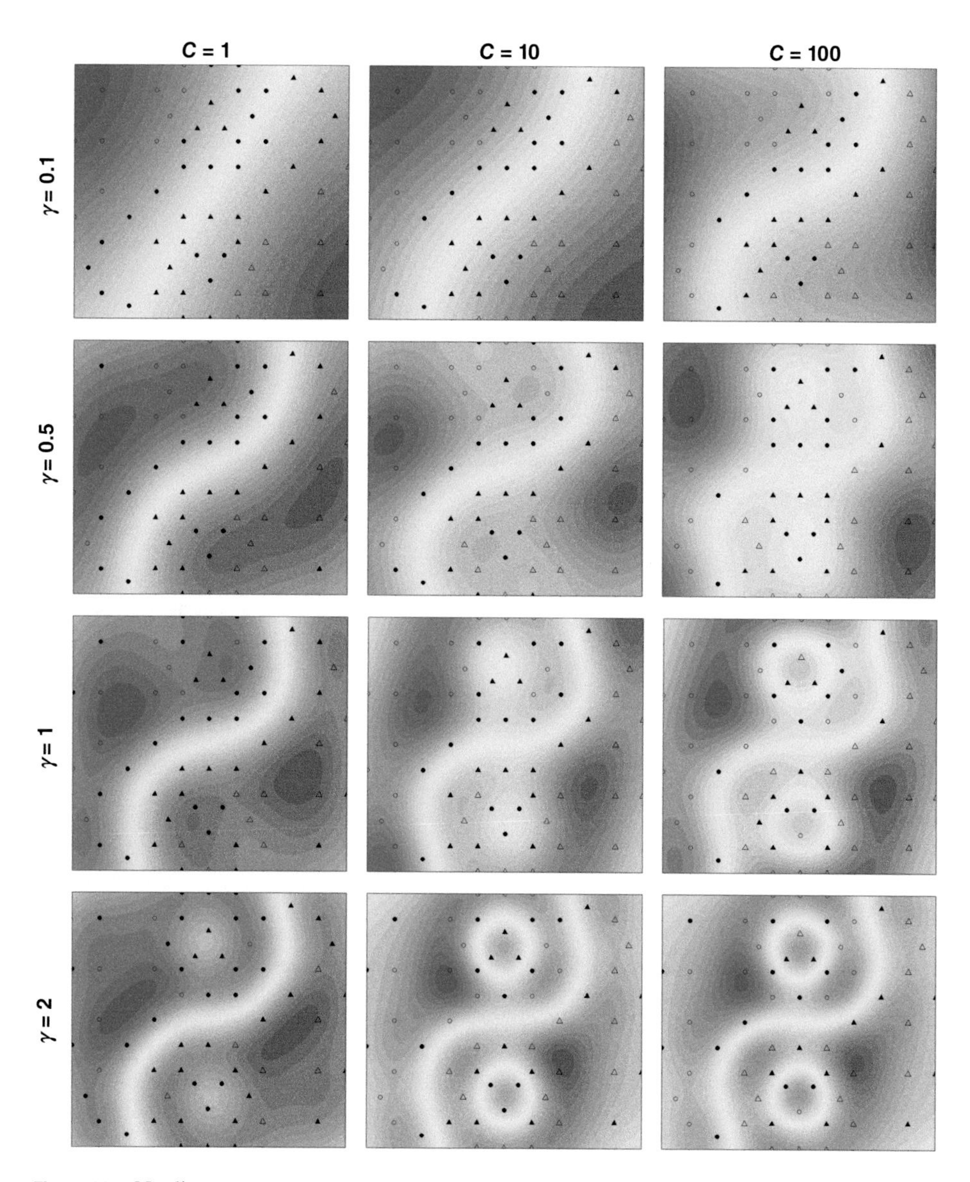

Figure 11.5 Nonlinear support vector classifier using the RFB kernel with different values of cost parameter C and kernel parameter γ. The training data is the same data used for experiments illustrated in Figure 11.3. White areas correspond to a separating hyperplane in the feature space of the kernel, but are of course nonlinear in the original data space. The color intensities of each class represent the probability of class membership.

11.5 Variable Importance

For the linear SVM classifier, vector $\boldsymbol{\beta}$ (which is estimated by $\hat{\boldsymbol{\beta}}$,

$$\hat{\boldsymbol{\beta}} = \sum_{i=1}^{N} \alpha_i y_i \mathbf{x}_i, \tag{11.28}$$

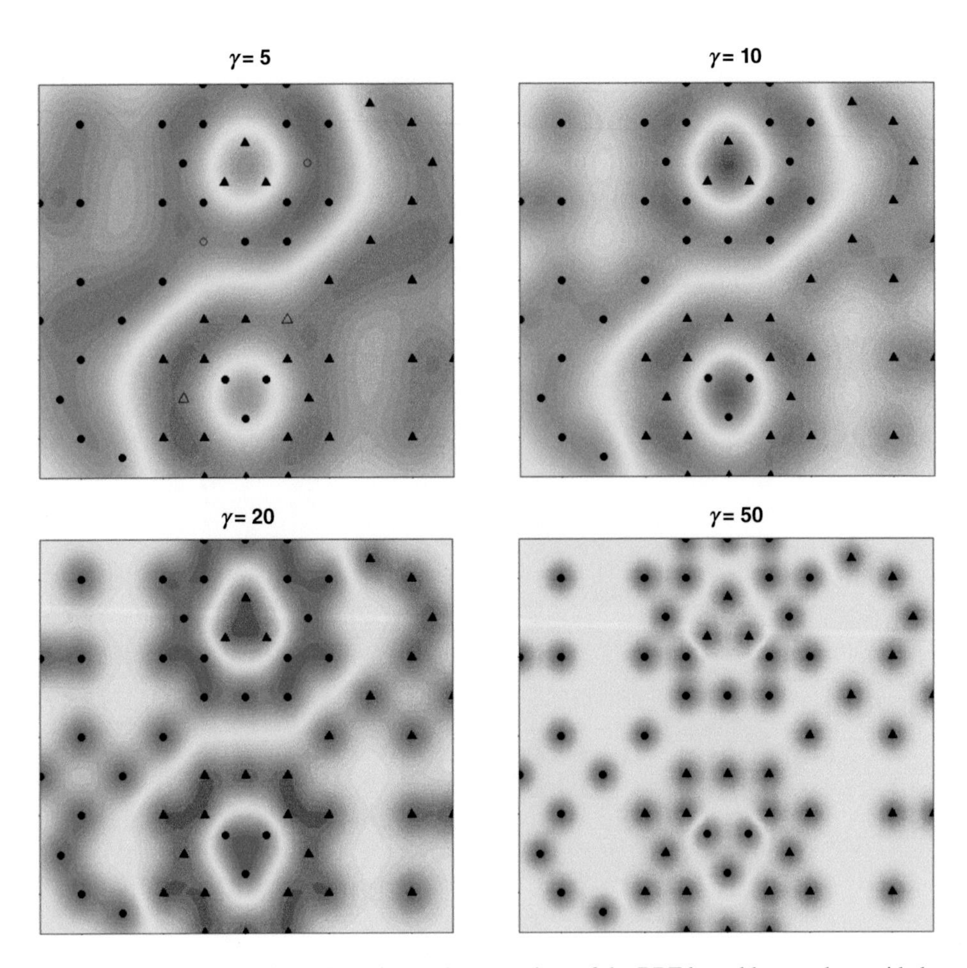

Figure 11.6 An example illustrating why too large γ values of the RBF kernel have to be avoided. The training data set here is the same as that used for Figures 11.3 and 11.5. Illustrated are four nonlinear SVM classifiers using the RBF kernel with $C = 1$ and four values of γ. When γ increases, the continuous class areas (as identified by each of the SVM classifiers) start to break into separate subareas, and eventually each of the training observations becomes an island, and that of course results in heavy overfitting of the training data. The start of this process is already visible for $\gamma = 5$; however, the white areas corresponding to the separating hyperplane in the feature space still properly represent the class boundaries, including those encircling the two enclaves. With further increase in γ, the hyperplane seizes more and more of the continuous class areas, until eventually only the "hills" or "valleys" representing the training observations are beyond the hyperplane.

see Section 11.2.3) defines the orientation of the optimal separating hyperplane. The elements of this vector, $\beta_k, k = 1, \ldots, p$, are the weights associated with p independent variables. The larger the absolute value of the weight, the more influence the variable has on the orientation of the separating hyperplane. Hence, the absolute value (or, alternatively, the squared value) of the weight may be used as the multivariate importance of a variable. This means that, during recursive feature elimination, we would remove, from the current set of p^* variables, those with the smallest values of their importance measure $\mathcal{I}(k)$,

$$\mathcal{I}(k) = |\beta_k|, k = 1, \ldots, p^*. \tag{11.29}$$

Consequently, removing a variable from the current set is equivalent to setting its weight to zero.[4] It is also worth mentioning that if a proper multivariate metric of variable importance is implemented during feature selection experiments, then the variable importance values will change with changes in the size and composition of the considered sets of variables (since for each of them a different optimal separating hyperplane will have to be identified).

For nonlinear SVMs, a change in the value of the kernel function after a variable is eliminated (this change estimates the change in model performance) may be used as the variable importance measure (Lal et al. 2006). This *sensitivity-analysis* approach stems from the concept of associating variable importance with the change in some objective, or cost, function – the change caused by removing the variable from the current set of variables.[5] For nonlinear SVMs, the cost function may be based on $\|\boldsymbol{\beta}\|^2$,

$$\|\boldsymbol{\beta}\|^2 = \sum_{i=1}^{N}\sum_{j=1}^{N} \alpha_i\alpha_j y_i y_j K(\mathbf{x}_i, \mathbf{x}_j), \tag{11.30}$$

and each iteration of recursive feature elimination would remove the variable whose elimination corresponds to the smallest change in $\|\boldsymbol{\beta}\|^2$. Consequently, the measure of variable importance can be calculated for each variable $k, k = 1, \ldots, p^*$ as

$$\begin{aligned} \mathcal{I}(k) &= \left| \|\boldsymbol{\beta}\|^2 - \|\boldsymbol{\beta}\|^2_{(p^*-k)} \right| \\ &= \left| \sum_{i=1}^{N}\sum_{j=1}^{N} \alpha_i\alpha_j y_i y_j K(\mathbf{x}_i, \mathbf{x}_j) - \sum_{i=1}^{N}\sum_{j=1}^{N} \alpha_i\alpha_j y_i y_j K(\mathbf{x}_{i,(p^*-k)}, \mathbf{x}_{j,(p^*-k)}) \right|, \end{aligned} \tag{11.31}$$

where $(p^* - k)$ refers to the current set of variables without variable k, and vector $\boldsymbol{\alpha}$ is associated with the optimal solution for a set of p^* variables (see optimization problem 11.23). For computational efficiency, it may be assumed that vector $\boldsymbol{\alpha}$ does not change when the variable is removed, and that variable importance $\mathcal{I}(k)$ can be estimated by calculating only the change in the value of the kernel function $K(\mathbf{x}_i, \mathbf{x}_j)$ caused by the removal of the variable (Guyon et al. 2002; Rakotomamonjy 2003; Lal et al. 2006).

[4] Other ways of defining multivariate variable importance for linear SVM have also been proposed, for example, the sign-consistency bagging method (Gómez-Verdejo et al. 2019). In this method, an ensemble of linear SVM classifiers is trained on bootstrap samples of the training data, and the variable importance is associated with the proportion of classifiers in which the variable weight coefficient has the same sign. The more consistent is the sign of the weight across the classifiers of the ensemble, the higher importance is assigned to the variable. The reasoning behind this approach is that the sign of weights is less sensitive to possible multicollinearities than their values.

[5] Using the absolute or squared value of the variable weight coefficient in linear SVM may also be justified by the sensitivity-analysis approach (Guyon et al. 2002).

11.6 Cost-Sensitive SVMs

The incorporation of misclassification costs into classification has been discussed in Chapter 4. One of the methods of doing this is to make such design changes into a cost-insensitive learning algorithm that will make it cost-sensitive. It turns out that this can be quite easily done to the SVM learning algorithm. The primal optimization problem for linear SVM, see (11.6), includes the regularization term

$$C\sum_{i=1}^{N}\xi_i, \tag{11.32}$$

which penalizes margin violations quantified by nonnegative slack variables $\xi_i, i=1, \ldots, N$, associated with the training observations $\mathbf{x}_i, i=1, \ldots, N$. The cost (or regularization) parameter C controls the tradeoff between maximizing the margin and minimizing its violations. This cost is, however, applied equally to all observations violating their class margin, regardless of the class of the training observation. To make the SVM learning algorithm sensitive to different class misclassification costs, it is enough to replace the single regularization term with two terms, each of them associated with one of the differentiated classes. This will give us the following primal optimization problem for cost-sensitive linear SVM classifiers (Fernandez et al. 2018):

$$\begin{aligned}\underset{\boldsymbol{\beta},\beta_0}{\text{minimize}}\ &\frac{1}{2}\|\boldsymbol{\beta}\|^2 + C^+\sum_{i|y_i=+1}\xi_i + C^-\sum_{i|y_i=-1}\xi_i\\ \text{subject to}\ & y_i\left(\beta_0 + \mathbf{x}_i^T\boldsymbol{\beta}\right) \geq 1-\xi_i,\\ &\xi_i \geq 0, \quad \text{for } i=1, \ldots, N,\end{aligned} \tag{11.33}$$

where C^+ and C^- are separate cost parameters for each of the classes, which may reflect the misclassification costs from the cost matrix (see Table 4.7). Observe also that each of the two regularization terms sums up the slack variables associated only with one of the classes, that is, associated with the training observations having either positive ($y_i = +1$) or negative ($y_i = -1$) value of the response variable.

Since Lagrange multipliers (see Section 11.2.2) are also associated with the training observations, the dual representation of the optimization problem for cost-sensitive linear SVM classifiers can be now presented as

$$\begin{aligned}\underset{\boldsymbol{\alpha}}{\text{maximize}}\ &\sum_{i=1}^{N}\alpha_i - \frac{1}{2}\sum_{i=1}^{N}\sum_{j=1}^{N}\alpha_i\alpha_j y_i y_j \mathbf{x}_i^T\mathbf{x}_j\\ \text{subject to}\ &\sum_{i=1}^{N}\alpha_i y_i = 0 \ \text{and}\\ &0 \leq \alpha_i \leq C^+ \ \text{if } y_i = +1,\\ &0 \leq \alpha_i \leq C^- \ \text{if } y_i = -1,\\ &i=1, \ldots, N,\end{aligned} \tag{11.34}$$

where Lagrange multipliers associated with training observations belonging to the positive and the negative classes will have different ranges of their values. The class with the higher misclassification cost will have fewer of its training observations misclassified, which can be interpreted as moving the optimal separating hyperplane away from this class.

Observe that the introduction of class-specific costs does not contradict the balancing of the impact of the positive and negative support vectors as specified in (11.13) by the condition requiring that $\sum \alpha_i y_i = 0$. Assume that we increase misclassification costs of the positive class, that is, $C^+ > C^-$. This moves class boundaries away from the positive class and thus allows for fewer positive observations to violate their class margin while simultaneously increasing the number of negative observations violating their class margin. However, all observations violating the class margins are now assigned different α_i values depending on their class: $\alpha_i = C^+$ for the positive class and $\alpha_i = C^-$ for the negative class, and since $C^+ > C^-$, the required balancing condition may still be satisfied.

Finally, as described in Section 11.3, the cost-sensitive linear SVM classifier can be extended to nonlinear cases by using a nonlinear kernel function $K(\mathbf{x}_i, \mathbf{x}_j)$. This will result in the following optimization problem for a nonlinear cost-sensitive SVM classifier:

$$\begin{aligned} \underset{\boldsymbol{\alpha}}{\text{maximize}} \sum_{i=1}^{N} \alpha_i - \frac{1}{2} \sum_{i=1}^{N} \sum_{j=1}^{N} \alpha_i \alpha_j y_i y_j K(\mathbf{x}_i, \mathbf{x}_j) & \\ \text{subject to } \sum_{i=1}^{N} \alpha_i y_i = 0 \text{ and} & \\ 0 \leq \alpha_i \leq C^+ \text{ if } y_i = +1, & \\ 0 \leq \alpha_i \leq C^- \text{ if } y_i = -1, & \\ i = 1, \ldots, N. & \end{aligned} \tag{11.35}$$

12 Discriminant Analysis

12.1 Introduction

Discriminant Analysis is a classical supervised learning algorithm for classification. It is a powerful method that should be included in the toolbox of any bioinformatics or data science professional. In addition to its classification capabilities, it also provides low-dimensional visualization of the classification space. The main assumptions under which discriminant analysis is performed include the independence of training observations, no singularities or severe multicollinearities, and no extreme outliers. *Usually*, we also assume multivariate normality of variables; however, this assumption is, at least theoretically, not necessary. Fisher derived the original version of discriminant analysis without making this assumption (Fisher 1936). The goal of his design was to find such a solution to a classification problem that maximally separates the differentiated classes by maximizing the ratio of the variation between classes to the variation within classes. In contrast, the goal of the Bayesian design of discriminant analysis is to minimize the probability of misclassification; this design assumes that the variables follow a multivariate normal distribution in each of the differentiated classes.

There is a lot of confusion surrounding discriminant analysis, not only about whether the assumption of multivariate normality is necessary or not but also about terminology and the relation between the two designs. To make it more transparent, we will refer to the original design as *Fisher's discriminant analysis* (FDA), and to the design using the Bayes approach, which assumes multivariate normality (Welch 1939), as *Gaussian* discriminant analysis. As already stated, the former maximizes the ratio of the variation between classes to that within classes, and the latter minimizes the probability of misclassification under the assumption of multivariate normality. Furthermore, Fisher's solution provides only linear boundaries between classes, while the Gaussian approach may identify either linear or quadratic boundaries. If the homogeneity (equality) of class covariance matrices is assumed, we have Gaussian *linear discriminant analysis* (LDA); if this assumption is not made, we have Gaussian *quadratic discriminant analysis* (QDA).

It is also worth stressing that both Fisher's approach and the Gaussian approach work well for multiclass classification problems, and that their good performance is attributed to the fact that simple – linear or quadratic – boundaries between classes are more likely to be supported by the available data than intricate alternatives; this is especially true for linear decision boundaries. Furthermore, Fisher's discriminant analysis and Gaussian

LDA lead to the same solution when the misclassification costs and the prior probabilities are the same for each class (Welch 1939; Hastie et al. 2009). Hence, both minimize the probability of misclassification when the differentiated populations are normally distributed (assumed or not).

Assuming that our classification problem involves J classes and p independent variables, calculations for discriminant analysis would require the inversion of a $p \times p$ matrix. This cannot be done when the number of variables p is greater than the number of observations N (for QDA, it would be the number of observations per class $N_j, j = 1, \ldots, J$).[1] That will, however, in no way decrease the usability of this classification method, as it should be obvious that when dealing with high-dimensional $p > N$ data (and especially $p \gg N$ data), the feature selection step should be performed first[2] before building the final classifier. Although any supervised and multivariate feature selection method can be used before performing discriminant analysis, it may be advantageous – in the context of classification with discriminant analysis – to use such a feature selection algorithm that will select an optimal subset of variables using an aligned criterion of maximizing the variation between classes in relation to the variation within classes. Furthermore, discriminant analysis can be used as a learning algorithm within a feature selection schema implementing either a forward or hybrid stepwise search (for example, within the extended version of the T^2-based feature selection algorithm described in Chapter 5).

12.2 Fisher's Discriminant Analysis

Fisher's discriminant analysis (FDA) identifies a set of discriminant functions (which are linear combinations of the independent variables) that will maximize the ratio of the variation between J classes, $J \geq 2$, to the variation within the classes (Fisher 1936, 1938). This means searching for such a projection that would maximally separate the class centers while simultaneously minimizing class variations. **No assumption about the distribution of variables is made** (so, in particular, there is no assumption that variables in each class have multivariate normal distribution). However, FDA makes an implicit assumption of the homogeneity of class covariance matrices.

Assume that in our training data we have p variables, J classes, and N observations. Each observation is represented by a $p \times 1$ vector $\mathbf{x}_i = \left[x_{1i}, \cdots, x_{pi}\right]^T$ and a class label y_i, $i = 1, \ldots, N$. Assume also that the class labels are $\mathcal{D}_1, \mathcal{D}_2, \ldots, \mathcal{D}_J$, and that the number of training observations in class j is $N_j, j = 1, \ldots, J$, where $\sum N_j = N$. The class centers (mean vectors for observations from each class) are

[1] Thus, LDA would be a better choice than QDA when we have a relatively small number of training observations (provided LDA's assumption of the homogeneity of class covariances is not severely violated).

[2] Some software implementations of LDA include feature selection as the first step of data analysis, before LDA is invoked.

$$\overline{\mathbf{x}}_j = \frac{1}{N_j} \sum_{i: y_i = \mathcal{D}_j} \mathbf{x}_i, \qquad j = 1, \ldots, J, \tag{12.1}$$

and the overall mean is

$$\overline{\overline{\mathbf{x}}} = \frac{1}{N} \sum_{i=1}^{N} \mathbf{x}_i. \tag{12.2}$$

We will first consider the simplest case of two-class Fisher's discriminant analysis (when $J = 2$) and then the general case of multiclass FDA (when $J \geq 2$).

12.2.1 Two-Class Fisher's Discriminant Analysis

When we only have two classes ($J = 2$), the class centers will always lie on a line. Thus, to maximally separate the classes, we will look for such a projection $\mathbf{u}$, for which a linear function $z = \mathbf{u}^T\mathbf{x}$ will project training observations from their original p-dimensional space onto such a line (one-dimensional space) that would maximize the distance between the projected class means and simultaneously minimize the variation of each class. Since the difference between the projected class means, $\overline{z}_1 = \mathbf{u}^T\overline{\mathbf{x}}_1$ and $\overline{z}_2 = \mathbf{u}^T\overline{\mathbf{x}}_2$, may be negative, we will use the squared distance,

$$\begin{aligned} (\overline{z}_1 - \overline{z}_2)^2 &= \left(\mathbf{u}^T\overline{\mathbf{x}}_1 - \mathbf{u}^T\overline{\mathbf{x}}_2\right)^2 \\ &= \mathbf{u}^T(\overline{\mathbf{x}}_1 - \overline{\mathbf{x}}_2)(\overline{\mathbf{x}}_1 - \overline{\mathbf{x}}_2)^T\mathbf{u}. \end{aligned} \tag{12.3}$$

Observing that the variation *between* the two classes is represented by a $p \times p$ between-class scatter matrix

$$\mathbf{B} = (\overline{\mathbf{x}}_1 - \overline{\mathbf{x}}_2)(\overline{\mathbf{x}}_1 - \overline{\mathbf{x}}_2)^T, \tag{12.4}$$

the distance between the projected class means can be presented as $\mathbf{u}^T\mathbf{B}\mathbf{u}$. Since the variation of class j can be represented by a $p \times p$ scatter matrix $\mathbf{W}_j$,[3]

$$\mathbf{W}_j = \sum_{i: y_i = \mathcal{D}_j} \left(\mathbf{x}_i - \overline{\mathbf{x}}_j\right)\left(\mathbf{x}_i - \overline{\mathbf{x}}_j\right)^T, \qquad j = 1, 2, \tag{12.5}$$

the variations of the projected classes are $\mathbf{u}^T\mathbf{W}_1\mathbf{u}$ and $\mathbf{u}^T\mathbf{W}_2\mathbf{u}$; hence, to minimize them, we may minimize

$$\mathbf{u}^T\mathbf{W}_1\mathbf{u} + \mathbf{u}^T\mathbf{W}_2\mathbf{u} = \mathbf{u}^T(\mathbf{W}_1 + \mathbf{W}_2)\mathbf{u}. \tag{12.6}$$

Observing that the variation *within* the two classes can be represented by a $p \times p$ within-class scatter matrix $\mathbf{W}$,

[3] Recall that variance is calculated as variation divided by its degrees of freedom, and observe that the matrix of variation of class j is $(N_j - 1)$ times the covariance matrix of the class, that is, $\mathbf{W}_j = \left(N_j - 1\right)\mathbf{S}_j$ (Duda et al. 2001).

$$\begin{aligned}\mathbf{W} &= \mathbf{W}_1 + \mathbf{W}_2 \\ &= \sum_{j=1}^{2} \sum_{i:y_i=\mathcal{D}_j} \left(\mathbf{x}_i - \overline{\mathbf{x}}_j\right)\left(\mathbf{x}_i - \overline{\mathbf{x}}_j\right)^T,\end{aligned} \tag{12.7}$$

the projected within-class variation will be $\mathbf{u}^T\mathbf{W}\mathbf{u}$. Consequently, to maximize (12.3) and simultaneously minimize (12.6), we may maximize their ratio,

$$\underset{\mathbf{u}}{\text{maximize}} \ \frac{\mathbf{u}^T\mathbf{B}\mathbf{u}}{\mathbf{u}^T\mathbf{W}\mathbf{u}}. \tag{12.8}$$

Since $\mathbf{u}$ is a vector and we are only interested in its *direction* (which defines the direction of the discriminant function z), and not in its magnitude, we may replace (12.8) with the equivalent optimization problem,

$$\begin{aligned}&\underset{\mathbf{u}}{\text{maximize}} \ \mathbf{u}^T\mathbf{B}\mathbf{u} \\ &\text{subject to} \ \mathbf{u}^T\mathbf{W}\mathbf{u} = 1.\end{aligned} \tag{12.9}$$

As done in a few other places, to solve such a constrained optimization problem, we may introduce a nonnegative Lagrange multiplier λ, represent the problem by the Lagrangian

$$L(\mathbf{u}, \lambda) = \mathbf{u}^T\mathbf{B}\mathbf{u} - \lambda\left(\mathbf{u}^T\mathbf{W}\mathbf{u} - 1\right), \tag{12.10}$$

and solve it by setting its partial derivative to zero,

$$\frac{\delta L}{\delta \mathbf{u}} = 2\mathbf{B}\mathbf{u} - 2\lambda\mathbf{W}\mathbf{u} = 0, \tag{12.11}$$

which gives us

$$\mathbf{B}\mathbf{u} = \lambda\mathbf{W}\mathbf{u}. \tag{12.12}$$

This is a generalized eigenproblem, which for the case of binary classification has one eigenvalue λ and one eigenvector associated with this eigenvalue. We may check this by rewriting (12.12) as a regular one-matrix eigenproblem (which can be done under the assumption that $\mathbf{W}$ is not singular),

$$\mathbf{W}^{-1}\mathbf{B}\mathbf{u} = \lambda\mathbf{u}. \tag{12.13}$$

Although $\mathbf{B}$ and $\mathbf{W}$ are $p \times p$ matrices, the rank of $\mathbf{B}$ is only one (as it is the outer product of two vectors); thus, matrix $\mathbf{W}^{-1}\mathbf{B}$ is also of rank one, which means that the solution to (12.13), as well as (12.12), is the single eigenvector of matrix $\mathbf{W}^{-1}\mathbf{B}$. This $p \times 1$ vector $\mathbf{u} = \left[u_1, \cdots, u_p\right]^T$ defines the direction of the *linear discriminant function* line z. This discriminant function is a linear combination of p independent variables, $z = u_1x_1 + \ldots + u_px_p$, and it projects any observation, say $\mathbf{x}_{new}$, from the original p-dimensional space onto the one-dimensional discriminatory space (see Figure 12.1).

The only additional thing needed for classification is to decide about the threshold, that is, the point on line z (between the projected class centers $\overline{z}_1 = \mathbf{u}^T\overline{\mathbf{x}}_1$ and $\overline{z}_2 = \mathbf{u}^T\overline{\mathbf{x}}_2$),

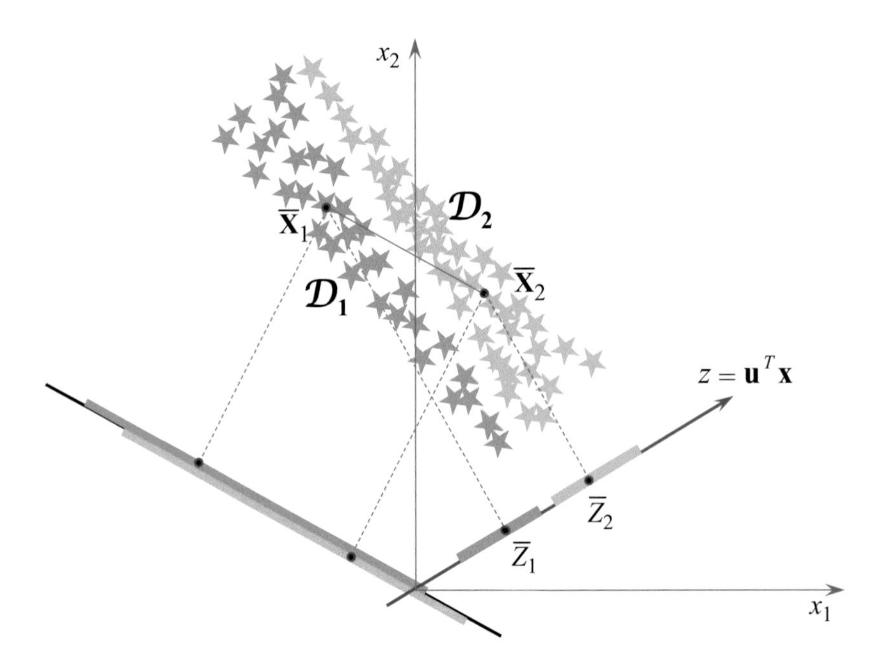

Figure 12.1 Visualization of two-class Fisher's discriminant analysis ($J = 2$) for a toy example with only two independent variables ($p = 2$). With two classes, there is only one linear discriminant function, $z = \mathbf{u}^T\mathbf{x}$, whose direction is defined by vector $\mathbf{u}$ (red line). This direction maximizes class separation – in this example the training observations belonging to classes $\mathcal{D}_1$ and $\mathcal{D}_2$ are completely separated after they are projected on line z. Recall that maximal class separation means maximizing the ratio of the between-class to within-class variation, which corresponds to minimal overlap between classes, rather than the maximal distance between class centers (the black line represents direction that maximizes the distance between class centers; projecting training observations on this line would result in a very heavy overlap between classes). For a geometrical interpretation of the FDA method, we may imagine starting with the line that connects class centers and then rotating this line until the overlap between classes is minimized. The same is true for multiclass FDA, except that we rotate the (J–1)-dimensional hyperplane that contains the J class centers.

which would define a boundary between the classes.[4] If the projection of the observation, $z_{new} = \mathbf{u}^T\mathbf{x}_{new}$, is less than the threshold, the observation would be classified to one class, otherwise to another. If the midpoint between class centers, $(\overline{z}_1 + \overline{z}_2)/2$, is selected as the threshold, the observation will be classified to the class whose projected center is closer to z_{new} (Fisher 1936). Another option is to assume that the *projected* classes are normally distributed and then use this assumption to calculate an optimal threshold (that would minimize the probability of misclassification). Recall that Fisher's linear discriminant function has been identified without making the assumption of multivariate normality. However, since this discriminant function, $z = \mathbf{u}^T\mathbf{x}$, is the sum

[4] Since FDA does not explicitly provide the threshold, it is sometimes described as a projection method rather than a classification method. However, considering that it identifies a discriminatory space that maximizes class separation, it seems equally appropriate to consider it as a classification method.

of many random variables, the central limit theorem provides some justification for making the assumption that the *projected* data is approximately normal in each class (Bishop 2006).

The main goal of this section is to describe FDA in a simple two-class situation, as an introduction to the multiclass case. However, if we were interested only in the two-class case, then we would not really need to solve eigenproblem (12.13). It has been shown (Duda et al. 2001; Rencher 2002) that, for a two-class case, we have

$$z = \mathbf{u}^T \mathbf{x} = (\overline{\mathbf{x}}_1 - \overline{\mathbf{x}}_2)^T \mathbf{W}^{-1} \mathbf{x}, \tag{12.14}$$

and, consequently, classification can be performed the same way as described earlier, but z_{new}, $\overline{z}_1$, and $\overline{z}_2$ would be calculated using the rightmost part of Equation (12.14).

12.2.2 Multiclass Fisher's Discriminant Analysis

The centers of two classes lie on a line; the centers of three classes on a two-dimensional plane; and the centers of J classes will always lie on a $(J - 1)$-dimensional hyperplane. Hence, in a multiclass case,[5] FDA will identify $(J - 1)$ linear discriminant functions that would project observations from their original p-dimensional space into a $(J - 1)$-dimensional space in which the classes will be maximally separated, that is, the ratio of between-class to within-class variations will be maximized. Let us first consider these variations for a multiclass case. The *variation within classes* is based on the distances between each observation in the class and its class center and is represented by a $p \times p$ scatter matrix $\mathbf{W}$,

$$\mathbf{W} = \sum_{j=1}^{J} \sum_{i : y_i = \mathcal{D}_j} (\mathbf{x}_i - \overline{\mathbf{x}}_j)(\mathbf{x}_i - \overline{\mathbf{x}}_j)^T. \tag{12.15}$$

Observe the matrix of the within-class variation $\mathbf{W} = (N - J)\mathbf{S}$, where $\mathbf{S}$ is the common covariance matrix (see Equation 12.38). This means that FDA is performed under the *implicit* assumption of homogeneity of class covariance matrices and thus results in linear boundaries between classes.

The *variation between classes* is based on the distances between class centers. However, instead of calculating all $\binom{J}{2}$ pairwise distances for J classes, we may exploit the fact that the total variation $\mathbf{T} = \sum_{i=1}^{N} (\mathbf{x}_i - \overline{\overline{\mathbf{x}}})(\mathbf{x}_i - \overline{\overline{\mathbf{x}}})^T = \mathbf{B} + \mathbf{W}$. Since both $\mathbf{T}$ and $\mathbf{W}$ are easy to calculate, we may compute $\mathbf{B}$ as $\mathbf{T} - \mathbf{W}$; as a result, the variation between classes may be calculated in a much easier way by only using the distances between each class center and the overall mean, and represented by the $p \times p$ scatter matrix $\mathbf{B}$,

$$\mathbf{B} = \sum_{j=1}^{J} N_j (\overline{\mathbf{x}}_j - \overline{\overline{\mathbf{x}}})(\overline{\mathbf{x}}_j - \overline{\overline{\mathbf{x}}})^T. \tag{12.16}$$

These three variations and their degrees of freedom are summarized in Table 12.1.

[5] The multiclass case may also be considered a general case with $J \geq 2$.

Table 12.1 Partitioning total variation into variation between classes and variation within classes, $\mathbf{T} = \mathbf{B} + \mathbf{W}$. Recall that variance is variation divided by its corresponding degrees of freedom. Thus, for example, the common within-class variance–covariance matrix $\mathbf{S} = \mathbf{W}/(N - J)$.

Source of variation	Variation (sum of squares)	Degrees of freedom
Between-class	$\mathbf{B} = \sum_{j=1}^{J} N_j(\overline{\mathbf{x}}_j - \overline{\overline{\mathbf{x}}})(\overline{\mathbf{x}}_j - \overline{\overline{\mathbf{x}}})^T$	$J - 1$
Within-class	$\mathbf{W} = \sum_{j=1}^{J} \sum_{i:y_i=\mathcal{D}_j} (\mathbf{x}_i - \overline{\mathbf{x}}_j)(\mathbf{x}_i - \overline{\mathbf{x}}_j)^T$	$N - J$
Total	$\mathbf{T} = \sum_{i=1}^{N} (\mathbf{x}_i - \overline{\overline{\mathbf{x}}})(\mathbf{x}_i - \overline{\overline{\mathbf{x}}})^T$	$N - 1$

The projection from the original p-dimensional space into a $(J - 1)$-dimensional space is represented by $\mathbf{z} = \mathbf{U}^T\mathbf{x}$, where $\mathbf{U}$ is now a $p \times (J - 1)$ matrix, which will be used to map p-dimensional observations $\mathbf{x}$ into their $(J - 1)$-dimensional representations $\mathbf{z}$. Thus, we search for the projection matrix $\mathbf{U}$ that maximizes the ratio of the projected between-class variation to the projected within-class variation, which will now be represented as[6]

$$\underset{\mathbf{U}}{\text{maximize}} \; \frac{tr(\mathbf{U}^T\mathbf{B}\mathbf{U})}{tr(\mathbf{U}^T\mathbf{W}\mathbf{U})}, \tag{12.17}$$

where $tr()$ denotes the trace of a matrix (that is, the sum of its diagonal elements). Following the same reasoning as before, we may rewrite (12.17) as a constrained optimization problem,

$$\begin{aligned} \underset{\mathbf{U}}{\text{maximize}} \; & tr(\mathbf{U}^T\mathbf{B}\mathbf{U}) \\ \text{subject to} \; & tr(\mathbf{U}^T\mathbf{W}\mathbf{U}) = 1, \end{aligned} \tag{12.18}$$

introduce a vector of nonnegative Lagrange multipliers $\boldsymbol{\lambda}$, represent the problem in the form of the Lagrangian

$$L(\mathbf{U}, \boldsymbol{\lambda}) = tr(\mathbf{U}^T\mathbf{B}\mathbf{U}) - \boldsymbol{\lambda}\left[tr(\mathbf{U}^T\mathbf{W}\mathbf{U}) - 1\right], \tag{12.19}$$

and, finally, solve it by setting its partial derivative $\delta L/\delta \mathbf{U}$ to zero. This will result in the generalized eigenproblem

$$\mathbf{B}\mathbf{U} = \boldsymbol{\lambda}\mathbf{W}\mathbf{U}. \tag{12.20}$$

[6] Observe that $\mathbf{U}^T\mathbf{B}\mathbf{U}$ is not a scalar now, as we have $J - 1$ projections; summing up the between-class variations in all $J - 1$ dimensions is equivalent to calculating the trace of $\mathbf{U}^T\mathbf{B}\mathbf{U}$. The same is true for within-class variations.

We can solve it directly in this form or, alternatively, if matrix $\mathbf{W}$ is nonsingular, in the form of a one-matrix eigenproblem $\mathbf{W}^{-1}\mathbf{B}\mathbf{U} = \boldsymbol{\lambda}\mathbf{U}$. Since the between-class scatter matrix $\mathbf{B}$ is the sum of J rank-one matrices, its rank, and thus the rank of matrix $\mathbf{W}^{-1}\mathbf{B}$, is $J - 1$.[7] Thus, as a solution to (12.20), we will have a $(J-1) \times (J-1)$ diagonal matrix $\boldsymbol{\lambda}$ of eigenvalues and a $p \times (J-1)$ matrix $\mathbf{U}$, whose columns will represent normalized eigenvectors $\mathbf{u}_1, \mathbf{u}_2, \ldots, \mathbf{u}_{(J-1)}$ associated with the eigenvalues,

$$\boldsymbol{\lambda} = \begin{bmatrix} \lambda_1 & 0 & \cdots & 0 \\ 0 & \lambda_2 & \cdots & 0 \\ \vdots & \vdots & \ddots & \vdots \\ 0 & 0 & \cdots & \lambda_{(J-1)} \end{bmatrix}, \quad \mathbf{U} = \begin{bmatrix} u_{11} & u_{12} & \cdots & u_{1(J-1)} \\ u_{21} & u_{22} & \cdots & u_{2(J-1)} \\ \vdots & \vdots & \ddots & \vdots \\ u_{p1} & u_{p2} & \cdots & u_{p(J-1)} \end{bmatrix}. \tag{12.21}$$

Therefore, the projection from the original p-dimensional data space to the $(J-1)$-dimensional space of the maximum class separation will be defined by $(J-1)$ *linear discriminant functions*,

$$\begin{aligned} z_1 &= \mathbf{u}_1^T \mathbf{x}, \\ z_2 &= \mathbf{u}_2^T \mathbf{x}, \\ &\cdots \\ z_{(J-1)} &= \mathbf{u}_{(J-1)}^T \mathbf{x}. \end{aligned} \tag{12.22}$$

Since the nonzero eigenvalues are ranked, $\lambda_1 > \lambda_2 > \ldots > \lambda_{J-1} > 0$, the single direction that best separates the classes is the one associated with the first discriminant function, z_1. Consequently, if we want to visualize the discriminatory space when $J > 4$, we would use the first two or three discriminant functions. The relative magnitude of class separation provided by each of the discriminant functions can be calculated as the proportion of its eigenvalue to the sum of all eigenvalues (Rencher 2002),

$$\frac{\lambda_k}{\sum_{l=1}^{J-1} \lambda_l}, k = 1, \ldots, J-1. \tag{12.23}$$

Recall from Chapter 5 the Lawley-Hotelling trace statistic T^2,

$$T^2 = tr\left(\mathbf{W}^{-1}\mathbf{B}\right), \tag{12.24}$$

which also provides a measure of class separation represented by the ratio of the variation between classes to the variation within classes. It has a direct relationship with Fisher's discriminant analysis[8] and can be used to evaluate discriminatory power of any set of variables or features. For example, the discriminatory power of feature k corresponding to discriminant function z_k is

[7] Theoretically, it is min(p, $J - 1$), though it is unlikely to have fewer variables than classes.

[8] In his second paper on FDA, Fisher (1938) discussed the relation between discriminant analysis and the T^2 statistic.

$$T^2(z_k) = \lambda_k, \quad k = 1, \ldots, J-1, \tag{12.25}$$

and the discriminatory power of all $(J-1)$ features is

$$T^2(z_1, \ldots, z_{J-1}) = \sum_{l=1}^{J-1} \lambda_l. \tag{12.26}$$

To classify a new observation $\mathbf{x}_{new}$, we would project it into the $(J-1)$-dimensional discriminatory space,

$$\mathbf{z}_{new} = \mathbf{U}^T \mathbf{x}_{new}, \tag{12.27}$$

and then assign it using some classification criterion; for example, we may assign the observation to class j, whose projected center, $\overline{\mathbf{z}}_j = \mathbf{U}^T \overline{\mathbf{x}}_j$, is closest to the projected observation. This common practice exploits the fact that in a $(J-1)$-dimensional space of uncorrelated features represented by the discriminant functions, the features have unit variances and zero covariances, which means that the within-class Mahalanobis distance[9] is the same as Euclidean distance (Ripley 1996). This also means that the class areas are represented by hyperspheres (rather than hyperellipsoids in the original p-dimensional space). Consequently, the *classification function* for Fisher's discriminant analysis may be written (Johnson and Wichern 2007) as

$$Class(\mathbf{x}_{new}) = \arg\min_j \sum_{k=1}^{J-1} \left[\mathbf{u}_k^T \left(\mathbf{x}_{new} - \overline{\mathbf{x}}_j\right)\right]^2 \quad j = 1, \ldots, J. \tag{12.28}$$

Although the FDA's discriminatory space defined by $(J-1)$ linear discriminant functions has been identified without assuming multivariate normality of classes in the original p-dimensional space, we may now – after transforming the data into this low-dimensional discriminatory space – assume multivariate normality of the $(J-1)$ features in each of the *projected* classes (justifying this, like for the binary case, by invoking the central limit theorem).

For two, three, or four differentiated classes $(J < 5)$, the discriminatory space, training data, and classification results can be graphically presented in one-, two-, or three-dimensional space (respectively) that includes 100 percent of the discriminatory information (see Figure 12.2). For $J \geq 5$, we can visualize a subspace of the discriminatory space using the first two or three discriminant functions.

12.3 Gaussian Discriminant Analysis

Fisher's discriminant analysis can be perceived as a nonparametric method employing the frequentist approach. In contrast, Gaussian discriminant analysis is based on a

[9] If the variances of the features were not equal (as it would be for the original p variables), then to properly measure the distances between the classified observation and class centers, Mahalanobis distance would need to be used as it takes into account such different variances.

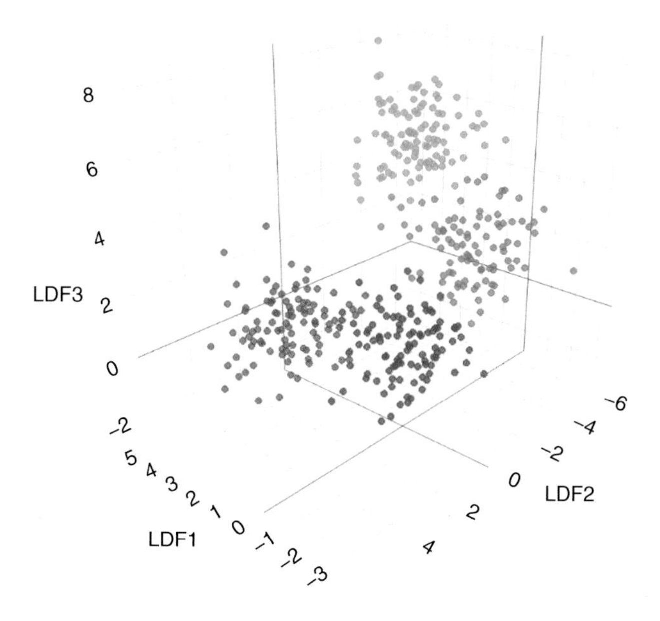

Figure 12.2 An example of the discriminatory space for four classes ($J = 4$). When four classes are differentiated, Fisher's discriminant analysis identifies three linear discriminant functions (LDF1, LDF2, LDF3); they define a three-dimensional discriminatory space, in which classes are maximally separated. This space includes 100 percent of the discriminatory information.

Bayesian approach and is a parametric method since, for each class, it makes the assumption of multivariate normal distribution (that is, Gaussian densities) of the independent variables (Welch 1939). These lead to a solution for which the probability of misclassification is minimized.

Assume that our target population includes J classes, say, J disease states labeled $\mathcal{D}_1, \mathcal{D}_2, \ldots, \mathcal{D}_J$, and that any observation, say, a patient under diagnosis, is represented by a $p \times 1$ vector $\mathbf{x}_i = [x_{1i}, \cdots, x_{pi}]^T$ corresponding to measured values of p variables (for example, gene expression levels). The core notion for a Bayesian classifier is the *posterior* probability $P(\mathcal{D}_j|\mathbf{x})$ of the patient belonging to class j (class with label $\mathcal{D}_j$) given that the vector representing the patient's results is $\mathbf{x}$. This posterior probability is calculated for each class j, $j = 1, \ldots, J$, using Bayes' rule,

$$P(\mathcal{D}_j|\mathbf{x}) = \frac{P(\mathbf{x}|\mathcal{D}_j)P(\mathcal{D}_j)}{\sum\limits_{m=1}^{J} P(\mathbf{x}|\mathcal{D}_m)P(\mathcal{D}_m)}, \tag{12.29}$$

where

- $P(\mathcal{D}_j)$ is the *prior* probability of class $\mathcal{D}_j$ (that is, the probability that a randomly selected patient belongs to class $\mathcal{D}_j$); denote it π_j and observe that $\sum_{j=1}^{J} \pi_j = 1$,
- $P(\mathbf{x}|\mathcal{D}_j)$ is the class-conditional density of the variables in $\mathbf{x}$ for class j, which – under the assumption of a multivariate normal distribution of the variables in

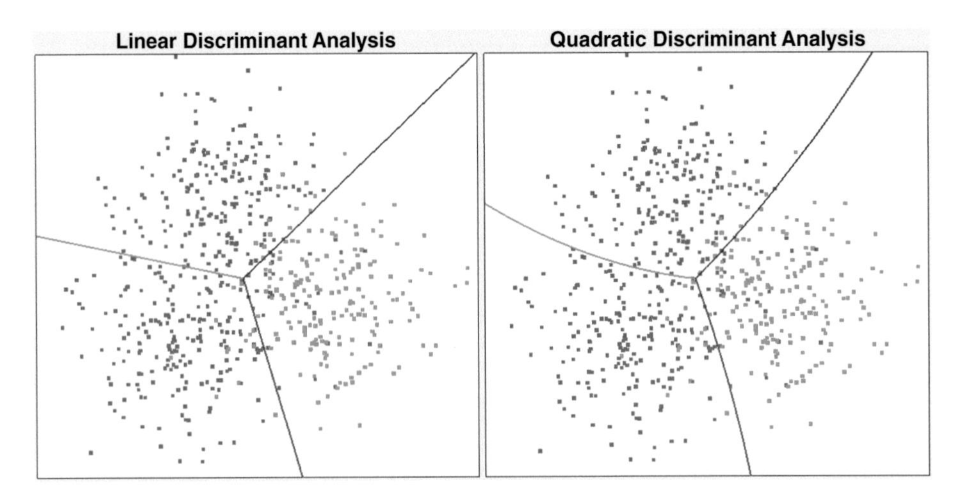

Figure 12.3 Visualization of the boundaries between three classes identified by LDA (left panel) and QDA (right panel).

each class, with the class mean $\boldsymbol{\mu}_j$ and the class covariance matrix $\boldsymbol{\Sigma}_j$ – will be represented by the Gaussian density function $f_j(\mathbf{x})$,[10]

$$f_j(\mathbf{x}) = \frac{1}{(2\pi)^{p/2}\left|\boldsymbol{\Sigma}_j\right|^{1/2}} e^{-\frac{1}{2}\left(\mathbf{x}-\boldsymbol{\mu}_j\right)^T \boldsymbol{\Sigma}_j^{-1}\left(\mathbf{x}-\boldsymbol{\mu}_j\right)}. \tag{12.30}$$

We can now rewrite Equation (12.29) for posterior probability of class $j, j = 1, \ldots, J$, as

$$P\left(\mathcal{D}_j|\mathbf{x}\right) = \frac{f_j(\mathbf{x})\pi_j}{\sum_{m=1}^{J} f_m(\mathbf{x})\pi_m}. \tag{12.31}$$

When classified based on this formula, the classified observation $\mathbf{x}$ will be assigned to class $\mathcal{D}_j$ with the largest posterior probability $P\left(\mathcal{D}_j|\mathbf{x}\right)$. If posterior probabilities $P\left(\mathcal{D}_j|\mathbf{x}\right), j = 1, \ldots, J$, are calculated under the assumption that all of the J class covariance matrices are equal (the homogeneity assumption), we will have Gaussian *linear* discriminant analysis, for which all boundaries between classes are linear (see the left panel of Figure 12.3 for a two-dimensional example). Otherwise, we will have Gaussian *quadratic* discriminant analysis (see the right panel of Figure 12.3).

12.3.1 Gaussian Linear Discriminant Analysis

Gaussian *linear discriminant analysis* (LDA), in addition to the multivariate normality assumption, makes the assumption of homogeneity of class covariances,

[10] Where $\left|\boldsymbol{\Sigma}_j\right|$ denotes the determinant of covariance matrix $\boldsymbol{\Sigma}_j$ and $\boldsymbol{\Sigma}_j^{-1}$ its inverse.

$$\boldsymbol{\Sigma}_1 = \boldsymbol{\Sigma}_2 = \ldots = \boldsymbol{\Sigma}_J = \boldsymbol{\Sigma}. \tag{12.32}$$

To compare posterior probabilities of J classes, we may first observe that the denominator of Equation (12.31) has the same value for all classes; thus, we need only to compare $f_j(\mathbf{x})\pi_j$ terms. Remembering the homogeneity of class covariances, from Equation (12.30) we have

$$f_j(\mathbf{x})\pi_j = \frac{1}{(2\pi)^{p/2}|\boldsymbol{\Sigma}|^{1/2}} e^{-\frac{1}{2}(\mathbf{x}-\boldsymbol{\mu}_j)^T \boldsymbol{\Sigma}^{-1}(\mathbf{x}-\boldsymbol{\mu}_j)}\pi_j. \tag{12.33}$$

However, the first term on the right side of this equation is now the same for all classes; thus, we may drop it. We may also take natural logarithm of the remaining part of Equation (12.33). As a result, finding the largest posterior probability $P(\mathcal{D}_j|\mathbf{x}), j = 1, \ldots, J$, is equivalent to finding the largest value of the following function (which can be considered a scaled posterior probability):

$$\begin{aligned} \delta_j(\mathbf{x}) &= -\frac{1}{2}(\mathbf{x}-\boldsymbol{\mu}_j)^T \boldsymbol{\Sigma}^{-1}(\mathbf{x}-\boldsymbol{\mu}_j) + \ln(\pi_j) \\ &= -\frac{1}{2}\mathbf{x}^T\boldsymbol{\Sigma}^{-1}\mathbf{x} + \mathbf{x}^T\boldsymbol{\Sigma}^{-1}\boldsymbol{\mu}_j - \frac{1}{2}\boldsymbol{\mu}_j{}^T\boldsymbol{\Sigma}^{-1}\boldsymbol{\mu}_j + \ln(\pi_j). \end{aligned} \tag{12.34}$$

Since the quadratic term $-\frac{1}{2}\mathbf{x}^T\boldsymbol{\Sigma}^{-1}\mathbf{x}$ is the same for all classes (the same covariance matrix and the same classified observation **x**), we may drop it and get a linear function in **x** for each class,

$$\delta_j(\mathbf{x}) = \mathbf{x}^T\boldsymbol{\Sigma}^{-1}\boldsymbol{\mu}_j - \frac{1}{2}\boldsymbol{\mu}_j{}^T\boldsymbol{\Sigma}^{-1}\boldsymbol{\mu}_j + \ln(\pi_j), \quad j = 1, \ldots, J, \tag{12.35}$$

and thus a linear decision boundary between any two of J classes.

All that is left now in order to design an LDA classifier is estimating population parameters from the training data. Assume that the training data include N observations, each of them represented by a vector $\mathbf{x}_i = [x_{1i}, \cdots, x_{pi}]^T$ and known to be from class y_i, $i = 1, \ldots, N$. Denote the class labels as $\mathcal{D}_1, \ldots, \mathcal{D}_J$, and the number of training observations in class j as $N_j, j = 1, \ldots, J$. If we do not know the prior probabilities of the classes in the target population, we estimate them from the proportions of training observations in each class,

$$\widehat{\pi}_j = \frac{N_j}{N}, \quad j = 1, \ldots, J. \tag{12.36}$$

The mean for each class is estimated as

$$\overline{\mathbf{x}}_j = \frac{1}{N_j}\sum_{i:y_i=\mathcal{D}_j}\mathbf{x}_i, \quad j = 1, \ldots, J, \tag{12.37}$$

and the common covariance matrix as

$$\mathbf{S} = \frac{1}{N-J}\sum_{j=1}^{J}\sum_{i:y_i=\mathcal{D}_j}(\mathbf{x}_i - \overline{\mathbf{x}}_j)(\mathbf{x}_i - \overline{\mathbf{x}}_j)^T. \tag{12.38}$$

Consequently, the *linear classification functions* are

$$\delta_j(\mathbf{x}) = \mathbf{x}^T\mathbf{S}^{-1}\overline{\mathbf{x}}_j - \frac{1}{2}\overline{\mathbf{x}}_j^T\mathbf{S}^{-1}\overline{\mathbf{x}}_j + \ln(\hat{\pi}_j), \quad j = 1, \ldots, J, \tag{12.39}$$

and the patient represented by vector $\mathbf{x}$ will be classified to class j for which the classification function $\delta_j(\mathbf{x})$ has the largest value, that is,

$$Class(\mathbf{x}) = \arg\max_j \delta_j(\mathbf{x}). \tag{12.40}$$

It is worth noting that if the prior probabilities are equal, that is, $\widehat{\pi}_1 = \widehat{\pi}_2 = \ldots = \widehat{\pi}_J$, then, in terms of classification results, the LDA classification rule (12.40) is equivalent to Fisher's classification rule (12.28) (Johnson and Wichern 2007).

We may also look at this LDA classification rule from a geometrical point of view. With the LDA's assumption of equal class covariance matrices, the class areas are represented by equal-size hyperellipsoids centered about their class means, with hyperplane boundaries between the classes. If the prior probabilities of the classes are the same, a patient represented by vector $\mathbf{x}$ will be classified to class j whose center is nearest to $\mathbf{x}$, when measured by the squared Mahalanobis distance $\left(\mathbf{x} - \overline{\mathbf{x}}_j\right)^T\mathbf{S}^{-1}\left(\mathbf{x} - \overline{\mathbf{x}}_j\right)$. In this case, a hyperplane boundary between any two classes crosses the line connecting the two class centers midway between the centers (although the hyperplane does not need to be orthogonal to this line, which would only be the case when the class areas are represented by hyperspheres). However, if the priors are not equal, the boundary will move toward the class with a lower prior probability (Duda et al. 2001).

Be aware that some authors call Functions (12.39) "linear *discriminant* functions". Yet, these J classification functions are very different from the (J – 1) Fisher's linear discriminant functions (described by 12.22), which:

- define a projection from a p-dimensional space of the original variables into a (J – 1)-dimensional discriminatory space, in which the classes are maximally separated,
- have coefficients that are eigenvectors of $\mathbf{W}^{-1}\mathbf{B}$,
- are ordered by their discriminatory power, and
- facilitate low-dimensional visualization of the discriminatory space.

Since Functions (12.39) have none of the above characteristics, and since their purpose is to allow classification in the p-dimensional space of the original variables, it seems more appropriate to call them linear *classification* functions (Rencher 2002; Huberty and Olejnik 2006).

12.3.2 Gaussian Quadratic Discriminant Analysis

Gaussian *quadratic discriminant analysis* (QDA) assumes multivariate normality, but does *not* make the assumption of homogeneity of class covariances. This means that instead of the LDA Equation (12.33), we will have

$$f_j(\mathbf{x})\pi_j = \frac{1}{(2\pi)^{p/2}|\mathbf{\Sigma}_j|^{1/2}} e^{-\frac{1}{2}(\mathbf{x}-\boldsymbol{\mu}_j)^T \mathbf{\Sigma}_j^{-1}(\mathbf{x}-\boldsymbol{\mu}_j)}\pi_j. \tag{12.41}$$

Now we may only drop the constant $1/(2\pi)^{p/2}$; hence, after taking natural logarithm, we will have the QDA version of a scaled posterior probability,

$$\delta_j(\mathbf{x}) = -\frac{1}{2}\ln|\mathbf{\Sigma}_j| - \frac{1}{2}(\mathbf{x}-\boldsymbol{\mu}_j)^T \mathbf{\Sigma}_j^{-1}(\mathbf{x}-\boldsymbol{\mu}_j) + \ln(\pi_j). \tag{12.42}$$

Instead of estimating the common covariance matrix, we will now need to estimate separate covariance matrices for each class,

$$\mathbf{S}_j = \frac{1}{N_j - 1}\sum_{i:y_i=\mathcal{D}_j}(\mathbf{x}_i - \bar{\mathbf{x}}_j)(\mathbf{x}_i - \bar{\mathbf{x}}_j)^T, \quad j = 1, \ldots, J, \tag{12.43}$$

and have the following *quadratic classification functions*:

$$\delta_j(\mathbf{x}) = -\frac{1}{2}\ln|\mathbf{S}_j| - \frac{1}{2}(\mathbf{x}-\bar{\mathbf{x}}_j)^T \mathbf{S}_j^{-1}(\mathbf{x}-\bar{\mathbf{x}}_j) + \ln(\hat{\pi}_j), \tag{12.44}$$

which define quadratic boundaries between classes; that is, the boundary between classes k and l, $k, l = 1, \ldots, J, k \neq l$, is defined by $\delta_k(\mathbf{x}) = \delta_l(\mathbf{x})$. The form of classification rule will be the same as described by Equation (12.40), but now observation $\mathbf{x}$ will be classified to class j corresponding to the largest value of the *quadratic* classification function.

12.3.3 Dimensionality Reduction in Gaussian Discriminant Analysis

In Fisher's version of discriminant analysis, dimensionality reduction is intrinsic; that is, $(J - 1)$ discriminant functions define a $(J - 1)$-dimensional discriminatory space, in which the ratio of between-to-within class variation is maximized, and in which classification is performed. This provides explicit dimensionality reduction from p to $J - 1$ (assuming, of course, that p is greater than $J - 1$). This, however, is far from obvious for Gaussian discriminant analysis. If we look at Gaussian classification functions – (12.39) for LDA and (12.44) for QDA – they are defined in a p-dimensional space of the original variables, and thus classification needs to be apparently performed in this space. Yet, we may recall that the centers of J classes will lie on a $(J - 1)$-dimensional hyperplane. Furthermore, even if the observation (say, patient) to classify is represented by a p-dimensional vector $\mathbf{x}$, and we want to calculate distances between $\mathbf{x}$ and the class centers, we may ignore distances that are orthogonal to this hyperplane, as they will be the same for each class. Hence, instead of performing classification in the original p-dimensional space, we may project the patient's vector $\mathbf{x}$ onto the $(J - 1)$-dimensional subspace of class centers and perform classification there.[11]

[11] It would involve sphering of data with respect to within-class variation $\mathbf{W}$ and then classifying the patient to the class with the nearest center; however, if prior class probabilities are not equal, $\ln(\pi_j)$ correction needs to be included in distance calculation (Hastie et al. 2009).

Furthermore, for LDA, we may also find a sequence of linear discriminant functions that will be ordered by their importance for classification and thus, similarly as it was natively available for FDA, provide visualization of the classification space (and the data) in fewer than $J-1$ dimensions (which would be particularly useful for $J > 4$). This approach is called reduced-rank LDA (Hastie et al. 2009) and will, surprisingly, result in the same discriminatory space as provided by the FDA, even if Fisher's analysis was done without the normality assumption. Consequently, LDA is equivalent to FDA. Moreover, although FDA does not make the multivariate normality assumption, if the classes *are* normally distributed, then the FDA solution will also minimize the probability of misclassification.

12.3.4 Regularized Discriminant Analysis

Estimates of class covariance matrices for QDA (see Equation 12.43) may become highly variable when the numbers of class observations $N_j, j=1, \ldots, J$, are relatively small, that is, not much greater that the number of variables p. This problem becomes more serious when the number of observations decreases, and when $N_j \leq p$, the covariance matrix $\mathbf{S}_j$ becomes singular and its inversion is either impossible or, even if numerically possible, the results would be unstable. This situation is similar to inverting a singular or multicollinear matrix $\mathbf{XX}^T$ in multiple regression (see Chapter 6), for which regularization via ridge regression (see Chapter 7) was one of possible solutions. Of course, instead of QDA we could use LDA, for which we need to estimate only one common covariance matrix, and thus the number of observations is counted over all classes. But what about a nonlinear solution in-between LDA and QDA? *Regularized discriminant analysis* (Friedman 1989) provides such a solution by shrinking class covariances $\mathbf{S}_j, j=1, \ldots, J$, toward the common covariance matrix $\mathbf{S}$. Regularized class covariances are calculated as

$$\mathbf{S}_j^{\text{Reg}}(\alpha) = \alpha\,\mathbf{S}_j + (1-\alpha)\mathbf{S}, \quad j=1, \ldots, J, \tag{12.45}$$

where $\alpha \in [0, 1]$ is the regularization parameter. With $\alpha = 0$, we have LDA; with $\alpha = 1$, we have QDA; and a continuum of solutions for α values between 0 and 1. An optimal value of α is usually decided by evaluating the performance of the classification model via cross-validation.

12.3.5 The Multivariate Normality Assumption

Since FDA and LDA will result in the same low-dimensional discriminatory space, and FDA does not assume multivariate normality of the variables in each class, a valid question is whether we should make this assumption. If we want to have a classifier that is optimal in the sense of minimizing the probability of misclassification, then the answer should be "yes." However, with a large number of independent variables, testing for their multivariate normality is a quite difficult task. Extensions of univariate tests would not be enough; for example, goodness-of-fit tests comparing observed and expected frequencies in a number of compartments would not work because for multivariate data, many, if not most, of such p-dimensional compartments will be empty even with a reasonably large number of training observations in a class. Hence, even if there are many tests for multivariate normality, we need to be aware that due to the sparseness of multivariate data, they may not be very powerful (Rencher 2002). Among the better-known tests are Mardia's test (Mardia 1970), Royston's test (Royston 1983), and Henze-Zirkler test (Henze and Zirkler 1990).

However, discriminant analysis is rather robust to some violations of the normality assumption, and thus, one option is to make the assumption without extensively checking its validity. If the variables in the classes are in fact multivariate normal, then we would have an optimal classifier.[12] Otherwise, if reliable estimates of the classifier's predictive power are sufficiently high, we still end up with a useful classification model. Hence, whether we build a discriminant analysis classifier with or without the normality assumption, it is of paramount importance to test it on a possibly large test data set, whose observations were never used during the predictive modeling process (which is, of course, true for any supervised learning algorithm, not discriminant analysis alone).

12.4 Partial Least Squares Discriminant Analysis

For discriminant analysis to work, we need to have more training observations than variables, and the more observations we have, the better it will work (well, the latter is true for any learning algorithm). For FDA and LDA, the total number of observations counts; whereas for QDA, the number of observations in the smallest class counts. For high-dimensional data, this means that discriminant analysis cannot be used as a learning algorithm within feature selection schemas implementing backward selection (for example, recursive feature elimination). However, it can be successfully used either after feature selection or within a feature selection schema using stepwise hybrid search. Since, in multivariate biomarker discovery, we are almost always interested in parsimonious biomarkers, these limitations of discriminant analysis are not really limiting its use. Consequently, there should be little reason for using this algorithm in situations when the number of training observations is smaller than (or similar to) the number of variables. Similarly, if high-dimensional data suffer from multicollinearity, properly designed multivariate feature selection should be performed instead of considering classification without the feature selection step. Nevertheless, one may perhaps imagine an unusual situation when the target variable really depends on very many independent variables (or, maybe, this was just assumed to be the case and no feature selection was performed), possibly with multicollinearities, and that there is a reason to use discriminant analysis in this situation. In such a case, the *partial least squares discriminant analysis* (PLSDA) could be the method of choice.

[12] Optimal from the point of view of minimizing the probability of misclassification.

PLSDA combines the partial least squares (PLS) approach to supervised dimensionality reduction with classification via discriminant analysis. The PLS part works the same way as for the partial least squares regression (PLSR), described in Chapter 6, except that instead of having a continuous target variable as in PLSR, classes need to be encoded into the target. For binary classification, classes can be encoded with values 0 and 1 of a single response variable, represented by vector $\mathbf{y}$ of N values – this corresponds to PLSR using the PLS1 algorithm. When there are more than two classes ($J > 2$), we need J dummy response variables (alternatively, J classes can be encoded by $J - 1$ dummy variables)[13] – in this situation, the PLS2 algorithms will be used. If we use J dummy response variables, they may be represented by a $J \times N$ matrix $\mathbf{Y}$,

$$\mathbf{Y} = \begin{bmatrix} y_{11} & y_{12} & \cdots & y_{1N} \\ y_{21} & y_{22} & \cdots & y_{2N} \\ \vdots & \vdots & \ddots & \vdots \\ y_{J1} & y_{J2} & \cdots & y_{JN} \end{bmatrix}, \tag{12.46}$$

whose rows represent J classes and columns represent N training observations. Each column will contain a single "1" value, corresponding to the class of this column's training observation (all other values in the column will be zero).[14]

The first stage of PLSDA is the same as in PLSR – a PLS algorithm (either PLS1 or PLS2) is used to identify the projection of a p-dimensional space of the original variables (possibly with multicollinearities) into an m-dimensional space of orthogonal PLS components, where theoretically $m \leq p$, but we are typically interested in $m \ll p$. As a result, we obtain a $p \times m$ matrix $\mathbf{C}$ whose columns represent m PLS components (for details, see Chapter 6). In the second stage, matrix $\mathbf{C}$ is used to project the training data into the space of the PLS components, and discriminant analysis (typically FDA or LDA) is performed in this m-dimensional space of those uncorrelated PLS components. An optimal number of PLS components m may be selected via cross-validation.

Note, however, that although the difficulties that discriminant analysis would have had with high-dimensional or multicollinear data have been circumvented, no *feature selection* has been performed as the PLS components are linear combinations of all original p variables. Since it is typical for high-dimensional data to include many noisy or irrelevant variables, solutions provided by PLSDA, if applied to such data, are likely to be suboptimal in terms of the predictive abilities of the PLSDA model as well as its interpretability. It is strongly recommended that, before deciding on using PLSDA, a serious consideration be given to multivariate feature selection.

[13] One of the classes will not have a dummy variable assigned, but will be represented by zero values of all $J - 1$ dummy variables.

[14] Software implementations of PLSDA usually perform the encoding of class information into dummy variables internally.

13 Neural Networks and Deep Learning

13.1 Introduction

One could say that in the context of multivariate biomarker discovery, which focuses on parsimonious biomarkers, there would be no need, or even reason, to cover artificial neural networks. It would be hard to dispute such an opinion. Neural networks, in this context, are niche algorithms that are unlikely to provide parsimonious and interpretable multivariate biomarkers developed from high-dimensional data. Nevertheless, we will describe them since those niche algorithms quite consistently outperform other learning algorithms in such applications as the classification of images or videos.[1] In particular, convolutional neural networks seem to be promising in the classification of medical images.

Conceptually, the terms *neural networks* and – the recently popular – *deep learning* are synonymous, with deep learning (or deep neural networks) referring to neural networks with many hidden layers. Consequently, with this relatively new and apparent subdivision of neural networks, those networks with only a few layers can be associated with *shallow* learning, although no number of layers has been defined as a boundary between deep and shallow networks. Often, deep neural networks may have tens of hidden layers, sometimes even hundreds.

Although the origins of neural networks (as well as of deep learning) can be traced back to the 1940s and 1950s (McCulloch and Pitts 1943; Hebb 1949; Rosenblatt 1958), the first period of their popularity in the machine learning context started in mid-1980s. Even if they were inspired by the architecture and learning process of the human brain, neural networks cannot be treated as models of human brain's activity or its information processing. Generally, the topology of a neural network consists of the input layer, a number of hidden layers, and the output layer (Bishop 1995; Ripley 1996). The input layer is where observations from the training data (during network training) or new observations (during testing or when the network is used for classification) are presented to the network. Each of the consecutive layers includes a number of neurons (processing units), which are connected to both the preceding and the next layer. Each connection is associated with a weight. During network training, the weights are initialized to random values and then incrementally adjusted after training observations are presented to the network and the output values are compared to the real values of the dependent variable; this adjustment process is called backpropagation

[1] Other areas of application of neural networks include computer vision, voice recognition, face recognition, machine translation, and natural language understanding.

(Werbos 1974; Rumelhart et al. 1986). Consequently, the classification abilities of a neural network are contained in the values of the weight parameters. Since there could be hundreds, thousands, or even millions (for example, in the case of deep learning) of those parameters, it is unrealistic to expect any meaningful interpretation of the results provided by neural network classifiers, which therefore makes them "black box" predictive models.

During the first period of their popularity, neural networks were most often built with only one to three hidden layers, as such solutions were deemed satisfactory for most applications; only in some situations were additional layers considered, and not too many of them due to the limitations of computational resources then available. In the mid-1990s, however, neural networks were overshadowed by the development of such learning algorithms as support vector machines and random forests. The current reincarnation of neural networks, which started after 2010, was mostly due to the significant increase in easily available computational power as well as the emergence of massive data sets. Since significant efforts in current neural network research are focused on networks with many hidden layers, the learning algorithm has been given the moniker "deep learning," even if, by itself, the concept of multilayer networks is nothing new. An advantage of using deep learning for image classification is that it works on the raw data (pixels or their representations), so no data preparation or feature selection is necessary before building a classifier. However, a straightforward development of deep neural networks with many hidden layers and often millions of weight parameters would be a good recipe for overfitting. Therefore, for such overparametrized networks to be able to provide generalizable classifiers, they have to be heavily regularized (Efron and Hastie 2016; Drori 2023).

13.2 Network Topology

The general topology of a neural network for classification is shown in Figure 13.1. The network includes the input layer, some number of hidden layers, and the output layer. All of the layers, with the exception of the input layer, are neuronal layers, that is, they are composed of a number of processing units called *neurons*. Each neuron consists of a linear summation unit (represented by an orange circle) and an activation unit (a green hexagon) that applies a nonlinear function, $f(\cdot)$ or $g(\cdot)$, to the output of the summation unit. The network is composed of the input layer and L neuronal layers, and the layers are indexed by $l, l=0, \ldots, L$, where $l=0$ corresponds to the input layer, $l=L$ the output layer, and the hidden layers are indexed from 1 to $L-1$. The input layer has p nodes, where p is the number of independent variables in the training data. Each hidden layer includes n_l neurons, where $l=1, \ldots, L-1$, and the output layer has J neurons, where J is the number of differentiated classes. Thus, in a fully connected network, there are $p \times n_1$ connections between the input layer and the first hidden layer, $n_{l-1} \times n_l$ connections between any two subsequent hidden layers (where $l=2, \ldots, L-1$), and $n_{L-1} \times J$ connections between the last hidden layer and the output layer. Each connection is associated with its own weight. Consequently, the weights associated with each neuronal layer may be represented by a matrix $\mathbf{W}_l, l=1, \ldots, L$. The linear (summation) units of a neuronal layer multiply the vector of signals from the preceding layer by the

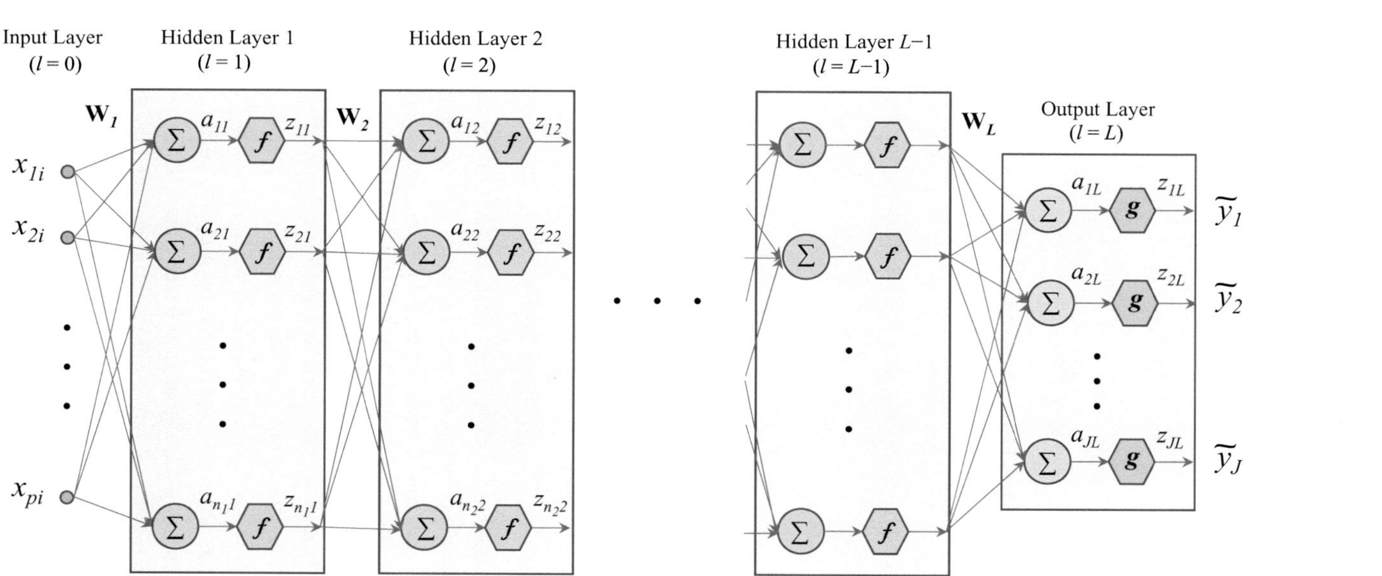

Figure 13.1 A topology of a feedforward neural network for classification. The network has $L + 1$ layers indexed by l where $l = 0, \ldots, L$. The input layer ($l = 0$) has p units corresponding to p independent variables. Each of the hidden layers ($l = 1, \ldots, L - 1$) includes n_l units called neurons. Each neuron is a combination of linear summation (an orange circle) and nonlinear activation (a green hexagon). Although all nonlinear activation functions in the hidden layers are denoted in this figure by f, it is possible for different hidden layers to use different activation functions. The output layer ($l = L$) consists of J neurons, where J is the number of differentiated classes. For multiclass classification, a commonly used activation function in the output layer ($\boldsymbol{g}$) is the softmax function, which ensures that the output class probabilities sum up to 1. In a fully connected network, each neuron of layer l, where $l > 0$, is connected to n_{l-1} outputs from preceding layer $l - 1$. Connection weights are the parameters of the neural network; they are initially set to random values and then learned from the training data during network training.
To design a neural network (whether shallow or deep), we need to decide on the number of hidden layers, number of neurons in each of them (which may be different), and nonlinear activation functions to be used in the network.

matrix of weights and produce activation vector $\mathbf{a}_l, l = 1, \ldots, L$. Each element of this vector is then the subject of a nonlinear activation operation function, $f(\cdot)$ or $g(\cdot)$, which as a result produces the layer's output vector $\mathbf{z}_l, l = 1, \ldots, L$.

For the first hidden layer, for which the signal is represented by a $p \times 1$ training vector $\mathbf{x}_i = [x_{1i}, \cdots, x_{pi}]^T$, and the weights by a $p \times n_1$ matrix $\mathbf{W}_1$, we have an $n_1 \times 1$ activation vector $\mathbf{a}_1 = [a_{11}, \cdots, a_{n_1 1}]^T$, where

$$\mathbf{a}_1 = \mathbf{W}_1{}^T \mathbf{x}_i, \tag{13.1}$$

and an $n_1 \times 1$ output vector $\mathbf{z}_1 = [z_{11}, \cdots, z_{n_1 1}]^T$, where

$$\mathbf{z}_1 = f(\mathbf{a}_1). \tag{13.2}$$

Then, for each subsequent hidden layer l, where $l = 2, \ldots, L-1$, the input signal is represented by the output of the preceding layer, that is, a $n_{l-1} \times 1$ vector $\mathbf{z}_{l-1} = [z_{1(l-1)}, \cdots, z_{n_{l-1}(l-1)}]^T$, and the weights by a $n_{l-1} \times n_l$ matrix $\mathbf{W}_l$. Combining linear and nonlinear processing of the signal in layer's neurons, the output signal from each of these layers is

$$\begin{aligned} \mathbf{z}_l &= f(\mathbf{a}_l) \\ &= f(\mathbf{W}_l{}^T \mathbf{z}_{l-1}), \quad l = 2, \ldots, L-1. \end{aligned} \tag{13.3}$$

Finally, the signal reaches the output layer L, whose weights are represented by a $n_{L-1} \times J$ matrix $\mathbf{W}_L$, and the network's output for J differentiated classes is

$$\begin{aligned} \tilde{\mathbf{y}} &= g(\mathbf{a}_L) \\ &= g(\mathbf{W}_L{}^T \mathbf{z}_{L-1}). \end{aligned} \tag{13.4}$$

13.2.1 Adding the Bias

It is quite common in neural network designs that, in addition to the signal from a preceding layer, each neuron has one more input – the bias. The bias is associated with a constant signal of 1 and is not connected to a preceding layer. But, like other weights, its connection weight is adjustable.

Let's consider a single neuron $r, r = 1, \ldots, n_1$ from the first hidden layer. Without bias, its activation value is $a_r = \mathbf{w}_r{}^T \mathbf{x}_i$, where $\mathbf{w}_r$ (the rth column of matrix $\mathbf{W}_1$) is the vector of weights of all connections entering the neuron. If we add the bias (associated with the constant signal $x_0 = 1$ and weight w_{0r}), then we have $a_r = \mathbf{w}_r{}^T \mathbf{x}_i + w_{0r}$.[2] For all neurons of layer 1, the bias terms are represented by a $n_1 \times 1$ vector of weights $\mathbf{w}_{0(1)}$ and a single "node" representing the signal $x_0 = 1$ (see Figure 13.2).

We can incorporate the bias signal into input vector $\mathbf{x}$, and the bias vector of weights into matrix $\mathbf{W}_1$. Similarly, for each of the subsequent layers, both the bias weight vector $\mathbf{w}_{0(l)}$ and the signal $z_{0(l-1)} = 1$ may be incorporated into their weight matrices and input vectors. This way, forward propagation can still use Equations (13.1)–(13.4); however, we now have:

[2] Hence, the weight associated with bias can be interpreted as the offset of the hyperplane $\mathbf{w}_r{}^T \mathbf{x}_i$.

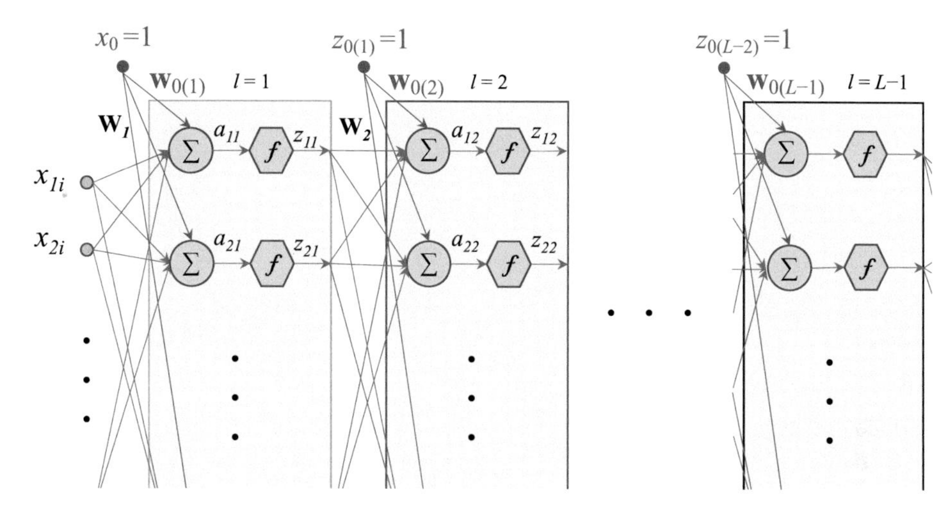

Figure 13.2 A part of the neural network from Figure 13.1, with the bias terms added (in red). The bias nodes are not connected to the previous layer and have their signals (bias activation) set to 1. Their weights (like all other weights of the network) are learned during training.

- $(p+1) \times 1$ training vector $\mathbf{x}_i = \left[x_0 = 1, x_{1i}, \cdots, x_{pi}\right]^T$,
- $(p+1) \times n_1$ matrix $\mathbf{W}_1$,

$$\mathbf{W}_1 = \begin{bmatrix} w_{01} & w_{02} & \cdots & w_{0n_1} \\ w_{11} & w_{12} & \cdots & w_{1n_1} \\ \vdots & \vdots & \ddots & \vdots \\ w_{p1} & w_{p2} & \cdots & w_{pn_1} \end{bmatrix}, \tag{13.5}$$

- $(n_{l-1}+1) \times 1$ vector $\mathbf{z}_{l-1} = \left[z_{0(l-1)} = 1, z_{1(l-1)}, \cdots, z_{n_{1l-1}(l-1)}\right]^T, l = 2, \ldots, L$,
- $(n_{l-1}+1) \times n_l$ matrix $\mathbf{W}_l$,

$$\mathbf{W}_l = \begin{bmatrix} w_{01} & w_{02} & \cdots & w_{0n_l} \\ w_{11} & w_{12} & \cdots & w_{1n_l} \\ \vdots & \vdots & \ddots & \vdots \\ w_{n_{l-1}1} & w_{n_{l-1}2} & \cdots & w_{n_{l-1}n_l} \end{bmatrix}, l = 2, \ldots, L-1, \text{and} \tag{13.6}$$

- $(n_{L-1}+1) \times J$ matrix $\mathbf{W}_L$,

$$\mathbf{W}_L = \begin{bmatrix} w_{01} & w_{02} & \cdots & w_{0J} \\ w_{11} & w_{12} & \cdots & w_{1J} \\ \vdots & \vdots & \ddots & \vdots \\ w_{n_{L-1}1} & w_{n_{L-1}2} & \cdots & w_{n_{L-1}J} \end{bmatrix}, \tag{13.7}$$

whose columns correspond to the weight vectors associated with J output neurons.[3]

[3] For simplicity, a layer index has been dropped from the elements of matrices $\mathbf{W}_1$, $\mathbf{W}_l$, and $\mathbf{W}_L$, since it is the same as the layer index of each of the matrices.

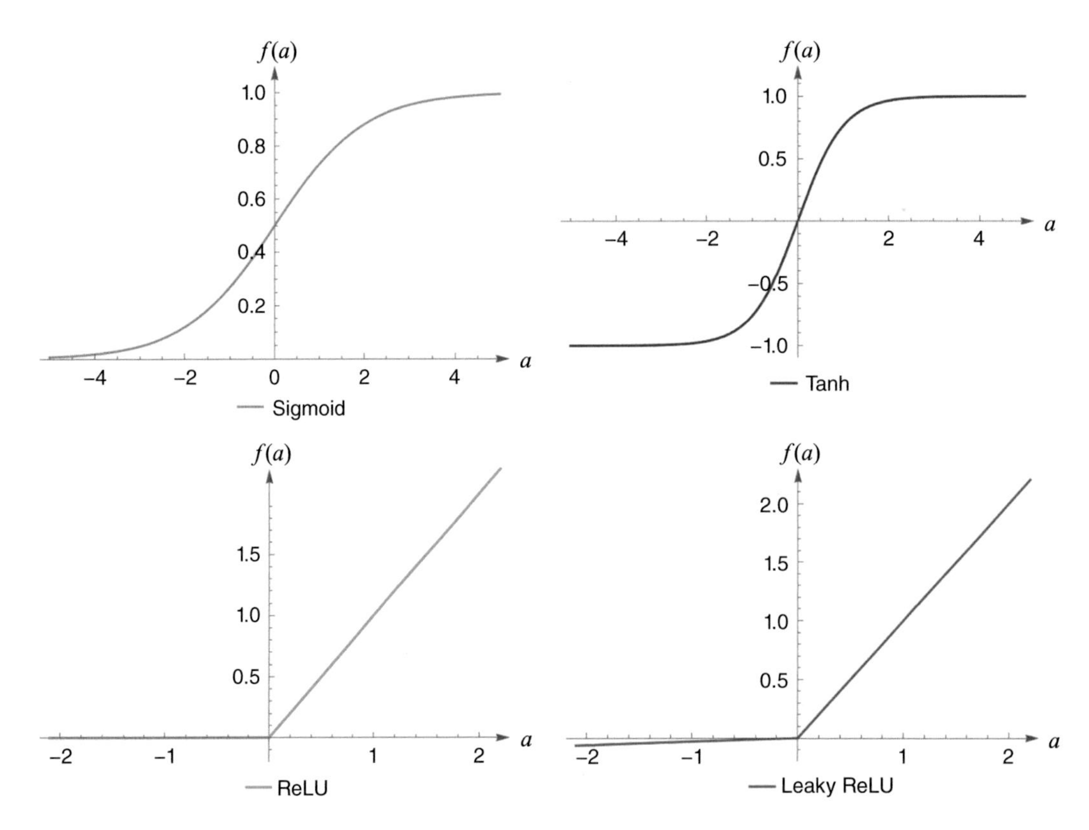

Figure 13.3 Nonlinear activation functions. The sigmoid function (orange curve) maps the activation signal to the output in the range (0, 1). The hyperbolic tangent function (blue curve) provides output in the range (–1, 1). The rectifier linear unit function (green curve) and its leaky version (red curve) are piecewise linear, and for positive activation signals their gradient is one.

13.2.2 Nonlinear Activation

A nonlinear activation function is the reason for neural networks to produce nonlinear models. Among commonly used activation functions are sigmoid, hyperbolic tangent (tanh), rectifier linear unit (ReLU), and leaky rectifier linear unit (leaky ReLU).

The *sigmoid* function, which maps the activation signal to the output in the range (0, 1), is defined as

$$f(a)_{sigmoid} = \frac{1}{1 + e^{-a}} \tag{13.8}$$

and is represented by the orange curve in Figure 13.3. It was originally used for binary classification. However, since its output is in the range (0, 1), it can be treated as a probability when used in the output layer. For multiclass classification, it is convenient to use its generalization – the *softmax* activation function, for which the output probabilities sum up to 1. For J classes, the softmax output for each class $j, j = 1, \ldots, J$, is defined as

$$g(a_j)_{softmax} = \frac{e^{a_j}}{\sum_{s=1}^{J} e^{a_s}}. \tag{13.9}$$

Similar in shape to the sigmoid is the *hyperbolic tangent* function,

$$f(a)_{tanh} = \frac{e^a - e^{-a}}{e^a + e^{-a}}. \tag{13.10}$$

Both the sigmoid and the hyperbolic tangent saturate for very small or very large values of the activation signal, but the latter translates this signal into the output in the range (–1, 1). In Figure 13.3, the hyperbolic tangent is represented by the blue curve.

Very popular in deep networks is the *ReLU* function,

$$\begin{aligned} f(a)_{ReLU} &= \begin{cases} a & \text{if } a \geq 0 \\ 0 & \text{otherwise} \end{cases} \\ &= \max(0, a), \end{aligned} \tag{13.11}$$

which is piecewise linear and outputs 0 for negative activations (see the green curve in Figure 13.3). Similar to the ReLU function, but not returning 0 for negative activation signals, is the *leaky ReLU* function,

$$f(a)_{Leaky\ ReLU} = \begin{cases} a & \text{if } a \geq 0 \\ \beta a & \text{if } a < 0, \end{cases} \tag{13.12}$$

where β is close to 0 and nonnegative.[4] The leaky ReLU function is represented by the red curve in Figure 13.3.

13.3 Backpropagation

Neural networks are trained by iteratively minimizing some loss function $\mathcal{L}$, which quantifies the difference between the output from the network and the true value of the response variable for training observations. When training starts, all weights of the network are initialized to random values. Then, training samples are presented to the network, usually in a random order, and for each of the samples, the output signal is calculated using forward propagation Equations (13.1)–(13.4). The error of prediction, represented by the value of loss function, is calculated for each of the network outputs, and the weights of network connections are updated in the direction of gradient descent of the loss function. This process is called *backpropagation*, and for it to work, the loss function has to be differentiable with respect to the network's weight parameters. The weights may be updated after all training observations are presented to the network (batch learning), after a subset of training data is presented (mini-batch learning), or

[4] A version of this function, in which β is treated as a tunable parameter (rather than a constant), is called parametric ReLU.

after each training observation is propagated through the network (stochastic learning). Nevertheless, presenting all training observations to the network constitutes an epoch, and training usually requires multiple epochs.

Denote the loss (or error) for training observation $\mathbf{x}_i, i = 1, \ldots, N$ by

$$\mathcal{L}_i(\mathbf{y}_i, \tilde{\mathbf{y}}_i), \tag{13.13}$$

where $\mathbf{y}_i$ is the target vector of network outputs (corresponding to true class label y_i for observation $\mathbf{x}_i$), $\tilde{\mathbf{y}}_i = F(\mathbf{x}_i, \mathbf{W})$ is the output observed after $\mathbf{x}_i$ is forward-propagated through the network, and $\mathbf{W}$ represents all connection weights of the network. To train the network, the error is backpropagated from the output layer to the hidden layer L–1 and then, one by one, through all preceding hidden layers. At each step, the weights of a layer are adjusted in the direction of gradient descent of the loss, in order to find a minimum of loss function – over all training data points – with respect to network weights,

$$\underset{\mathbf{W}}{\text{minimize}} \frac{1}{N} \sum_{i=1}^{N} \mathcal{L}_i(\mathbf{y}_i, F(\mathbf{x}_i, \mathbf{W})). \tag{13.14}$$

Since loss functions are not necessarily convex with respect to the elements of $\mathbf{W}$, we are typically searching for a good local minimum (Efron and Hastie 2016; Drori 2023). This may be accomplished by separately minimizing individual losses $\mathcal{L}_i(\mathbf{y}_i, F(\mathbf{x}_i, \mathbf{W}))$ associated with each training observation $\mathbf{x}_i$. Each of the training observations is first forward-propagated through the network, and the outputs of the network calculated using current network weights $\mathbf{W}$. Then, the current error is backpropagated and network weights $\mathbf{W}$ adjusted.

13.3.1 Backpropagation through the Output Layer

Let's first take a closer look at error backpropagation from the output layer L to the hidden layer L–1. Each of the weights connecting neuron $j, j = 1, \ldots, J$, of the output layer with the outputs of the pervious layer (L–1) needs to be adjusted by the negative derivative of loss function (gradient descent) with respect to the weight,

$$w_{rj} \leftarrow w_{rj} - \gamma \Delta w_{rj}, \quad r = 0, \ldots, n_{L-1}, \tag{13.15}$$

where

$$\begin{aligned} \Delta w_{rj} &= \frac{\partial \mathcal{L}_i}{\partial w_{rj}} \\ &= \frac{\partial \mathcal{L}_i}{\partial a_{jL}} \cdot \frac{\partial a_{jL}}{\partial w_{rj}} \\ &= \frac{\partial \mathcal{L}_i}{\partial a_{jL}} \cdot z_{r(L-1)}, \end{aligned} \tag{13.16}$$

and $\gamma > 0$ is the learning rate parameter that controls the magnitude of weight changes during each update. This learning rate is usually decreased from epoch to epoch in order to achieve a reasonable compromise between the time necessary to train the network and the ability to identify good local minima.

Let's assume that the loss is calculated as

$$\begin{aligned}\mathcal{L}_i(\mathbf{y}_i, \tilde{\mathbf{y}}_i) &= \frac{1}{2}\sum_{j=1}^{J}\left(y_{ij} - \tilde{y}_{ij}\right)^2 \\ &= \frac{1}{2}\sum_{j=1}^{J}\varepsilon_j^2,\end{aligned} \tag{13.17}$$

where ε_j represents the error at output $\tilde{y}_j$ of the network,[5] and observe that in Equation (13.16),

$$\delta_j = \frac{\partial \mathcal{L}_i}{\partial a_{jL}} \tag{13.18}$$

represents the contribution of neuron j to the error of output prediction and thus quantifies the local gradient of the error at neuron j. Continuing with the chain rule for derivatives, we have

$$\begin{aligned}\delta_j &= \frac{\partial \mathcal{L}_i}{\partial \varepsilon_j}\cdot\frac{\partial \varepsilon_j}{\partial \tilde{y}_j}\cdot\frac{\partial \tilde{y}_j}{\partial a_{jL}} \\ &= \varepsilon_j\cdot(-1)\cdot g'\left(a_{jL}\right),\end{aligned} \tag{13.19}$$

and consequently

$$\begin{aligned}\Delta w_{rj} &= \delta_j\cdot z_{r(L-1)} \\ &= -\varepsilon_j g'\left(a_{jL}\right) z_{r(L-1)},\end{aligned} \tag{13.20}$$

and

$$w_{rj} \leftarrow w_{rj} + \gamma\varepsilon_j g'\left(a_{jL}\right) z_{r(L-1)}, \quad r = 0, \ldots, n_{L-1}. \tag{13.21}$$

13.3.2 Backpropagation through Any Number of Hidden Layers

Once the weights of connections to the output layer are adjusted, and the local gradients for each output neuron calculated, they will be backpropagated to the last hidden layer ($l = L - 1$), and then this process is repeated for each remaining hidden layer until all connection weights of the network are adjusted.

For each hidden layer $l, l = L - 1, \ldots, 1$, assume that

- $r = 0, \ldots, n_{l-1}$ indexes neurons of layer $l - 1$, with $r = 0$ corresponding to the bias,
- $s = 1, \ldots, n_l$ indexes neurons of the current layer l, and

[5] Another popular loss function for classification is cross-entropy loss $\mathcal{L}_i(\mathbf{y}_i, \tilde{\mathbf{y}}_i) = -\sum_{j=1}^{J} y_{ij} \ln \tilde{y}_{ij}$.

- $t = 1, \ldots, n_{l+1}$ indexes neurons of layer $l + 1$.

Then, for each connection weight w_{rs} between layers $l - 1$ and l, we have

$$w_{rs} \leftarrow w_{rs} - \gamma \Delta w_{rs}, \tag{13.22}$$

and

$$\Delta w_{rs} = \delta_s \cdot z_{r(l-1)}, \tag{13.23}$$

where δ_s is the local gradient of the error at neuron s of the current layer l. This local gradient depends upon all local gradients of the layer's $l + 1$ neurons, to which neuron s is connected, and is calculated as

$$\delta_s = f'(a_{sl}) \sum_{t=1}^{n_{l+1}} \delta_t w_{st}. \tag{13.24}$$

Observe that for $l = 1$, layer $l - 1 = 0$ is the input layer, and hence $f'(a_{sl})$ represents the value of one of the independent variables, $x_{ki}, k = 1, \ldots, p$, of training observation $\mathbf{x}_i$.

13.3.3 Derivatives of Nonlinear Activation Functions

For backpropagation to work properly, nonlinear activation functions implemented in a network need to be differentiable. Let's then look at the derivatives of activation functions described in Section 13.2.2.

The derivative of sigmoid,

$$\begin{aligned} \frac{d\left(f(a)_{sigmoid}\right)}{da} &= \frac{e^{-a}}{(1 + e^{-a})^2} \\ &= f(a)_{sigmoid}\left(1 - f(a)_{sigmoid}\right), \end{aligned} \tag{13.25}$$

is represented by the orange curve in Figure 13.4. Although sigmoid is differentiable in the entire range of activation values, its gradient vanishes for large negative or positive values of a. Furthermore, the maximum value of this gradient is 0.25, which means that repetitive multiplications during backpropagation decrease the magnitude of weight adjustments, which may result in slow training of the network.

The derivative of the hyperbolic tangent function,

$$\begin{aligned} \frac{d\left(f(a)_{tanh}\right)}{da} &= \frac{4}{(e^a + e^{-a})^2} \\ &= 1 - f^2(a)_{tanh}, \end{aligned} \tag{13.26}$$

is graphed as the blue curve in Figure 13.4. Even if the maximum value of this gradient is 1 and thus, for smaller values of the activation signal, training may be faster than for the sigmoid, the gradient of the hyperbolic tangent vanishes quickly when the absolute value of a increases.

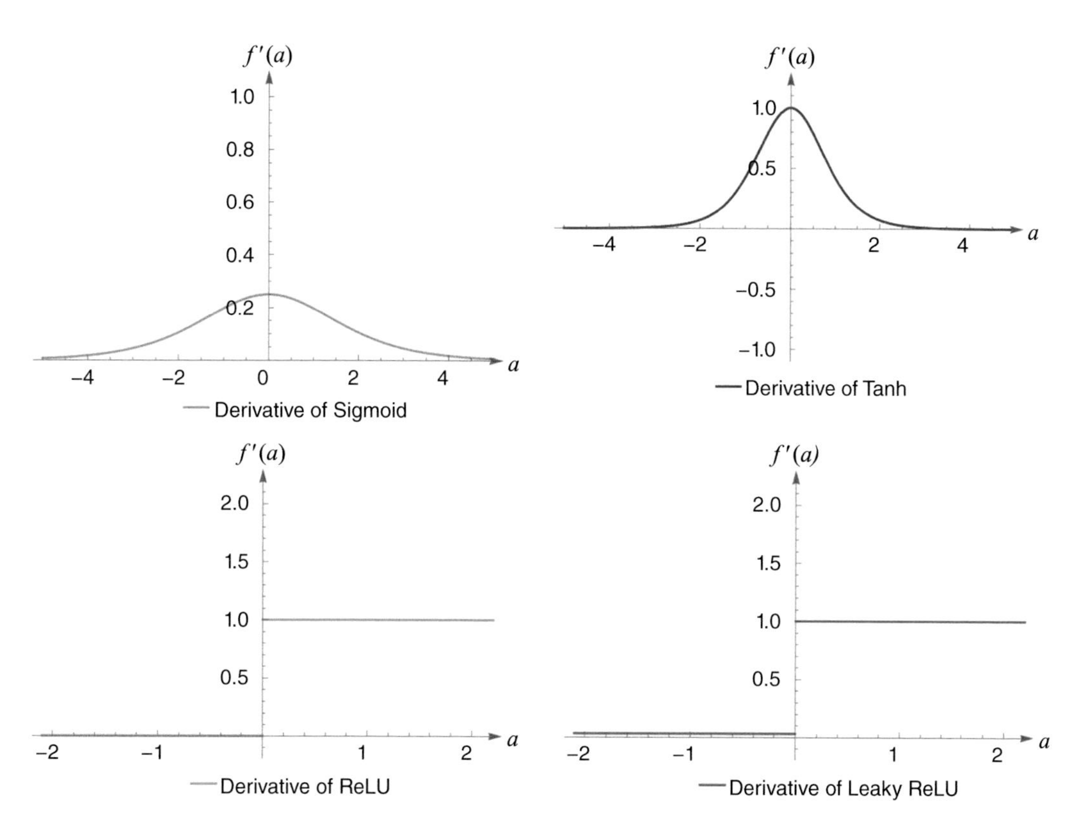

Figure 13.4 Derivatives of the nonlinear activation functions from Figure 13.3. Both the sigmoid (orange curve) and the hyperbolic tangent (blue curve) are differentiable in the entire range of activation values. However, their gradients vanish for large negative and large positive values of a. Both the ReLU (green curve) and the leaky ReLU (red curve) are not differentiable at $a = 0$, but they have constant gradients for positive and negative activations (1 and 0 for ReLU, and 1 and β for the leaky ReLU).

The derivative of the ReLU function,

$$\frac{d\left(f(a)_{ReLU}\right)}{da} = \begin{cases} 1 & \text{if } a > 0 \\ 0 & \text{if } a < 0, \\ undefined & \text{if } a = 0 \end{cases} \tag{13.27}$$

is represented by the green curve. Even if ReLU is not differentiable when the activation signal is exactly 0,[6] it is commonly used in deep networks; as long as the activation signal is positive, ReLU's gradient will not vanish, no matter how many hidden layers are included in the network.[7] Furthermore, since the ReLU output as well as its gradient are very simple to compute, training of ReLU-based networks may be efficient even for deep networks using large training sets.

[6] In software implementations, the nonexistent gradient of ReLU for $a = 0$ is typically set to 0.

[7] Only when all ReLU output signals in a layer are equal to 0 will its gradient vanish.

The derivative of the leaky ReLU function,

$$\frac{f(a)_{Leaky\ ReLU}}{da} = \begin{cases} 1 & \text{if } a > 0 \\ \beta & \text{if } a < 0, \\ undefined & \text{if } a = 0 \end{cases} \tag{13.28}$$

is represented by the red curve in Figure 13.4. Its nonzero gradient for negative activations resolves the vanishing gradient issue of ReLU.

13.4 Classification of Medical Images: Deep Convolutional Networks

In the area of biomedical applications, deep neural networks seem to have a great potential in the classification of medical images (Liu et al. 2019; Aggarwal et al. 2021; Korot et al. 2021). Such networks are typically designed as convolutional feedforward neural networks, which are characterized by sparse connectivity and weight sharing, and whose main components are convolutional and pooling layers. Training data consist of images, and each image can be represented by a $q \times q$ input matrix of pixels, plus the third dimension of their color components, that is, the depth of the image (which for RGB images is 3).

A *convolutional* layer applies a number of *filters* (or kernels), which are much smaller $v \times v$ matrices[8] ($v \ll q$)[9] designed to recognize specific local features of the image (for example, lines or edges). Each filter slides over the input matrix (horizontally, then vertically), and for each of its overlapping positions, convolution of input and filter data is performed (that is, summing up the results of pointwise multiplications of the image and filter pixels). For each of the filters, this procedure will generate a *feature map*, which is either a $q \times q$ or smaller matrix[10] of convolution results.[11] Consequently, the output from the convolutional layer is an array of images – feature maps. Usually, the ReLU activation function is then applied to the output of the convolutional layer, which replaces all "negative pixels" with zeroes.[12]

A *pooling* layer aggregates information from neighboring features by replacing small $u \times u$ (for example, 2×2) nonoverlapping segments of each feature map with a single value (typically the maximum value of the segment). This will reduce both redundancy

[8] The depth of each filter is the same as the color depth of the input image, so both the image and a filter can be treated as 3D boxes.

[9] Typical v values are 3 or 5. Since filters are smaller than the input image, their application will result in a network with sparse connectivity, which – in the context of a fully connected network – would mean that many connection weights are set to 0 (Goodfellow et al. 2016).

[10] Edge padding (that is, supplementing the borders of an input matrix with columns and rows of zeroes), as well as sliding with the stride of one, would result in q horizontal and q vertical positions of the filter. Otherwise, a feature map would be smaller than the input matrix.

[11] Each segment of an input image overlapping with a filter is connected to one neuron of the feature map associated with this filter (sparsity). Furthermore, since a filter detects the same feature independently of its location in the input image, neurons of the feature map share a set of connection weights.

[12] An additional advantage of using ReLU in this setting is that having the gradient equal to either 0 or 1 enables efficient training.

(likely to occur within such segments) and the size of the images presented to the next layer. Typically, a convolutional network consists of some number of convolutional as well as pooling layers (which can be seen as a feature detection part of the network), followed by one or more "regular," fully connected layers (acting as a classification part of the network) with the softmax activation function in the output layer.

With many layers of a deep network and images with large numbers of pixels, such networks may have huge numbers of parameters. Although convolutional networks reduce this number via sparse connectivity and weight sharing (blocks of connection weights having the same value), overfitting resulting from overparameterization is a real danger for any deep network.

13.5 Deep Learning: Overfitting and Regularization

One of the common ways to avoid overfitting (which is especially important for over-parametrized deep networks) is regularization. It could be done in many ways, including adding a regularization term to the loss function, via dropout, or via distortion and augmentation of training images. Although regularization may be important for any model, it is especially important for deep neural networks, whose large number of layers, as well as large and possibly very different numbers of neurons in hidden layers, could result, during training, in quickly increasing or vanishing gradients.

13.5.1 Regularization of the Loss Function

A common way to decrease the complexity of a neural network model is penalizing large values of its connection weights. This can be done by adding a regularization (or penalty) term to the loss function and

$$\underset{\mathbf{W}}{\text{minimize}} \ \frac{1}{N}\sum_{i=1}^{N} \mathcal{L}_i(\mathbf{y}_i, F(\mathbf{x}_i, \mathbf{W})) + \lambda G(\mathbf{W}), \tag{13.29}$$

where $\lambda > 0$ is a tunable penalty coefficient and $G(\mathbf{W})$ is a function of network connection weights. One such popular penalty term implements L_2 regularization,

$$G(\mathbf{W}) = \sum_{l=1}^{L}\sum_{r=1}^{n_{l-1}}\sum_{s=1}^{n_l} (w_{rs})^2, \tag{13.30}$$

which, in the context of deep learning, is also known as *weight decay*. Being a sum of squares of the model parameters[13] (as in ridge regression), this penalty term is represented by a smooth function and thus can work very well with backpropagation.

Other popular penalty terms correspond to lasso (L_1 regularization) and elastic net (both L_1 and L_2 regularization) penalties. In L_1 regularization,

[13] The weights of bias nodes are usually not penalized.

$$G(\mathbf{W}) = \sum_{l=1}^{L} \sum_{r=1}^{n_{l-1}} \sum_{s=1}^{n_l} |w_{rs}|, \tag{13.31}$$

the penalty term sums up the absolute values of weight parameters and – as in lasso regression (see Chapter 7) – sets some of the weight parameters to 0, which will increase the sparsity of a neural network. In elastic net type penalty, the penalty term from Formula (13.29) will be replaced by two penalty terms,

$$\begin{aligned} \lambda G(\mathbf{W}) &= \lambda_1 G_1(\mathbf{W}) + \lambda_2 G_2(\mathbf{W}) \\ &= \lambda_1 \sum_{l=1}^{L} \sum_{r=1}^{n_{l-1}} \sum_{s=1}^{n_l} (w_{rs})^2 + \lambda_2 \sum_{l=1}^{L} \sum_{r=1}^{n_{l-1}} \sum_{s=1}^{n_l} |w_{rs}|, \end{aligned} \tag{13.32}$$

where $\lambda_1 \geq 0$ is the ridge penalty coefficient and $\lambda_2 \geq 0$ is the lasso penalty coefficient. The lasso term will promote sparsity, while the ridge term (with quadratic penalty) will have a strong impact on large weight coefficients.

13.5.2 Dropout

Another method for the regularization of deep neural network models is dropout (Hinton et al. 2012; Srivastava et al. 2014). In this approach, randomly selected neurons (and all their connections) are temporarily removed from the network during training.[14] Usually one hidden layer (but sometimes more than one) is dedicated as a dropout layer, and only its neurons may be dropped. Each neuron of a dropout layer has some probability (for example, 0.5) of being removed, and thus at different training epochs (or batches), different sets of neurons are dropped. This will limit the development of codependency between neurons and also thwart the chances of simply memorizing – and relying on – some specific information content of the training data.[15] In effect, the network has to learn more robust characteristics of each class. Furthermore, removing different sets of neurons means that training is performed on a number of networks with a different topology, and that the trained network is the result of averaging them. In a way, this resembles bagging, except that, instead of aggregating information from an ensemble of independent classifiers (see, for example, Chapter 10), a single network classifier is developed as a result of aggregating information of a set (ensemble) of different, but parameter sharing, subnetworks that are subsequently considered during training (Heaton 2015; Goodfellow et al. 2016; Fan et al. 2020). Furthermore, if dropout is also extended to the input layer (Srivastava et al. 2014), then the network will also be trained on different subsamples of the training data, which makes the process even more similar to bagging.

[14] This can be done by multiplying all outputs from the neuron by 0.

[15] For example, if a neuron learns that a specific feature is associated with a particular class of images, then dropping this neuron will force the network to identify another feature characteristic of this class (Goodfellow et al. 2016).

Similar to regularization of loss function, dropout does not include bias neurons. Once the network is trained, all its neurons are used for classification.

13.5.3 Augmentation of the Training Data

Augmentation and distortion of training data is yet another regularization approach, which is highly effective for the classification of images. This method of improving generalization will increase the size of training data by adding images created by rotating, flipping, cropping, scaling, or distorting the original training images. However, modifications have to be class-invariant, which here means that for a human it would be obvious that both the original image and all its modifications belong to the same class (Goodfellow et al. 2016).

Part IV

Biomarker Discovery via Multistage Signal Enhancement and Identification of Essential Patterns

14 Multistage Signal Enhancement

In this part, we will present an approach to multivariate biomarker discovery that, although including all elements of the biomarker discovery process described in previous chapters, focuses on the multistage enhancement of predictive information (the signal) by the sequential elimination of the variables that either represent various types of noise or are unlikely to be associated with real population patterns. The main advantage of this approach is that it not only provides a parsimonious multivariate biomarker but also associates each variable of the biomarker with a group of variables representing the same essential biological pattern. Combining these two characteristics of the method allows for the identification of generalizable biomarkers that have the best chance for meaningful biomedical interpretation.[1]

14.1 Introduction

As mentioned in Chapter 2, when we deal with high-dimensional and sparse $p \gg N$ biomedical data, then the real predictive patterns (which would allow for a highly accurate prediction of the response variable for new observations – say, patients – from the target population) are buried under billions of random patterns that are present in the training data by chance, but do not represent patterns that exist in the target patient population. The general idea for the identification of real patterns is conceptually quite simple: remove noise and enhance the signal, then remove more (or a different kind of) noise and enhance the signal further, and continue the process until we are left with the variables that are most likely to represent real population patterns. Thus, at each step, some variables are eliminated, and for those still retained in the training data, the probability of their (multivariate) association with real predictive patterns increases with each step.

14.2 Removing Variables Representing Experimental Noise

We assume that the training data used in the experiment have already been preprocessed, and that the preprocessing could include (among other procedures) signal background

[1] The method presented in this part is a generalization and simplification of the method described in Dziuda (2010).

correction, adjustment for batch effects, and determination of experimental noise level (such preprocessing depends on the type of data). At this step, variables representing experimental noise will be removed. One way of doing this is to remove those variables that have all of their values below the experimental noise level that was determined (or estimated) for the training data.[2] However, we may also want to utilize the noise level to remove variables with an extremely low variance of values. Hence, removing variables whose amplitude of values across observations is below the noise level would combine those two criteria. For example, for gene expression data typically involving more than 20,000 variables, it is not uncommon for many of them to be not expressed at a biologically meaningful level (in a particular tissue), and thus it is again not uncommon that some 40 percent, or even more, of the original variables will be eliminated at this single step.

14.3 Removing Variables with Unreliable Measurements

This step can be performed on the data generated by those high-throughput technologies that associate each measurement with some indicator of its reliability. For example, some low-level preprocessing procedures for gene expression microarray data provide – with the expression level calculated for each probe – a detection call representing the probability that the calculated expression level was reliably determined. Thus, if such reliability information is available, we may decide to retain only those variables that have at least some threshold proportion (for example, 30 percent) of reliable measurements across observations (or across at least one class in the case of classification). All other variables will be removed, as they represent noise associated with the unreliability of their measurement.

14.4 Determining an Optimal Size of the Biomarker and Removing Variables with No Multivariate Importance for Prediction

At this signal enhancement step, we target noise associated with a lack of predictive information. In particular, those variables not selected into any optimal-size multivariate biomarker with reasonable predictive power will be removed. Therefore, this step starts with the determination of the optimal size of a multivariate biomarker. Subsequently, biomarkers of the optimal size are selected from the training data and their predictive power is evaluated. After identifying the first biomarker with a satisfactory predictive power, its variables are removed from the training data and set aside. Then, the next biomarker is identified; if it also has satisfactory predictive power, its variables are also

[2] For more aggressive noise filtering, we may opt for retaining only the variables for which the proportion of their values that are above the noise level is above some threshold. For regression, this proportion would be calculated across all training observations. For classification, however, we would retain those variables that have at least the threshold proportion of their values greater than the noise level in at least one of the differentiated classes.

set aside. This process continues until no multivariate biomarker of the optimal size and satisfactory predictive power can be identified. The variables still in the training data are considered to have no multivariate importance for prediction.

14.4.1 Determining the Optimal Size of a Multivariate Biomarker

We start with the training data containing all independent variables, except those removed earlier for representing either experimental noise or unreliable measurements. To determine the optimal size of a biomarker, we may, for example, perform multiple feature selection experiments, each utilizing a different bootstrap sample of the training data and performing feature selection for a selected set of considered sizes.[3] Averaging of results from hundreds or even thousands of such parallel feature selection experiments will allow for the identification of the optimal size of a biomarker. Recall, from Chapter 5, that the optimality of a biomarker's size should preferably be based on a combined criterion of minimizing the size (parsimony) and maximizing the biomarker's predictive abilities. For example, upon identifying the size corresponding to the maximum performance (determined via testing predictive models or via some performance metrics not involving building predictive models), we may opt for a more parsimonious size, which still provides high-enough performance (using, for example, the one-standard-error method or simultaneous evaluation of several performance metrics, both described in Chapter 3). This way the chance of overfitting is decreased.

Details of feature selection experiments depend on their design and choice of algorithms. For example, if each of the parallel feature selection experiments implements recursive feature elimination, with nonlinear support vector machines used as the learning algorithm, we may need to limit the number of parallel experiments to some 200–300 in order to keep the processing time reasonable. However, if feature selection experiments implement a stepwise hybrid selection, with or without involving a learning algorithm, we may efficiently perform 1,000 or more such parallel experiments (see Chapter 5). Furthermore, if the evaluation of subsets considered by the feature selection experiments is done by utilizing a learning algorithm that assumes independence of observations, then bootstrapping with replacement may not be used, instead bootstrapping without replacement needs to be implemented (see Section 4.1.3.1).

So far, this step may appear similar to the feature selection approaches described in Chapter 5. However, now, we are not yet interested in selecting variables of an optimal biomarker – we are aggregating the results of multiple feature selection experiments to decide *only* on the optimal size of a multivariate biomarker. The reason is that before selecting variables of an optimal biomarker, we want to perform additional steps of noise reduction (such as the removal of variables that have unsatisfactory multivariate importance), as well as identify and then consider groups of variables with similar patterns of their values across observations.

[3] Since a decision on the optimal size of a multivariate biomarker is, to some extent, arbitrary, an alternative yet simpler method may be used to choose a potentially optimal size, and then multiple feature selection experiments will be performed to only validate this choice (see, for example, Study 1 in Chapter 16).

14.4.2 Removing Variables with No Multivariate Importance for Prediction

Once the optimal size of a multivariate biomarker is determined, we will split the independent variables into two groups: those with potential multivariate importance and those without. The latter would represent yet another kind of noise and are removed. To achieve this, we perform multivariate feature selection to identify the first biomarker of the optimal size and evaluate its predictive abilities. Of course (assuming that the training data include sufficient predictive information relevant to our goal), there may be other biomarkers having the same optimal size and similar predictive abilities as the one identified first. Therefore, the variables of the first biomarker (again assuming that this biomarker was evaluated as having satisfactory predictive power) are set aside, and the training data are used to identify the next biomarker of the optimal size. If it also has satisfactory predictive power, its variables are likewise set aside (into a pool of variables having potential importance for prediction), and the training data, without all of the variables currently set aside, are used to identify the next biomarker of the optimal size. Then, as we proceed, subsequent biomarkers with satisfactory predictive power are identified, the pool of potentially important variables increases, the number of variables in the consecutive versions of training data decreases, and we stop when no additional biomarker of the optimal size and satisfactory predictive power can be identified. Those variables that are still included in the latest iteration of training data are considered noise, with no important multivariate predictive information. Consequently, the training data for further analysis will include only the pool of potentially important variables.

14.4.3 The Pool of Potentially Important Variables

It is quite likely that the *pool of potentially important variables* includes only a relatively small percentage of the original number of independent variables (for example, few hundreds), while those deemed as representing noise may number in thousands. Since the majority of the original variables have been, so far, removed after being deemed as representing different kinds of noise, the pool should constitute a good representation of predictive information contained in the entire original training data. We may want to verify this assumption. A reasonable way of doing this is to perform another round of multiple feature selection experiments, similar to those described in Section 14.4.1, but this time using only pool variables, and stopping each experiment when the optimal number of variables is reached. For example, if 1,000 parallel experiments implementing stepwise hybrid feature selection had been performed in order to determine the optimal size of a biomarker, we may now run another 1,000 bootstrap-based experiments (this time using only pool variables) and compare – for optimal cardinality – the averaged OOB estimates of predictive power of the two ensembles. They should be similar; it is even possible that the estimates based on the pool of potentially important variables will be higher than those based on all the original variables – this may have been due to a significantly decreased noise level in pool variables.

15 Essential Patterns, Essential Variables, and Interpretable Biomarkers

15.1 Introduction

One could assume that now – after identifying the pool of potentially important variables, and after removing a majority of original variables by associating them with various kinds of noise – we are ready to choose the best set of variables to be included in the final multivariate biomarker of the optimal size. However, variables in typical high-dimensional biomedical data are far from being independent of each other; groups of coregulated variables with similar patterns (across observations) may represent the same biological processes and similar predictive information. The identification of such patterns may be very important for the biological interpretation of the final multivariate biomarker. Thus, instead of jumping into final variable selection, we want to include – in our biomarker discovery process – considerations of such interconnections among the variables present in the pool of potentially important variables. To do this, we will start with identifying groups of variables that have similar patterns of values across observations. An investigation into such patterns will lead to the identification of essential patterns, their essential variables, and eventually multivariate biomarkers that not only are parsimonious and generalizable but also have the best chance for a meaningful biological interpretation.

15.2 Groups of Variables with Similar Patterns

Finding groups of variables with similar patterns of values across observations requires a definition of similarity between variables. Observe that each variable (from the pool) can be represented by a point in an N-dimensional space of N observations (say, patients) – the point defined by an N-dimensional vector of the values of the variable across N patients. The commonly used metrics of similarity between such points are based on their distance in an N-dimensional space; examples of such metrics are the Euclidean and Mahalanobis distances (see Section 15.5.1). Using such metrics would help in grouping variables represented by points that are close to each other, which translates to similar magnitudes of their values and similar directions of their vectors. However, for such data as gene or protein expression levels, we may prefer to group variables only by the similarity of the shape of their expression patterns across observations, rather than by the magnitude of their expression – this would translate to having their vectors pointing in similar directions regardless of their length. Such grouping can be done by using correlation between variables as a metric of their distance (see Section 15.5.1).

To find groups of variables with similar patterns, we may use such clustering methods as *hierarchical clustering* or *self-organizing maps* (SOM).[1] A short description of these methods is provided in Section 15.5. When using either of these methods, we may decide on the number of clusters (and other parameters of clustering) most appropriate to our data. In hierarchical clustering, we may cut the resulting dendrogram (a graphical representation of cluster hierarchy) at a height providing a specific number of clusters. For SOM clustering, we may perform it several times with different topologies (such as rectangular or hexagonal grid) and different numbers of clusters and choose the most plausible results. Using SOM has some advantages: (i) its implementations usually provide graphical representations of patterns grouped into each cluster; (ii) clusters similar to each other are close on the grid of SOM results. In any case, as a result of such clustering, we identify a number of clusters representing groups of variables with a similar pattern of changes in their values across observations.

15.3 Essential Patterns

Recall that in order to evaluate the predictive performance of a pool of potentially important variables, we generated an ensemble of predictive models based on parallel feature selection experiments, with each experiment resulting in a biomarker of the optimal size (see Section 14.4.3). In an example described there, we performed 1,000 bootstrap-based experiments resulting in an ensemble of 1,000 predictive models whose performance can be estimated by predicting the value of the response variable from their respective OOB samples. Now we can use the same predictive models to investigate how the variables belonging to each of the identified clusters are distributed among biomarkers used by these models. The assumption is that variables from the clusters associated with the most important biological processes underlying our predictive goal should be used more often than the variables from clusters of lesser importance. Thus, we hypothesize that clusters whose variables are used most often by the predictive models of the ensemble represent patterns that are essential for prediction. Although the other patterns can hardly be equated to noise, limiting further analysis to the ***essential patterns*** should further enhance the signal and increase the chances that the final multivariate biomarker will represent a real population pattern associated with the biological processes underlying changes in the investigated response variable.

[1] Observe that clustering, as an unsupervised approach, may not be used for any task with supervised goals, such as feature selection or dimensionality reduction in a biomarker discovery study. Here, however, clustering is applied to a pool of potentially important variables, being identified as associated with the discriminatory information relevant to the supervised goal of the study. Since grouping variables based on their expression patterns is an unsupervised task, clustering is here a proper approach.

15.4 Essential Variables and Interpretable Biomarkers

Once the essential patterns have been identified, we may look at their variables, since they may not be necessarily equally important for prediction. As was done for patterns, we now look at the distribution of each of the essential pattern variables among the same – previously built and used – ensemble of predictive models. The variables that were selected into many predictive models are rather unlikely to be selected by chance. They will constitute a set of essential variables that are most likely to represent real patterns underlying the investigated biological phenomena. Consequently, if we now limit our considerations only to the ***essential variables of the essential patterns*** and use them for performing feature selection to identify the final multivariate biomarker, this biomarker will not only be parsimonious but also have the best chance of being generalizable and interpretable.[2]

15.5 Ancillary Information

15.5.1 Hierarchical Clustering: A Snapshot

Hierarchical clustering is an unsupervised method of grouping objects (either observations or variables) into a hierarchy of nested clusters. This hierarchy may be graphically represented by a *dendrogram*, a tree-resembling structure visualizing distances between objects and their clusters (see Figure 15.1). In order to perform hierarchical clustering, two metrics of distance (or dissimilarity) need to be defined – one for measuring distances between objects, and the other for calculating distances between clusters. Typical metrics of distance between objects include the Euclidean distance, Mahalanobis distance, and correlation distance. When we cluster p variables from a pool of potentially important variables, each variable can be represented by an $N \times 1$ vector $\mathbf{v}_k = [v_{1k}, \cdots, v_{Nk}]^T$, where $k = 1, \ldots, p$ and N is the number of patients (or observations) in the training data. Euclidean distance between two variables, $\mathbf{v}_k$ and $\mathbf{v}_l$, where $k, l = 1, \ldots, p, k \neq l$, is defined as

$$distance_{Euclidean} = \sqrt{(\mathbf{v}_k - \mathbf{v}_l)^T (\mathbf{v}_k - \mathbf{v}_l)}. \tag{15.1}$$

Mahalanobis distance is defined as

$$distance_{Mahalanobis} = \sqrt{(\mathbf{v}_k - \mathbf{v}_l)^T \mathbf{S}^{-1} (\mathbf{v}_k - \mathbf{v}_l)}, \tag{15.2}$$

where $\mathbf{S}^{-1}$ is an $N \times N$ inverted variance–covariance matrix (recall that the two variables are treated here as N-dimensional objects).

[2] The approach to multivariate biomarker discovery described in Chapters 14 and 15 is used by the study presented in Chapter 16.

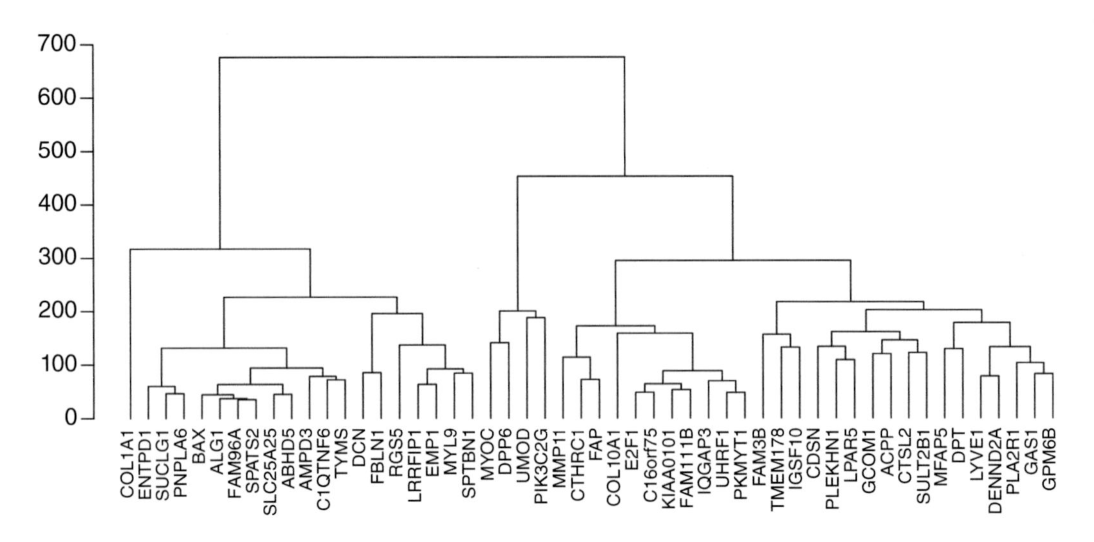

Figure 15.1 An example of a dendrogram. Fifty-two variables have been clustered using the agglomerative hierarchical clustering method with the Euclidean distances between variables and the complete linkage between clusters. The distance between any two objects (variables or clusters) is represented on the dendrogram by the height at which the objects are connected.

Correlation distance is defined as

$$distance_{correlation} = \frac{1 - Cor(\mathbf{v}_k, \mathbf{v}_l)}{2}, \tag{15.3}$$

where $Cor(\mathbf{v}_k, \mathbf{v}_l)$ denotes the Pearson correlation between $\mathbf{v}_k$ and $\mathbf{v}_l$.

The distance between clusters can be represented by the distance between the objects of two clusters that are: (i) the nearest neighbors (single linkage); (ii) the farthest neighbors (complete linkage); (iii) the vectors representing the centers of the clusters (centroid linkage); or (iv) by calculating all pairwise distances between objects of two clusters and taking their average value (average linkage).

Hierarchical clustering can be performed using either an *agglomerative* (bottom-up) or a *divisive* (top-down) approach. Assuming that we cluster M objects, the agglomerative approach starts by treating each of the objects as its own one-element cluster. Then, pairwise distances between these clusters are calculated and the two clusters that are closest to each other are merged into a higher-level cluster. This process is iteratively repeated until we end up with a single cluster joining all objects and their subgroups. The divisive approach initially treats all objects as a single cluster and then, at each iteration, splits one of the currently defined clusters into two subgroups in a way that maximizes the distance between the two newly created subgroups. This process ends when we have M clusters, with each including only one object. Each of these approaches results in a (most likely different) nested structure of clusters. A dendrogram is an informative way of visualizing such structures. The height at which objects or clusters are merged represents their dissimilarity, that is, objects or clusters that are more similar are merged at lower heights. To identify a particular number of clusters, all that we need to do is cut the dendrogram using a horizontal line placed at such a height that the line crosses that particular number of vertical lines of the dendrogram.

15.5.2 Self-Organizing Maps: A Snapshot

The self-organizing maps (SOM) algorithm (Kohonen 1982) is used to project high-dimensional data onto a two- or three-dimensional map that consists of a number of nodes organized in a square, rectangular, or hexagonal grid. Each node represents a cluster of objects, and the topological proximity of the nodes corresponds to the similarity between their clusters. To find a good and possibly most informative representation of the data, we may run SOM several times with different numbers of nodes (or even different topologies of the grid) and different settings of its parameters.

At first, the map is initialized by assigning each node a random vector representing its cluster's center[3] (those vectors have the same dimensionality as that of the data to be clustered). Then, an object from the data set is randomly selected and assigned to the node whose center is closest to the vector representing the object (using, for example, Euclidean or correlation distance in the original space). This node is called the "winner," and its cluster's center is adjusted toward the object. The extent of this adjustment is regulated by the *learning rate*, a parameter. However, not only the winner's center is adjusted. Each node is associated with a set of neighbors, which are grid nodes whose centers are not farther from the winner's center than a specified *threshold*, another parameter. Their centers are also adjusted toward the currently considered data object, though those adjustments are smaller than that for the winner and are defined by the *neighborhood function* determining the extent of adjustment upon the distance of a neighbor from the winner. Then, all of this is repeated for another object randomly selected from the data. When all data objects are presented to the algorithm, the first iteration ends.

The entire clustering process usually involves hundreds or thousands of iterations. After each iteration, the learning rate and the threshold parameter (defining the size of local neighborhood) are decreased. This means that initially the algorithm focuses on global relationships between clusters, that adjustments to their centers are relatively large, and that later those relations are more and more fine-tuned by decreasing the magnitude of adjustments and by focusing on smaller and smaller local neighborhoods. The clustering ends either when a predetermined number of iterations is completed or when there are no changes in the assignment of objects to the grid nodes.

The SOM algorithm described above, in which data objects are presented to the grid one at a time, is called the "online" version of the algorithm (its original version). There is also the "batch" version of SOM algorithm (Kohonen 1995), in which all data objects (at each iteration) are presented to the grid simultaneously, and each cluster's center is updated to the weighted average of those data objects that map to a cluster's neighbor (Ripley 1996). Although the details of the two versions differ, the general idea of adjusting centers in local neighborhoods is the same for both.

[3] Alternatively, nodes may be initialized by assigning randomly selected objects from the data.

Part V

Multivariate Biomarker Discovery Studies

16 Biomarker Discovery Study 1:

Searching for Essential Gene Expression Patterns and Multivariate Biomarkers That Are Common for Multiple Types of Cancers

16.1 Introduction

The goal of this study is to search for parsimonious multivariate gene expression biomarkers that could possibly be used for the early detection of multiple types of cancers. The study implements the biomarker discovery approach described in Chapters 14 and 15, that is, identifying generalizable and interpretable multivariate biomarkers via multistage signal elevation (and simultaneous sequential elimination of various kinds of noise) and searching for essential gene expression patterns that are common for multiple cancer types. If such patterns exist and are identified, they may lead to significant improvements in early cancer diagnosis, which is the most important factor of successful treatment.

Cancer is the second leading cause of mortality in the USA (slightly trailing heart disease) and quite consistently accounts for over 500,000 deaths per year (Hoyert and Xu 2012; Rana et al. 2021; Ahmad et al. 2022). We can report significant improvements in the diagnosis and treatment of many cancer types; for example, the survival rate for pediatric acute lymphocytic leukemia (ALL) has increased from less than 10 percent in the 1960s to about 90 percent in the 2000s and 2010s (Hunger et al. 2012; American Cancer Society 2023). However, for some types of cancer, we still fail a significant proportion of patients; for example, the mortality rate for pediatric acute myelogenous leukemia (AML) is still above 30 percent (American Cancer Society 2023). The lowest overall survival rates are: 11 percent for pancreatic cancers, 20 percent for liver and esophageal cancers, and 22 percent for lung cancers (Siegel et al. 2022).

There are many factors influencing the outcome of cancer treatment (for example, the ability to predict the risk of relapse, patient's sensitivity to radiation, or individual response to a particular therapy), but, as indicated by survival statistics, the single most important factor of successful treatment and long-term survival is early diagnosis. Unfortunately, too often patients are diagnosed with cancer only when symptoms occur, and the disease may already be in its advanced stage. Although we have some genetic or genomic biomarkers for the early detection of some types of cancer (for example, BRCA1 and BRCA2 for breast cancer, PSA for prostate cancer, or CA125 for ovarian cancer), many of them have a too high false positive rate to be considered very reliable.

Other, more traditional methods (like biopsy or even colonoscopy) are often too invasive to be adequate for routine population screening.

Although there are many types of cancer with their distinct phenotypic characteristics, there are also some common biological phenomena (such as uncontrolled growth, resisting cell death, or the capability for local invasion and distant metastasis) underlying all of them (Hanahan and Weinberg 2011). Therefore, it should be possible to identify characteristic molecular signatures (biomarkers) that are associated with biological processes common to many – or even all – cancer types. Any such biomarker would be a multivariate biomarker (consisting of a set of genes, proteins, or metabolites) representing a multivariate pattern that is characteristic for the considered tumor types. The identification of such biomarkers may constitute a breakthrough event in early cancer diagnosis, especially if the implementation of multicancer tests is minimally invasive and requires, for example, only blood samples.

The early detection of multiple cancer types is recently an area of very active research whose goal is to design multicancer early detection (MCED) tests that would use blood samples to screen for many cancers (Etzioni et al. 2022).[1] From the point of view of the phases of biomarker development, MCED studies are still in their early phase (as defined by Pepe et al. 2002). Retrospective MCED studies have reported multicancer sensitivity from about 40 percent to about 80 percent (with lower sensitivities for stage I and higher for stage III tumors), and specificity reaching 98–99 percent, which is significantly higher than for such tests as PSA or mammogram (Etzioni et al. 2022). A prospective study involving over 10,000 women with no history of cancer has reported MCED blood test sensitivity of 27.1 percent, specificity of 98.9 percent, and positive predictive value of 19.4 percent (Lennon et al. 2020). Recall from Chapter 4 that for population screening tests, it is crucial to have a high specificity (especially if the follow-up tests to differentiate between the true positive and false positive are invasive).

The MCED biomarker discovery study presented in this chapter is a retrospective study utilizing RNA-Seq data from the Cancer Genome Atlas. The objectives of the study are: (i) to identify characteristic gene expression patterns that are common for multiple cancer types, and (ii) to identify a parsimonious multivariate biomarker representing those patterns.

16.2 Data

16.2.1 RNA-Seq Data from The Cancer Genome Atlas

The Cancer Genome Atlas (TCGA) repository was used as the source of the data analyzed in this project.[2] The TCGA project started in 2005 and is supervised by the

[1] It is worth noting that apart from obvious benefits of early cancer diagnosis, there are also risks and costs associated with overdiagnosis. Overdiagnosis refers to detecting inconsequential cancers that "*would never cause symptoms or death during a given patient's lifetime*" (Srivastava et al. 2019). Unnecessary treatments following overdiagnosis, including biopsy, surgery, radiation, or chemotherapy, may cause physical and psychological harm, even leading to patient's death.

[2] **Acknowledgment**: The data utilized in this study have been generated by the TCGA Research Network: www.cancer.gov/tcga.

Table 16.1 Five tumor types selected for analysis. Each of them is represented by at least 500 tumor samples, and they represent cancers of five different tissues.

Study	Tumor	Normal	Total	Tissue	Disease
BRCA	1,102	113	1,215	Breast	Breast invasive carcinoma
KIRC	534	72	606	Kidney	Kidney renal clear cell carcinoma
HNSC	521	43	564	Head and neck	Head & neck squamous cell carcinoma
THCA	513	59	572	Thyroid	Thyroid carcinoma
LUAD	513	58	571	Lung	Lung adenocarcinoma

National Cancer Institute and the National Human Genome Research Institute. The goal of the project was to generate and catalog high-throughput genomic and proteomic data for multiple types of cancer. The TCGA data can be accessed via the Genomic Data Commons (GDC) Data Portal.[3] For this study, we were interested in the *Next-Generation Sequencing* (NGS)-based RNA-Seq gene expression data.[4] Although data sets for more than thirty types of cancer are available in TCGA, many have a relatively small number of tumor samples, or very few or no matched normal samples. Our selection criteria for the data sets to be included in this study were: (i) at least 500 tumor samples, (ii) a reasonable number of matched normal samples, preferably fifty or more, (iii) representation of cancers of different tissues.[5] Based on these criteria, five data sets were selected; basic information on the five *downloaded* data sets[6] is summarized in Table 16.1. Together, the selected data included RNA-Seq gene expression data for 3,528 patients (3,183 tumor samples and 345 normal ones), each sample represented by the gene expression levels of 20,530 genes. Thus, the cumulative gene expression matrix included about 74.5 million data points.

For the five selected data sets, possible batch effects associated with *BatchID*, *TSS* (Tissue Source Site), and *ShipDate* were evaluated and quantified using the MBatch

[3] https://portal.gdc.cancer.gov.

[4] Next-Generation Sequencing is an umbrella term used to describe high-throughput sequencing technologies that have emerged in the mid-2000s and are constantly being developed and improved since then. Before NGS, the Sanger method (Sanger et al. 1997) was commonly used for genome sequencing (for example, in the Human Genome Project). Although Sanger sequencing features a lower error rate, new NGS technologies are more scalable and much less expensive (in terms of the cost per megabase of sequence). The TCGA RNA-Seq data used in this study were generated using the Illumina HiSeq 2000 NGS sequencing platform.

[5] Two lung cancer data sets (LUAD and LUSC) passed the quantitative selection criteria; the one with slightly greater numbers of tumor as well as normal samples was selected to satisfy the "different tissue" criterion.

[6] An adjustment will be done later in order to have a balanced representation of the selected tumor types.

Omic Browser from MD Anderson Cancer Center.[7] None of the five data sets required any batch correction.[8]

The NGS technology quantifies a gene's expression level by counting the sequences (identified in an analyzed biological sample) that map to a particular gene. To compare gene expression levels obtained from different biological samples, the expression data from these biological samples have already been preprocessed and normalized to the same value of their upper quartiles.[9] The downloaded data were also already $\log_2$-transformed.

16.2.2 Training and Test Data Sets

Before starting any analysis, the data have to be split into training and test data sets. Stratified random selection was applied and 80 percent of samples associated with each of the five tumor types was selected into the training set. The remaining 20 percent of the data was set aside as the test set to be used only to evaluate the final multivariate biomarker identified in the study. To have a more balanced design (not overrepresenting one of the tumor types among the tumor samples), only half of the initially assigned BRCA tumor training samples were randomly selected and left in the final training set. Consequently, the training data set includes 2,387 samples; 2,108 of them are tumor samples and 279 are normal. The set-aside test set (holdout set) includes 700 samples – 634 tumor samples and 66 normal ones. Details of the training and test sets are provided in Table 16.2.

16.2.3 Removing Variables Representing Experimental Noise

The variables in the training data represent genes (more precisely gene expression levels). Many genes in a tissue are not expressed at biologically significant levels – the variables associated with such genes can be seen as experimental noise, which may have detrimental effects on the analysis. They should be identified and eliminated. For these data, the noise level has been estimated to be 30 counts per million (CPM)[10], where CPM represents the number of sequences identified for a particular gene per 1 million sequences identified in a particular biological sample. Consequently, the training data set was first noise-filtered to include only the genes with the gene expression levels greater than 30 CPM in at least 20 percent of samples of at least

[7] https://bioinformatics.mdanderson.org/MQA.

[8] Apart from visualizations of batch effects (such as via PCA and hierarchical clustering), the MBatch Omic Browser quantifies the effects by calculating – for each of the considered batch factors – a value for the Dispersion Separability Criterion (DSC). The DSC represents the ratio of between-batch to within-batch variation and is calculated as $DSC = \sqrt{tr(\mathbf{S}_B)}/\sqrt{tr(\mathbf{S}_W)}$, where $\mathbf{S}_B$ and $\mathbf{S}_W$ are matrices of between-batch and within-batch variation, respectively. DSC values above 1 usually indicate strong batch effects, while for values between 0.5 and 1, batch effects may need to be considered. All DSC values for the five selected data sets were below 0.5.

[9] The TCGA RNA-Seq data were quantified in Fragments Per Kilobase of transcript per Million mapped reads (FPKM), then subjected to upper quartile normalization (UQ-FPKM), and then $\log_2$-transformed.

[10] Based on the average value of the four-percent-trimmed CPM mean across samples.

Table 16.2 Training and test data sets. Only the training set including gene expression data for 2,387 samples is used for biomarker discovery analysis. The test set with remainder of 700 samples is set aside and will only be used to test the final multivariate biomarker and classifier.

Study	Training data set			Test data set		
	Tumor	Normal	Total	Tumor	Normal	Total
BRCA	441	91	532	220	22	242
HNSC	417	35	452	104	8	112
KIRC	428	58	486	106	14	120
LUAD	411	47	458	102	11	113
THCA	411	48	459	102	11	113
Total	2,108	279	2,387	634	66	700

one of the ten considered subclasses of samples (five tissue types with tumor and normal samples for each of them). After removing experimental noise, the training data set contains gene expression data of 2,387 biological samples and 11,346 genes.

16.2.4 Low-Dimensional Visualization of the Training Data

As part of the exploratory data analysis (EDA), the training data were visualized in the space of the first two principal components.[11] The results (see Figure 16.1) indicate that the unsupervised grouping of training observations is aligned with their tumor types. On this two-dimensional visualization, there are clearly four groups associated with tumor types – four instead of five, since the BRCA and LUAD observations heavily overlap. However, when the training observations are projected onto a three-dimensional space defined by the first three principal components (see Figure 16.2), all five tumor types become well separated. This is what was expected for this heterogeneous data set – that the primary factor in grouping of training observations would be tumor type (rather than, for example, *BatchID*).

When the training observations – on a two-dimensional principal component analysis (PCA) plot – are annotated by their class (tumor or normal), the normal observations seem to be assembled on the boundaries of the groups (see Figure 16.3). Recall that PCA is blind to the class variable;[12] hence, if normal and tumor observations are reasonably separated in this unsupervised visualization, then supervised methods should be able to find a projection of the training data, in which the classes are very well separated.

[11] Principal component analysis is described in Section 16.8.1. Recall that principal components are very useful for low-dimensional data visualization during EDA; however, they may not be used (as a replacement for the original variables) for predictive modeling (see Chapter 2).

[12] Whether the class (target) variable is defined by the diagnosis (tumor vs. normal) or five tumor/tissue types.

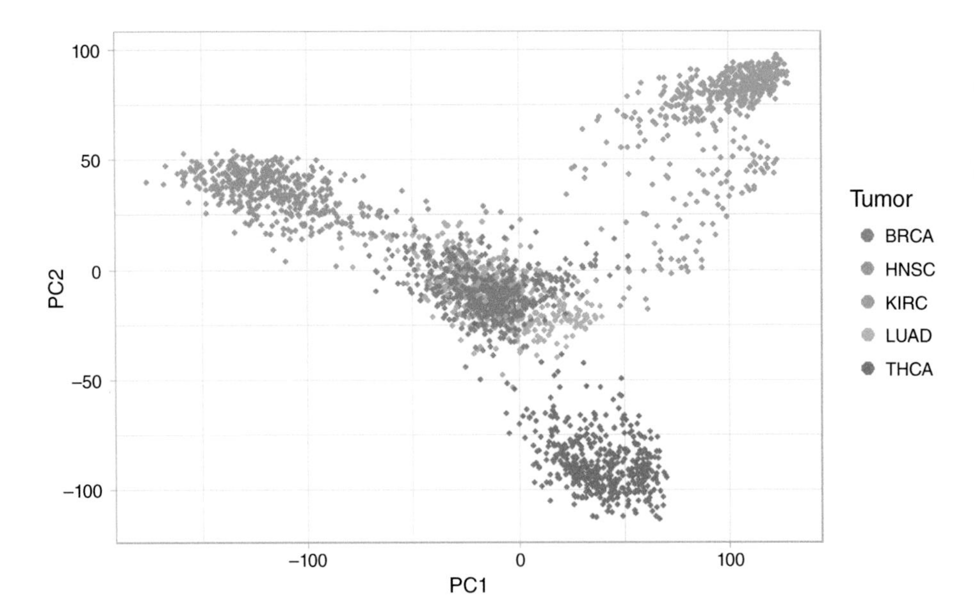

Figure 16.1 PCA visualization of the training data in the space of the first two principal components (which together explain 34.4 percent of the variation in the data represented by the 11,346 independent variables). The observations are annotated (and colored) by their tumor types. Three tumor types (HNSC, KIRC, and THCA) form three groups that are completely separated from each other, as well as reasonably separated from the fourth group, which includes the heavily overlapping observations from the two remaining tumor types (BRCA and LUAD).

16.3 Determining an Optimal Size of Multivariate Biomarker

In this study, Fisher's discriminant analysis (FDA) was used as the learning algorithm (see Chapter 12), and the T^2-driven stepwise hybrid search method was used for feature selection (see Section 5.3.3 of Chapter 5).[13] To decide on the optimal size of the multivariate biomarker, we started with evaluating the T^2 discriminatory power and the distribution of training observations in the FDA's discriminatory space, for subsets including one to thirty variables. This was done by performing a T^2-driven hybrid feature selection that stops when the current subset includes thirty variables (recall that we are interested in a parsimonious multivariate biomarker). The results indicate that sets of ten to fifteen variables may have satisfactory discriminatory power, as well as may provide reasonable separation of the classes. Figure 16.4 shows the FDA's discriminatory space,

[13] Experiments for this study were performed using the *MbMD* biomarker discovery software (www .multivariatebiomarkers.com). However, any software platform that implements discriminant analysis (either Fisher's or Gaussian) and stepwise hybrid feature selection can be used instead.

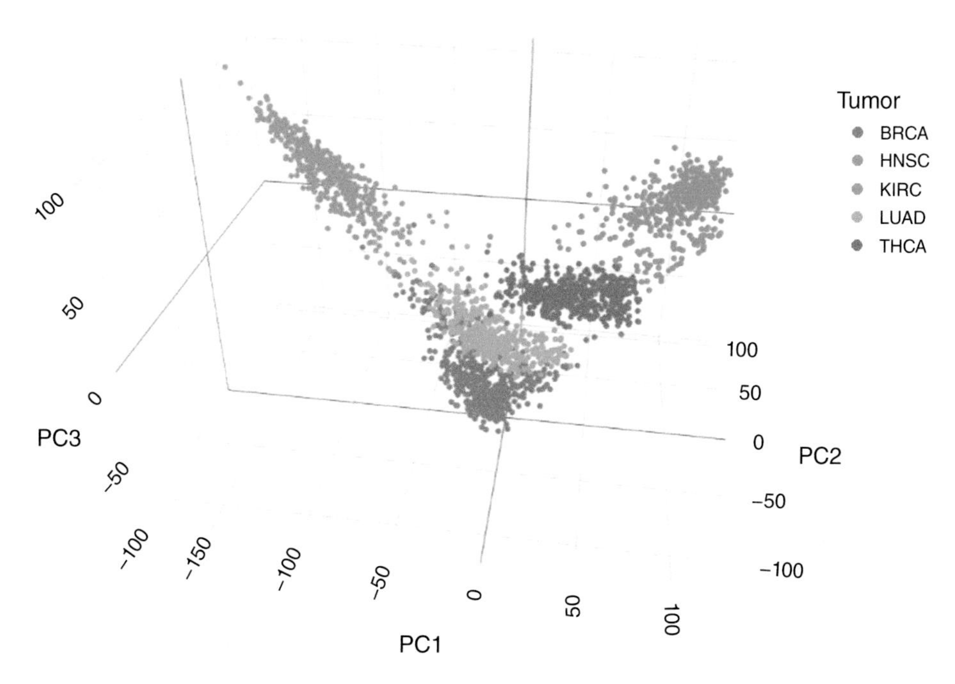

Figure 16.2 Three-dimensional visualization of the training data in the space of the first three principal components (which cumulatively explain 43.2 percent of the data variation). The observation data points are colored by their tumor types. Adding the third principal component dimension reveals that BRCA and LUAD are also well separated and –consequently – that the primary grouping of the training observations is well aligned with their tumor types.

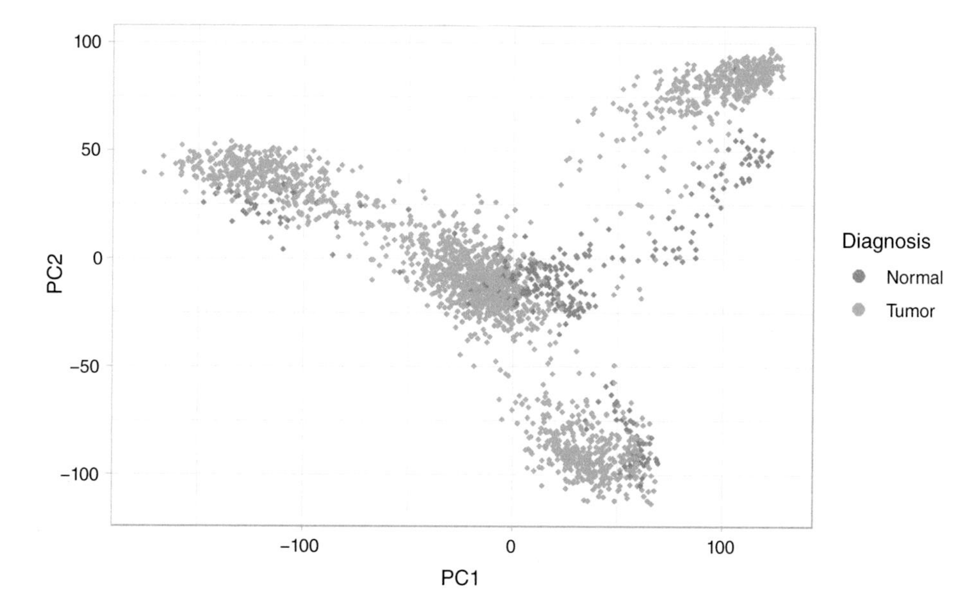

Figure 16.3 PCA visualization of the training data, with the observations annotated by their diagnosis class (tumor or normal). The normal observations (plotted in red) are gathering mostly along the boundaries of the groups associated with their tumor types.

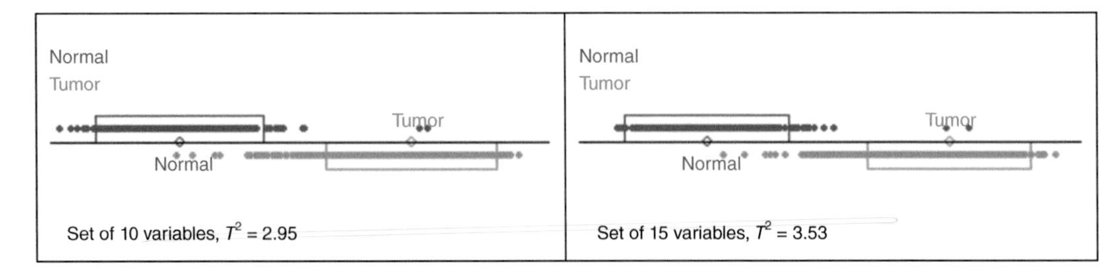

Figure 16.4 Discriminatory space of the FDA models using sets of ten variables (the left panel) and fifteen variables (the right panel); the sets were identified by a T^2-driven stepwise hybrid feature selection performed on the training data including 2,387 observations and 11,346 variables (genes). The separation of the tumor and normal classes, as well as their T^2 measures, may suggest satisfactory discriminatory power. Since, for two classes, FDA's discriminatory space is one-dimensional, the projections of the 2,387 training observations would be clamped together on the line defined by the FDA's linear discriminant function. For a better visual distinction between the classes, observations from each class are plotted with a different vertical offset. The horizontal dimensions and placements of the boxes show the areas of 95 percent probability for each class. Only a few observations of each class are projected to the area of the other class. For the FDA model built on the set of ten variables, 98.5 percent of the training observations are reclassified correctly into their own class, and 95.1 percent of the observations are projected within a 95 percent area of their class. For the model using fifteen variables, these numbers are 98.8% and 95.1%, respectively.

as well as the distribution of projected training observations for the FDA models built on the identified sets of ten and fifteen variables. Recall that with two differentiated classes (here tumor and normal), the discriminatory space is one-dimensional (that is, it is represented by a single linear discriminant function that maximally separates the classes). Though some of the training observations are projected into the area of the other class, there are 2,387 observations here, and only a very small percentage of them is reclassified incorrectly – for example, for the FDA model based on the ten-variable subset (the left panel of Figure 16.4), 98.5 percent of training observations are reclassified correctly, and 95.1 percent are within the marked 95 percent probability area of their own class (where we would expect to have 95 percent of the class samples).

Recall that, so far, all of these results were based on the T^2 measures of discriminatory power and on the reclassification of training data observations and, as such, constitute only necessary, but not satisfactory, conditions for selecting one of these set sizes. To more realistically estimate the performance of a classification model based on a specific number of variables, we will use out-of-bag (OOB) estimates of performance metrics averaged over a large number of bootstrap-based feature selection experiments. Since the results for the sets of ten and fifteen genes were very similar, and since we are always interested in parsimonious biomarkers, we will focus (or, at least, focus first) on the set size of ten. Of course, we cannot just build a classifier on whatever ten genes were selected into the set of ten identified during the initial hybrid feature selection, as this would lead to internal validation (no more reliable than the reclassification of training data already performed).

Table 16.3 OOB estimates of the performance of an average ten-gene biomarker. The estimates are based on aggregating information from 1,000 parallel feature selection experiments (each using its own bootstrap subsample of the training data), resulting in 1,000 different ten-gene biomarkers and 1,000 FDA classifiers built on those multivariate biomarkers.

Performance metric	OOB estimate (training data with 11,346 variables)
Average T^2	2.97
(Min–Max) T^2	(2.57–3.47)
Accuracy	98.2%
Tumor sensitivity	98.3%
Specificity	97.7%

To properly – and with high confidence – estimate the average classification performance of any set of ten variables selected from these training data via multivariate feature selection, we performed 1,000 parallel feature selection experiments, with each based on its own bootstrap sample randomly selected from the entire training data containing 2,387 patients and all 11,346 variables. Since we were using the stepwise hybrid feature selection with T^2 as well as Fisher's discriminant analysis, and since both of them assume independence of observations, we used a bootstrap method *without* replacement. Each of the 1,000 bootstrap samples was generated by stratified random sampling, selecting 80 percent of observations from each class. The remaining 20 percent of observations constituted the OOB sample associated with the bootstrap sample. Each of the 1,000 feature selection experiments ended when its own optimal set of ten variables was identified. Then, each of these 1,000 different optimal sets of ten variables was used to build an FDA classification model, and then each of the 1,000 models was tested on the OOB sample from its own feature selection experiment. Averaging performance measures over those 1,000 classification models provided the following OOB estimates for the performance of our future (and final) optimal biomarker of ten variables: 98.2 percent accuracy, 98.3 percent tumor sensitivity, and 97.7 percent specificity. The results are summarized in Table 16.3. Since OOB estimates based on many independent feature selection experiments can be considered quite reliable, we will search for a generalizable and interpretable multivariate biomarker containing ten genes.

16.4 Identifying the Pool of Potentially Important Variables

The goal of this stage of elevating signal (and reducing noise) was to identify those independent variables (genes) that have a potential multivariate importance and, thus, remove those without. Since this is only one of the several stages of this multistage approach to elevating signal, we decided that at this stage we would be slightly less aggressive in eliminating noise and would even keep genes whose multivariate importance was rather marginal (if they are really unimportant, they will be eliminated in the subsequent stages focusing on the essential patterns). Hence, we evaluated the multivariate importance of the variables by selecting them into fifteen-gene biomarkers, even if we previously decided that our final (and optimal) biomarker will include only ten genes.[14]

Consequently, we started with our entire training data, including 11,346 variables, and identified the first biomarker of fifteen genes (again using T^2-driven stepwise hybrid feature selection). We evaluated its predictive ability (represented here by the value of its T^2 metric), set aside those fifteen genes, and thus initialized the pool. Then, we used the training data without those fifteen variables and identified the next biomarker of fifteen genes. It also had a satisfactory discriminatory power (with satisfactory, assumed for this experiment, being $T^2 \geq 2.0$), and its genes were added to the pool (and removed from the training data). This process was continued; subsequent biomarkers with satisfactory discriminatory power were identified; their genes increased the pool and were removed from the training data. The process stopped when no additional fifteen-gene biomarker with satisfactory discriminatory power could be identified. As a result, thirty-three fifteen-gene biomarkers with $T^2 \geq 2.0$ were identified. Hence, the pool of potentially important variables included 495 genes.

This means that out of 11,346 genes included in the original training data, 10,851 were deemed as not having satisfactory multivariate importance and thus removed as representing yet another type of noise. Although we could be quite confident about the quality of these results, we nevertheless checked whether those 495 genes selected into the pool had virtually the same predictive power as the entire training data of 11,346 genes. To accomplish this, we performed another 1,000 parallel feature selection experiments, with each based on its own bootstrap sample; however, this time the training data included only 495 genes from the pool. Consequently, the 1,000 feature selection experiments identified 1,000 different ten-gene biomarkers, which were used to build 1,000 FDA classifiers. Those classifiers were then tested on the OOB samples from their feature selection experiments. As before, performance measures were averaged over the 1,000 classification models. This resulted in the following OOB estimates: 98.2 percent accuracy, 98.3 percent tumor sensitivity, and 97.4 percent specificity. The results are summarized in Table 16.4. Since these results are almost identical to those from the feature selection experiments based on all 11,346 variables, we were confident that the 495 variables of the pool captured important discriminatory information, and that the remaining 10,851 variables can be removed as they represented noise (with this type of noise corresponding to no multivariate importance).

[14] Another advantage of utilizing, at this stage, larger sets of genes (larger than the already decided optimal size of ten) is the possibility for additional optimization. If feature selection is continued after identifying a set of ten genes, then not only are new genes added into the set but some of the originally selected ten genes may be replaced. This may result in a better set of the first ten genes (that is, with a higher discriminatory power) included in the fifteen-gene biomarker.

Table 16.4 OOB estimates of the performance of an average ten-gene biomarker when the training data included only 495 variables from the pool of potentially important variables. The estimates were based on aggregating information from 1,000 parallel feature selection experiments (each using its own bootstrap subsample of the training data), resulting in 1,000 different ten-gene biomarkers and 1,000 FDA classifiers built on those biomarkers.

Performance metric	OOB estimate (training data with 495 pool variables)
Average T^2	2.97
(Min–Max) T^2	(2.54–3.39)
Accuracy	98.2%
Tumor sensitivity	98.3%
Specificity	97.4%

16.5 Identifying Essential Patterns

After identifying the pool of 495 potentially important genes (and after removing a majority of the original 11,346 genes by associating them with various types of noise), we *could* perform a final feature selection to identify an optimal biomarker of ten genes. However, we wanted to further increase chances that the genes selected into the final biomarker represent the real biological processes associated with groups of genes with similar expression patterns across the observations. Furthermore, associating the biomarker genes with such expression patterns may be very important for the biological interpretation of the multivariate biomarker. Therefore, we started with the identification of groups of genes having similar expression patterns across observations. Then, by analyzing the distribution of those patterns among the 1,000 biomarkers identified by the previously performed feature selection experiments, the essential patterns were identified.

16.5.1 Groups of Variables with Similar Expression Patterns

To identify groups of genes with similar expression patterns, we clustered the 495 genes selected into the pool of potentially important variables. The self-organizing maps (SOM) clustering algorithm (see Section 15.5.2) was used, with a 4 × 4 rectangular topology and with the average linkage used for calculating distances between clusters. To calculate distances between genes (within clusters), the correlation distance was used, since we wanted to group the genes with a similar *shape* of their expression level values across observations, rather than grouping them by similar magnitudes of their expression (see Section 15.2). By design of this experiment, sixteen clusters were identified, with the numbers of genes in each ranging from nine to ninety. This suggested that some of these cluster patterns may include distinct subpatterns that were combined together due to being significantly different from the other fifteen patterns. We did not investigate such potential subpatterns any further, since nothing would prevent the final feature selection algorithm from selecting more than one gene from any essential pattern, if such a selection would maximize the discriminatory power of the multivariate biomarker. The expression patterns of the genes included in each of the

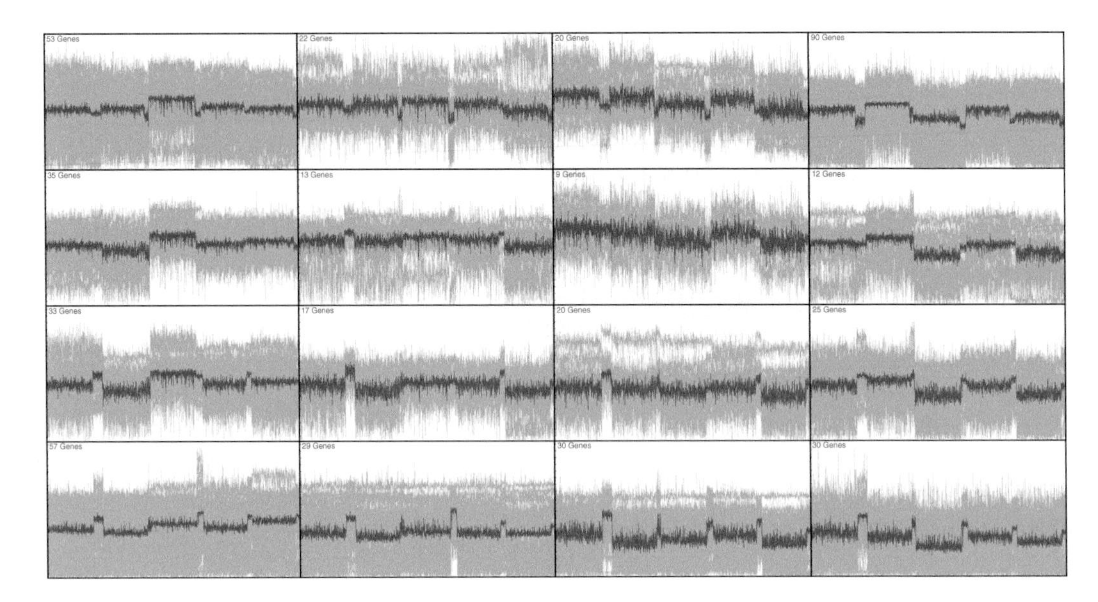

Figure 16.5 Results of SOM clustering of the 495 genes constituting the pool of potentially important variables. Clustering was performed using a 4 × 4 rectangular grid topology. The correlation distance was used to quantify dissimilarities between genes, and the average linkage was used to calculate distances between clusters. Observe that the gene expression patterns of clusters that are topologically close are more similar to each other than to the patterns of the more distant clusters.

sixteen clusters across all 2,387 observations are presented in Figure 16.5. After identifying the sixteen patterns, the next stage of signal elevation was to identify the essential patterns.

16.5.2 Essential Gene Expression Patterns

To select from the sixteen identified gene expression patterns those that would be deemed essential, we utilized the results of the parallel feature selection experiments performed previously on the training data including only the 495 genes constituting the pool of potentially important variables. Recall that those 1,000 bootstrap-based feature selection experiments resulted in 1,000 different ten-gene biomarkers and 1,000 FDA classifiers built on those biomarkers.

Assuming that the clusters whose genes are most frequently selected into those 1,000 bootstrap-based classifiers are most likely to represent real (essential) biological patterns, we investigated the distribution of the genes of the sixteen clusters among the biomarkers utilized by those classifiers. Table 16.5 shows the results of this investigation. Some of the clusters had their genes used by the 1,000 classifiers as few as fifty times, while four clusters had their genes used more than a thousand times (meaning that some classifiers used more than one gene from the same cluster). However, due to the very different cluster sizes, essential patterns were identified by considering the *average* number of times the genes from a cluster were used across the 1,000 classifiers.

Table 16.5 Results of SOM clustering of the 495 genes selected into the pool of potentially important variables, and results of investigating the distribution of the genes of the identified clusters among the biomarkers (and classifiers) identified by the previously performed 1,000 parallel feature selection experiments. Genes of six of the sixteen patterns are most commonly selected into the classifiers of the ensemble. These six patterns (including 255 genes) are deemed to represent the essential gene expression patterns associated with discrimination between tumor and control classes.

Results of SOM clustering of 495 genes		Distribution across 1,000 ten-gene biomarkers (resulting from 1,000 feature selection experiments on the pool of 495 genes)	
Pattern ID (cluster)	**Number of genes**	**Number of times pattern genes are used**	**Average use per gene**
12	25	1,342	53.7
3	20	891	44.6
16	30	1,289	43.0
13	57	1,225	21.5
9	33	676	20.5
4	90	1,827	20.3
8	12	213	17.8
5	35	555	15.9
1	53	822	15.5
2	22	316	14.4
14	29	383	13.2
11	20	148	7.4
6	13	84	6.5
7	9	50	5.6
10	17	69	4.1
15	30	99	3.3

Six clusters having their genes used, on average, by at least twenty classifiers were designated as representing the essential gene expression patterns (see Table 16.5). These six clusters included 255 genes.

16.6 Essential Variables of the Essential Patterns

After identifying the six essential gene expression patterns, we looked at their genes. The 255 genes included in these patterns are not necessarily equally important for cancer diagnosis. Thus, we now look at their *individual* distribution across the same 1,000 ten-gene biomarkers and 1,000 classifiers from the same feature selection experiments. The genes selected into at least ten classifiers are considered *essential* genes as they are least likely to be selected by chance. There are fifty-two such genes, and their distribution results are summarized in Table 16.6. Observe that some of these essential

Table 16.6 Essential genes of essential patterns: fifty-two (out of 255 genes included in the six essential gene expression patterns) were selected into at least ten out of 1,000 different ten-gene biomarkers identified by the 1,000 parallel feature selection experiments performed on the training data including only the 495 genes of the pool of potentially important variables.

	Gene	Pattern ID (cluster)	No. of times the gene was used in 1,000 biomarkers		Gene	Pattern ID (cluster)	No. of times the gene was used in 1,000 biomarkers
1	MYOC	13	727	27	CTHRC1	3	51
2	GCOM1	16	690	28	FAM96A	13	49
3	DCN	12	605	29	CDSN	4	47
4	COL10A1	4	550	30	DPT	16	38
5	GAS1	12	515	31	FAM3B	16	35
6	MMP11	3	498	32	UHRF1	4	34
7	KIAA0101	4	411	33	SUCLG1	13	32
8	LYVE1	9	288	34	IQGAP3	4	32
9	COL1A1	3	261	35	LPAR5	4	29
10	PLEKHN1	4	255	36	PNPLA6	13	27
11	UMOD	13	192	37	CTSL2	4	25
12	FAM111B	4	175	38	SULT2B1	4	24
13	ENTPD1	9	151	39	C16orf75	4	22
14	GPM6B	16	133	40	C1QTNF6	3	22
15	SLC25A25	16	126	41	TYMS	4	19
16	ABHD5	16	106	42	ALG1	3	19
17	TMEM178	13	106	43	FAP	3	19
18	PIK3C2G	16	88	44	LRRFIP1	12	18
19	DENND2A	9	85	45	PKMYT1	4	17
20	E2F1	4	77	46	DPP6	13	17
21	ACPP	12	65	47	SPTBN1	13	16
22	MFAP5	12	60	48	SPATS2	4	15
23	FBLN1	12	59	49	PLA2R1	16	14
24	RGS5	9	58	50	AMPD3	4	14
25	MYL9	9	56	51	BAX	4	14
26	IGSF10	16	54	52	EMP1	12	12

genes were selected into several hundreds of the 1,000 classifiers, and recall that each classifier was built on a different random subset of the training data.

16.7 Building and Testing the Final Multivariate Biomarker

After the multistage processing of signal elevation and noise reduction, the fifty-two essential genes were identified, which are the ones having the best chance to represent

Table 16.7 The ten genes selected into the final multivariate biomarker. They represent all six essential gene expression patterns. The order in the table reflects the order in which the genes were included in the final biomarker during stepwise hybrid selection.

Gene symbol	Pattern ID	Gene description
DCN	12	Decorin
COL10A1	4	Collagen type X alpha 1 chain
COL1A1	3	Collagen type I alpha 1 chain
GCOM1	16	GRINL1A complex locus 1
GAS1	12	Growth arrest specific 1
KIAA0101	4	PCNA clamp associated factor
MYOC	13	Myocilin
RGS5	9	Regulator of G protein signaling 5
PLEKHN1	4	Pleckstrin homology domain containing N1
TMEM178	13	Transmembrane protein 178A

real gene expression patterns associated with the biological processes implicit in, and common to, the investigated five tumor types. Therefore, if we now perform heuristic feature selection based only on these essential genes, the thereby identified parsimonious ten-gene (optimal size) multivariate biomarker will have the best chance of being well generalizable. Furthermore, by having the genes selected into the biomarker associated with their essential patterns, our final multivariate biomarker will also have the best chance for plausible biological interpretation.

16.7.1 The Final Multivariate Biomarker of Optimal Size

Using again the T^2-driven stepwise hybrid search, feature selection was performed on the training data including only the fifty-two essential genes. A ten-gene biomarker with a discriminatory power of $T^2 = 3.05$ was identified. This biomarker taps into all six essential patterns; the genes of the final biomarker (and their pattern associations) are presented in Table 16.7. Figure 16.6 shows a heat map visualization of the biomarker – expression levels of the ten genes of the biomarker across 2,387 training observations. The genes were clustered via hierarchical clustering using correlation distance and average linkage.

Based on this final multivariate biomarker, a Fisher's discriminant analysis classifier was built to be eventually used for the classification of new patients, or any patients not included in the training data.[15]

[15] Since only two classes are differentiated here, there is only one linear discriminant function that defines a one-dimensional Fisher's discriminatory space. Although this discriminatory space was identified without making the multivariate normality assumption of the original independent variables, we may now invoke the central limit theorem and assume that the feature representing the discriminant function is normally distributed in each of the *projected* classes (see Chapter 12). This allows to estimate the probability of a classified new observation's membership in either class; the observation is assigned to the class with the highest probability.

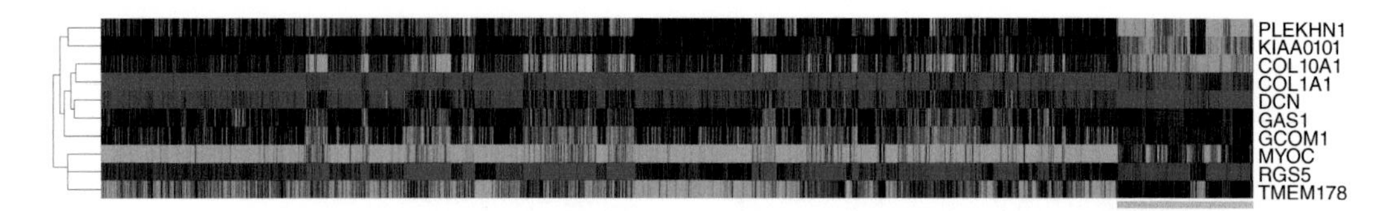

Figure 16.6 A heat map visualizing gene expression level of the ten genes selected into the final multivariate biomarker. The genes are clustered by their correlation distance (the dendrogram on the left side of the image represents the results of hierarchical clustering with the average linkage). The horizontal dimension corresponds to 2,387 observations in the training data. A quite distinct block on the right side of the heat map (marked by an orange bar below the map) corresponds to all 279 normal observations in the training data. Color intensities represent gene expressions levels – green corresponds to lower expression levels, and red to higher ones (with black corresponding to expression values around the median).

It is important to understand that our final multivariate biomarker is just one of the potential multivariate biomarkers that could be identified for this classification problem. Using different training data, or the same training data with different modeling approaches and different learning algorithms, or even having all of the above the same but with different parameters, would likely lead to a different optimal multivariate biomarker with a set of genes that may or may not overlap with those in our final biomarker. However, this neither invalidates the results nor indicates their intrinsic instability. If the procedures used for the identification of such biomarkers were sound, each of those biomarkers would represent the same, or similar, set of biological processes associated with the class differences, and each may be expected to provide quite similar predictive abilities for the classification of new observations.

Recall that each gene selected into our optimal biomarker represents one of the essential patterns – a group (or cluster) of genes with similar expression patterns across the training observations. Thus, our solution should be considered from the perspective of three levels: (i) the genes selected into the optimal biomarker, (ii) the essential patterns those genes belong to, and (iii) biological processes associated with those essential patterns. This means that apparently different results – that is, different possible sets of variables – may be equivalent when considered from a higher level they tap into. Consequently, this would define a gradation of variability: while a gene from our optimal biomarker could possibly be replaced by another gene from the same pattern, the variability in the identified essential patterns would be much smaller, and the variability in biological processes associated with those patterns – as well as involved in the etiology of disease state – would be either very small or none (if we can reliably identify a set of such processes). All this leads to the very likely equivalence of apparently different results.

16.7.2 Testing the Final Biomarker and Classifier

Recall that 20 percent of the original data were set aside before analysis. These test data were never used in any step of the analysis. Now, after completing the analysis, we will

Table 16.8 Confusion matrix resulting from testing the final biomarker (and the classifier) on the holdout test set including 700 observations. Of the 634 tumor test samples, 625 were properly classified into the tumor class (true positives), and out of the sixty-six normal test samples, sixty-four were classified into the normal class (true negatives). Consequently, of the 700 test observations, 689 were classified into their true class.

		Predicted Class	
		Tumor	Normal
True Class	Tumor	625	9
	Normal	2	64

Table 16.9 Classification performance of the final multivariate biomarker as measured by testing it on the 700 observations of the holdout test data set, which was not used during any step in the biomarker discovery analysis.

Performance metric	Results of testing the final ten-gene biomarker on the test data set (never seen during the analysis)
Accuracy	98.4%
Tumor sensitivity	98.6%
Specificity	97.0%

use this holdout test data set to validate our final ten-gene biomarker. When the 700 patients of the test data were classified by the FDA classifier implementing the final biomarker, the biomarker showed 98.6 percent tumor sensitivity, 97 percent specificity, and 98.4 percent accuracy. The confusion matrix produced by the test is provided in Table 16.8, and the performance of the biomarker is summarized in Table 16.9.

By consequently implementing the multistage approach to multivariate biomarker discovery (described in Chapters 14 and 15), each of the many steps in the presented study elevated the discriminatory signal, as well as removed a different type of noise (thus recursively increasing the signal-to-noise ratio). All this led to the identification of essential genes of essential gene expression patterns. Those genes and patterns are most likely to represent real biological phenomena underlying the differentiation between the heterogeneous *Tumor* class (representing here five different cancer diseases of five different tissues) and the *Normal* class. The result is a generalizable parsimonious multivariate biomarker not only with a high predictive power but also whose genes are associated with a larger set of genes representing distinct gene expression patterns. This association should aid in plausible biological interpretations of the final multivariate biomarker and its classification abilities.

16.8 Ancillary Information

16.8.1 Principal Component Analysis: A Snapshot

Although *principal component analysis* (PCA), an unsupervised learning algorithm, may not be used to drive a biomarker discovery project (as this could severely compromise the quality of the results and would amount to a misconception discussed in Chapter 2), this technique may be very useful during exploratory data analysis. Data visualization in a low-dimensional space of the first two or three principal components (PCs) can elucidate some internal data structures and relations among observations or among variables. In this study, PCA was used to gain an insight into the grouping of observations included in the training data. Recall that PCA is blind to the response variable (as well as to any meta information associated with each observation)[16], so the directions of PCs are identified without considering the class of each observation (Tumor or Normal) or its tumor type. However, once we completed the PCA and used its results to visualize the data in the space of the first two or three principal components, we can then add such meta information to the plot by drawing the projection of each observation using the color assigned to its class (or subclass) – see Figures 16.1–16.3.

Principal component analysis is an unsupervised dimensionality reduction method that can sequentially identify such orthogonal directions (principal components) that maximize data variation. Assuming the data include p independent variables and N observations, and that each observation is represented by a $p \times 1$ vector $\mathbf{x}_i = \left[x_{1i}, \cdots, x_{pi}\right]^T$, principal components can be identified by finding solutions to the eigenproblem,

$$\mathbf{Se} = \lambda \mathbf{e} \tag{16.1}$$

where $\mathbf{S}$ is a $p \times p$ data covariance matrix, while scalar λ and $p \times 1$ vector $\mathbf{e}$ represent the eigenvalue–eigenvector pair associated with each principal component (that is, each solution of the eigenproblem).[17] Although there are min(N – 1, p) distinct solutions to this problem, for the visualization purposes, we usually are interested only in the first two principal components (sometimes in the first three, for a three-dimensional visualization).

Each principal component is a linear combination of all original variables,

$$\mathrm{PC}_k = \mathbf{e}_k^T \mathbf{x}, \quad k = 1, \ldots, \min(N - 1, p), \tag{16.2}$$

[16] For example, *BatchID*, *TSS*, and *ShipDate*, which can be used to annotate observations on PCA visualizations offered by the MBatch Omic Browser of MD Anderson Cancer Center (see Section 16.2.1).

[17] For the gene expression data used in this study, as well as for any high-dimensional biomedical data, in which variables are measured with the same unit, the original variables should not be standardized before performing PCA. However, principal components derived from the data covariance matrix are sensitive to the measurement scales of the variables; therefore, for different applications wherein variables are measured with different units, we should either standardize them first or perform PCA using the data correlation matrix.

with eigenvector $\mathbf{e}_k$ defining the direction of PC_k in the original p-dimensional space, and eigenvalue λ_k corresponding to data variance in this direction. The first PC is associated with the direction of the largest data variation (hence it has the largest eigenvalue). The second PC identifies the direction that is orthogonal to the first PC and has the largest variation out of all such orthogonal directions. Then, each subsequent PC (if we identify more than two) will represent the direction of maximum data variation, under the constraint that this direction has to be orthogonal to all previously identified principal components.

To visualize the data in a low-dimensional space of the first m principal components, for each of the original observations, we will calculate its m-dimensional projection,

$$\mathbf{z}_i = \mathbf{E}_m^T \mathbf{x}_i, \tag{16.3}$$

where $\mathbf{E}_m$ is a $p \times m$ matrix composed of the first m eigenvectors. To find out the proportion of total data variance represented by low-dimensional visualization, we will calculate the *proportion of variance explained* (PVE) by the first m PCs,

$$\mathrm{PVE}(m) = \frac{\sum_{j=1}^{m} \lambda_j}{\sum_{k=1}^{\min(N-1,p)} \lambda_k}. \tag{16.4}$$

For example, Figures 16.1 and 16.3 provide two-dimensional visualization of the training data including 2,387 observations and 11,346 independent variables; hence there are 2,386 distinct principal components. However, just the first two of them represent 34.4 percent of data variance, while adding the third PC (see Figure 16.2) increases PVE to 43.2 percent, which was sufficient to conclude that the unsupervised grouping of training observations aligned with their tumor types.

16.8.2 Heat Maps: A Snapshot

A heat map is a graphical visualization of multivariate data in which values of variables are represented by different colors and color intensities. In bioinformatics, it is common to use heat maps to show the results of clustering of both variables and observations (though nonclustered data can also be visualized this way). The variables and observations can be clustered either independently or simultaneously. In the former case, hierarchical clustering (see Section 15.5.1) is usually used for this purpose. The latter is called two-way clustering, and its typical goal is to identify two-dimensional nonoverlapping blocks of variables and observations with similar (possibly homogeneous) patterns of values.[18] There are various algorithms for two-way clustering (for example, Hartigan 1972; Tibshirani et al. 1999), but the general concept is to start with the entire

[18] There are also two-way clustering algorithms that allow for overlapping blocks (for example, Lazzeroni and Owen 2002).

variables–times–observations matrix and recursively split it into smaller rectangular blocks in a way that minimizes the total within-block variation.

When both variables and observations are clustered (either separately or simultaneously), the heat map includes two dendrograms (see, for example, Figure 17.6). Sometimes, however, we may want to use heat map visualization when only variables are clustered. For example, Figure 16.6 shows a heat map for clustering variables of a ten-gene biomarker, while observations are ordered by their class (tumor and normal) and then by subclass (five tumor types). In either case, the colors and their intensities correspond to values of variables. The heat maps in Chapter 16 and 17 use the green–black–red continuous color scheme in which shades of green correspond to lower variable values (here gene expression levels); shades of red, higher variable values; and black, values close to the median. Typical software implementations of heat maps provide interactive visualizations in which placing the cursor in any spot on a heat map will display detailed information about the variable and the observation associated with that spot. However, in the static (printed) versions of heat maps presented in Chapters 16 and 17, only variables (genes) are labeled; there were too many training observations (2,387 in Study 1, and 339 in Study 2) to include their labels in an informative way.

17 Biomarker Discovery Study 2: *Multivariate Biomarkers for Liver Cancer*

17.1 Introduction

In this study, we will search for parsimonious multivariate biomarkers for liver cancer, a type of cancer with one of the lowest overall survival rates – about 20 percent (Siegel et al. 2022). As in Study 1, the data set for the present study comes from *The Cancer Genome Atlas* (TCGA) repository and includes the *Next-Generation Sequencing* (NGS)-based RNA-Seq gene expression data. Furthermore, this study will also take advantage of aggregating information from many parallel multivariate feature selection experiments. However, these will be the only similarities between the two studies; other aspects of the studies will be different.

The TCGA-LIHC liver cancer (liver hepatocellular carcinoma) data set includes only 423 biological samples, and thus is much smaller than the data set used in Study 1. Furthermore, including only data for one type of cancer, this set is much more homogeneous than the data set used in Study 1. We will use these differences as an opportunity to illustrate a different technique of performing biomarker discovery. The small size of the data set will allow for efficient use of R (instead of the specialized biomarker discovery software used for Study 1). Furthermore, data homogeneity will allow for a much simpler design of the study – recursive feature elimination (RFE) will be used as the main method for parallel feature selection experiments. RFE will be coupled with two learning algorithms most commonly used for this purpose – first random forests (RF) and then support vector machines (SVM).[1] Since the liver cancer data set includes 373 tumor observations and only fifty normal observations, and since both RF and SVM algorithms are sensitive to class imbalances (unlike Fisher's discriminant analysis used in Study 1), we will also illustrate a way of dealing with this challenge via a "technical" rebalancing of the training data.[2] Moreover, a very small number of normal observations will pose another challenge. If we split the data into training and test sets, the test set will have very few normal observations. This will result in a very large granularity in the evaluation of specificity based on testing the final

[1] Since both RF and SVM can work with $p \gg N$ data, and since they can internally calculate variable importance, they are well suited to be used within the RFE wrapper method, which always starts with building classifiers based on all independent variables (for more information on RFE, see Chapter 5).

[2] "Technical" qualification means that rebalancing of training data to improve the performance of classifiers is not explicitly related to misclassification costs (depending on the purpose of the biomarker, it may or may not be aligned with the potential misclassification costs; for more on this issue, see Chapter 4).

classifier(s) on the test data. Thus, the final test results would need to be interpreted in the context of out-of-bag (OOB) estimates, which will have much smaller granularity.

17.2 Data

17.2.1 RNA-Seq Liver Cancer Data from The Cancer Genome Atlas

The Cancer Genome Atlas repository was the source of liver cancer data (TCGA-LIHC) analyzed in this project.[3] There are many ways of accessing and retrieving data from the TCGA repository. We downloaded the data directly into R (R Core Team 2022) using Bioconductor's RTCGA package and its associated RNA-Seq data retrieval R script (Kosinski 2022).

```
if (!requireNamespace("BiocManager", quietly=TRUE))
  install.packages("BiocManager")

BiocManager::install("RTCGA")
library(RTCGA)
library(dplyr)

# Load newest RNA-Seq TCGA-LIHC data
dir.create( 'LIHC')
downloadTCGA( cancerTypes = 'LIHC',
              dataSet =
'Merge_rnaseqv2__illuminahiseq_rnaseqv2__unc_edu__Level_3__RSEM_genes_
normalized__data.Level',
              destDir = 'LIHC',
              date = NULL )

# shortening directory name
list.files( 'LIHC/') %>%
  file.path( 'LIHC',.) %>%
  file.rename( to = substr(.,start=1,stop=28))
# reading data
list.files( 'LIHC/') %>%
  file.path( 'LIHC',.) -> directory

directory %>%
  list.files %>%
  file.path( directory,.) %>%
  grep( pattern = 'illuminahiseq', x = ., value = TRUE) -> pathRNASeq
```

[3] **Acknowledgment**: The data utilized in this study have been generated by the TCGA Research Network: www.cancer.gov/tcga.

```
readTCGA(path = pathRNASeq, dataType = 'rnaseq') -> lihc
rownames(lihc) <- lihc[,1]     # sample names
lihc[,1] <- NULL
```

The downloaded RNA-Seq data included 423 samples (373 tumor observations and fifty normal ones) and 20,531 variables (gene expression levels). Data representing each of the biological samples were already preprocessed and normalized to the same value of the third quartile, but not log transformed. As in Study 1, the MBatch Omic Browser from MD Anderson Cancer Center was used for evaluating possible batch effects. The values of dispersion separability criterion (DSC) for *BatchID*, *TSS* (Tissue Source Site), and *ShipDate* were all small (much less than 0.5), and thus no batch correction was required.[4]

17.2.2 Training and Test Sets

With the data including only fifty normal observations, there is no perfect way of balancing the two contradictory aspects of splitting the data into training and test sets: (i) the larger the test set, the better the estimates of the classification model's performance, and (ii) the larger the training set, the higher the quality of the model expected. Theoretically, one could consider limiting the validation of the final biomarker and the classification model to resampling techniques; for example, properly performed bootstrapping should provide OOB estimates of the model performance that are as reliable as those based on the test data of the same size as the training data (Breiman 2001). However, as a golden rule, we always recommend having a test set that was never seen or used during the entire process of predictive modeling. Consequently, before performing any analysis, the data were split into training and test sets. Using stratified random selection, 80 percent of observations from each class were assigned to the training data set, and the remaining 20 percent to the holdout set, which was then set aside to be used only to test the final classifier(s). The `createDataPartition` function of the `caret` package was used for this purpose (Kuhn and Johnson 2013; Kuhn 2022).

```
# Retrieve Class from sample code
Class = c()  # class of each sample (each row)
for (i in 1:nrow(lihc)) {
    sample <- rownames(lihc)[i]
    TCGA_code <- substr(sample, 14, 15 )
    if(TCGA_code<10) Class[i] <- "tumor"
    else             Class[i] <- "normal"
}

Class <- as.factor(Class)
lihc$Class <- Class
```

[4] The data in the two classes were also properly matched by gender (DSC = 0.123). For a description of dispersion separability criterion, see Chapter 16.

Table 17.1 Training and test data sets. Only the training set including gene expression data from 339 samples will be used during biomarker discovery analysis. The test set with eighty-four samples will be set aside and used only to test the final multivariate biomarkers and classifiers.

Study	Training data set			Test data set		
	Tumor	Normal	Total	Tumor	Normal	Total
LIHC	299	40	339	74	10	84

```
#   Stratified selection into Train and Test
library(caret)
set.seed(369)
inTrain <- createDataPartition(lihc$Class,
                               p = 0.8,
                               list = FALSE)
trainData <- lihc[ inTrain, ]
testData  <- lihc[-inTrain, ]

Diagnosis <- as.factor(trainData$Class)
Test_Class <- as.factor(testData$Class)
Train <- trainData
Test <- testData
Test$Class <- NULL
```

The training data set includes 339 observations; 299 of them are tumor samples and forty normal. The holdout set includes eighty-four observations; seventy-four tumor samples and ten normal (see Table 17.1). With only ten normal observations in the test set, the estimate of specificity will have a large granularity of 10 percent. This means, for example, that specificity evaluated on the test set could be equal to 90 percent or 100 percent, but no value between the two would be possible.

17.2.3 Removing Variables Representing Experimental Noise

Since many genes in a tissue are not expressed at biologically significant levels, we will remove the genes deemed to represent experimental noise from the training data. We will use the same noise level estimate as used in Study 1, 30 counts per million (CPM), where CPM represents the number of sequences associated with a particular gene per 1 million sequences identified in a particular biological sample. We will also use the same filtering criterion – that is, the genes with expression level greater than 30 CPM in at least 20 percent of samples of at least one of the two classes will be retained.

```
# Filtering Train Data
# Keeping genes with at least 20% values >30 CPM in at least one class
```

```
CPM <- Train
CPM$Class <- NULL

totalCount <- rowSums(CPM)              # total count per sample
CPM <- CPM/totalCount*1000000           # replace expression with CPM
CPM$Class <- Train$Class

CPM_Tumor  <- subset(CPM, Class == "tumor")
CPM_Normal <- subset(CPM, Class == "normal")
CPM_Tumor$Class  <- NULL
CPM_Normal$Class <- NULL

# tumor class - identify genes to retain
CPM_Tumor <- CPM_Tumor[, colSums(CPM_Tumor >30)/nrow(CPM_Tumor) >= 0.2]
genes_T <- colnames(CPM_Tumor)

# normal class - identify genes to retain
CPM_Normal <- CPM_Normal[, colSums(CPM_Normal >30)/nrow(CPM_Normal) >= 0.2]
genes_N <- colnames(CPM_Normal)

genes <- unique(c(genes_T, genes_N))
retain <- colnames(Train)%in%genes
Train <- Train[,retain]

# log2 the data
Train <- log2(Train+1)
Test <- log2(Test+1)
```

With experimental noise removed, the training data set now consists of gene expression data from 339 biological samples and 6,548 genes. The final step of data preparation was $\log_2$ transformation of all expression values. Since the variance of gene expression data has a tendency to increase with an increase in expression level (heteroscedasticity of variance), it is a common practice to apply some kind of transformation that would stabilize variance across the range of expression values; $\log_2$ transformation is one such method.[5]

[5] Stabilizing the variance of gene expression data is very important for parametric learning algorithms that make the assumption of homoscedasticity; it is, however, not crucial for nonparametric methods such as random forests.

17.2.4 Low-Dimensional Visualization of the Training Data

As a part of exploratory data analysis, the training data were visualized in the space of the first three principal components[6] using the interactive 3D graphics generated by the `plot_ly` function of the `plotly` package (Sievert 2020).

```
# 3D interactive visualization of the training data
pca <- prcomp(Train)

pcaData <- as.data.frame(pca$x[, 1:3])
pcaData <- cbind(pcaData, Diagnosis) # add diagnosis to df
colnames(pcaData) <- c("PC1", "PC2", "PC3", "Diagnosis")

library(plotly)
plot_ly(data = pcaData, x = ~PC1, y = ~PC2, z = ~PC3,
        type = "scatter3d",
        mode = "markers",
        marker = list(size=3),
        color = ~Diagnosis,
        colors = c("#F8766D", "#00BFC4")) %>%
  layout(legend = list(itemsizing='constant',
         font = list(size=12),
         title = list(text='<b> Diagnosis </b>')))

# Proportion of variance explained
pr.var <- pca$sdev^2
PVE <- pr.var / sum(pr.var)
head(PVE)
```

On the plot (see Figure 17.1), the 339 training observations are color-annotated by their class (tumor or normal). The normal observations seem to be grouped on the boundary of the tumor class. Such a separation by the unsupervised analysis – which is blind to the class variable – allows us to expect that the supervised learning algorithms used in this study should be able to identify a parsimonious multivariate biomarker with at least reasonable predictive power.

17.3 Feature Selection Experiments with Recursive Feature Elimination and Random Forests

To properly estimate the average classification performance of multivariate biomarkers consisting of a different number of variables – in order to select the optimal size and

[6] Principal component analysis is described in Section 16.8.1.

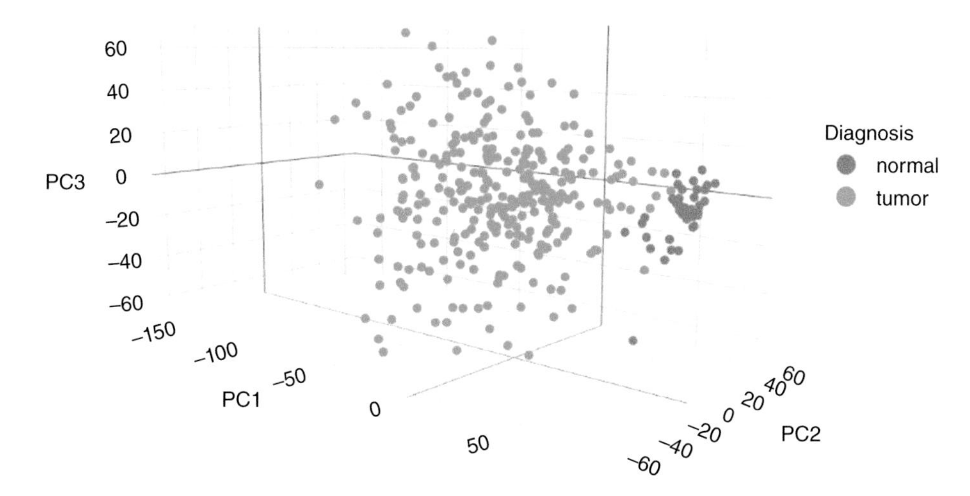

Figure 17.1 3D visualization of the training data in the space of the first three principal components (which cumulatively explain 36.6 percent of the variation in the data represented by the 6,548 independent variables). The observation data points are colored by their diagnosis class (tumor or normal). The normal observations (plotted in red) are gathering mostly on a boundary of the tumor observations.

composition of the final biomarker – 500 parallel feature selection experiments were performed. Each of them implemented the recursive feature elimination (RFE) method coupled with the random forests (RF) learning algorithm – the RF-RFE approach.[7] Each of the 500 feature selection experiments was based on a different random subsample of the training data. The default resampling technique in the `rfe` function of the `caret` package (used in these experiments) is bootstrapping implementing Efron's bootstrap with replacement. However, since the random forests learning algorithm is sensitive to heavy class imbalances, and since the ratio of tumor-to-normal observations in the liver cancer training data is almost exactly 7.5:1, a stratified resampling method was used instead to rebalance the classes.

The severe underrepresentation of normal training observations may result in RF classifiers with relatively low specificity. To boost specificity via rebalancing class proportions, we performed stratified resampling with different proportions of observations selected from each class. However, in order to evaluate the effectiveness of this "technical" balancing tactic, we also performed the baseline experiment using random sampling without balancing.

17.3.1 Resampling without Balancing (Substudy 1)

For resampling without balancing, the repeated cross-validation method was used. Fifty repetitions of 10-fold cross-validation were performed, with each of the 500 random subsamples of the training data including 90 percent of observations from each class

[7] Recursive feature elimination is described in Chapter 5, and random forests for classification in Chapter 10.

(thirty-six normal and 270 tumor observations). The `createMultiFolds` function of the `caret` package was used to generate the `index` object with 500 entries listing observations selected into each subsample.

```
# Create 500 resampling sets
# default 10-fold CV repeated 50 times
set.seed(369)
index <- createMultiFolds(Diagnosis, times = 50)
```

17.3.2 Resampling with Rebalancing Class Proportions (Substudy 2)

In order to boost specificity, the ratio of tumor-to-normal observations was decreased to 2:1. Another 500 resamples of training data were generated, each including a random selection of 90 percent of normal training observations and 24 percent of tumor observations (thirty-six and seventy-two observations, respectively). Although the same index structure was used, the 500 subsamples were generated independently of each other, and thus no *K*-fold cross-validation was involved.

```
# 500 independent resampling sets
# each with 90% control and 24% tumor observations
# loaded into index$Foldxx.Repyy
index_Tumor   <- which(Diagnosis == "tumor")
index_Control <- which(Diagnosis == "normal")

index <- list()
set.seed(369)
for (rep in 1:50){
  for(fold in 1:10){
    Balanced_Tumor   <- sample(index_Tumor,
                               size = round(0.24*length(index_Tumor)),
                               replace = FALSE)
    Balanced_Control <- sample(index_Control,
                               size = floor(0.9*length(index_Control)),
                               replace = FALSE)

    order <- 10*(rep-1)+fold
    if(fold < 10) list_name <- paste("Fold0", fold, ".Rep", rep, sep="")
    else          list_name <- paste("Fold",  fold, ".Rep", rep, sep="")

    index[order] <- list(sort(c(Balanced_Tumor, Balanced_Control)))
    names(index)[order] <- list_name
  }
}
```

17.3.3 Performing Feature Selection Experiments

By defining two different sets of 500 random subsamples of the training data, setups for two different substudies were created – one that does not balance the proportion of classes (Substudy 1), and the other that does (Substudy 2). Since both of them utilized the same index structure to store information about the generated training subsamples, the same R script can be used for each to perform feature selection experiments.

For both substudies, each of its 500 recursive feature elimination experiments starts with building a random forest classifier on all 6,548 variables and its own subsample of 108 training observations.[8] The observations not selected into the subsample (here we call them the out-of-bag, or OOB, observations)[9] are used to test the classifier and record the values of its performance metrics – AUC (the area under the ROC curve), sensitivity, specificity, and accuracy. Then, variable importance is calculated for all 6,548 variables, and those with least importance are removed. For each of the subsequent RFE iterations, only the variables still remaining in the training subsample are used, and the entire process is repeated: a new random forest classifier is built, tested on the OOB data, variable importance is recalculated, and the least important variables are removed.

The number of variables removed at each of the RFE iterations is defined by the predetermined cardinalities of the considered subsets. As described in Chapter 5, at each of the early iterations (steps) of the RFE process, many variables can be eliminated and then – after reaching relatively low cardinalities – fewer variables are eliminated at each step. When the cardinality decreases to the range of reasonably parsimonious biomarkers, only one variable is eliminated at a time. For these substudies, the following cardinalities were considered: 6,548, 4000, 2000, 1000, 500, 250, 150, 75, 50, and then every subset size between thirty and two. The main function used in the following R script, which is based on Kuhn and Johnson (2013), is `rfe` from the `caret` package.

```
# Multiple RFE runs on the randomly generated subsamples
#
# custom function for selecting optimal set of variables
selectVar_custom <- function(y, size){
  library(dplyr)
```

[8] Each random forest classifier built during the 500 RFE experiments consisted of 500 trees, and the default number of variables was randomly selected at each node of each tree (see Chapter 10). A preliminary run of the random forests algorithm indicated that OOB error rates stabilize – for this training data set – between 400 and 500 trees.

[9] Recall that for the feature selection experiments with rebalancing of class proportions, independent random subsampling was performed for each of the 500 RF-RFE experiments. This sampling procedure was equivalent to stratified bootstrapping without replacement, in which stratification was performed in the way that resulted in the predefined 2:1 ratio of tumor-to-normal observations in each of the bootstrap samples. The observations not selected into a bootstrap sample defined, of course, the out-of-bag (OOB) sample associated with the bootstrap sample. The subsampling without balancing was based on repeated 10-fold cross-validation, rather than bootstrapping; so, technically, the observations in the "test" fold of each RF-RFE experiment (of Substudy 1) were not OOB observations. Yet, for the simplicity of comparing and discussing the results of the two approaches, we will also refer to them as an OOB sample.

```
  finalImp <-
    dplyr::filter(y, Variables == size) %>%
    dplyr::select(c(Overall, var)) %>%
    group_by(var) %>%
    summarize(frequency = n()) %>%
    arrange(desc(frequency))
  as.character(finalImp$var[1:size])
}

# custom rf functions
custom_rfFuncs <- rfFuncs
custom_rfFuncs$summary <- function(...)
                          c(twoClassSummary(...), defaultSummary(...))
custom_rfFuncs$selectVar <- selectVar_custom

# RFE with Random Forests
cardinality <- c(2:30, 50, 75, 150, 250, 500, 1000, 2000, 4000)

ctrl <- rfeControl(method = "repeatedcv",
                   number = 10,     # folds - ignored when index is used
                   repeats = 50,    # repeats - ignored when index is used
                                    # but both are reported in rfe results
                   verbose = TRUE,
                   functions = custom_rfFuncs,
                   returnResamp = "all",
                   rerank = TRUE, # recalculate var importance for each size
                   index = index
)

set.seed(369)
RF_RFE <- rfe(x = Train,
              y = Diagnosis,
              sizes = cardinality,
              metric = "ROC",
              rfeControl = ctrl,
              ntree = 500
)
RF_RFE
```

17.3.4 Results of Feature Selection Experiments

The main results of feature selection experiments are OOB estimates of the average performance of subsets of the considered cardinalities. For each of those subset sizes, AUC, sensitivity, specificity, and accuracy were calculated by averaging the values of

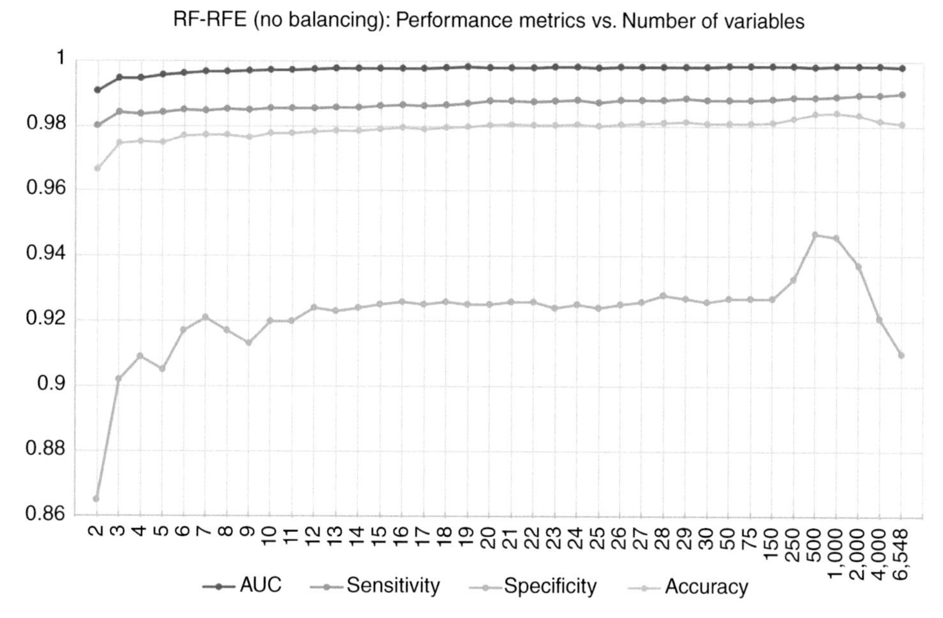

Figure 17.2 OOB performance metrics averaged over 500 RF-RFE feature selection experiments performed on random subsamples of the training data **with no balancing** (Substudy 1). Each subsample included 90 percent of randomly selected training observations from each class (270 tumor and 36 normal observations). Each point on the graph represents the value of a performance metric – for a specific cardinality – calculated by averaging the values of the metric across the 500 RF-RFE feature selection experiments. Due to random forests' sensitivity to the heavy class imbalance, specificity is significantly lower than sensitivity. In the range of subset sizes of 10–30, where the specificity is quite stable, the average specificity is about 0.925 (with the minimum and maximum values in this range of 0.920 and 0.928, respectively). In the same range of sizes, the average sensitivity to tumor is about 0.987 (with the minimum and maximum values of 0.985 and 0.989).

each of these performance metrics across the 500 RF-RFE feature selection experiments. The results of Substudy 1 (with no balancing) are visualized in Figure 17.2, while Figure 17.3 shows the results of Substudy 2, in which class proportions were rebalanced.

The OOB results of RF-RFE experiments without balancing (see Figure 17.2) confirmed the random forests learning algorithm's sensitivity to a severe class imbalance. As expected, the algorithm's focus on the overall performance resulted in specificity – associated with the severely underrepresented normal class – being significantly lower than sensitivity to the tumor class. Specificity is quite stable for cardinalities between ten and thirty, with the average value about 0.925. This value is much lower than the average sensitivity in the same range, which is about 0.987.

When specificity was boosted in RF-RFE experiments with class rebalancing (Substudy 2), its average value, in the same cardinality range of ten to thirty, increased to about 0.988 (see Figure 17.3). The price for this 6.9 percent increase in specificity was a 1.1 percent decrease in average sensitivity (to about 0.977). A summary of

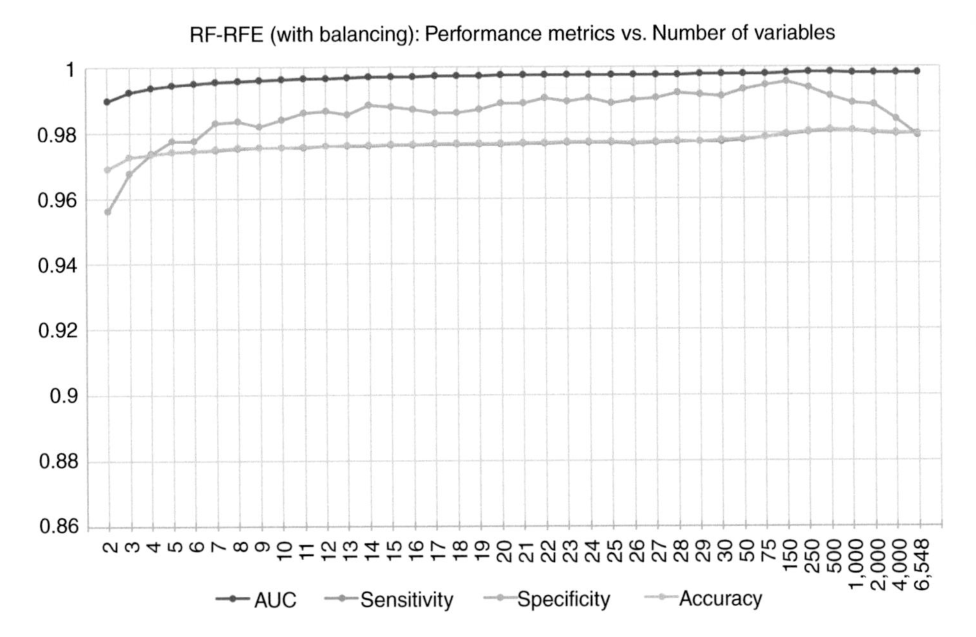

Figure 17.3 OOB performance metrics averaged over 500 RF-RFE feature selection experiments performed on random subsamples of the training data **with rebalancing** the ratio of tumor to normal observations from 7.5:1 in the training data to 2:1 in each subsample generated for Substudy 2. Hence, each randomly selected subsample included 90 percent of training observation from the normal class (36 normal observations) and only 24 percent of the tumor training observations (72 tumor observations). In the result of this rebalancing, the average specificity in the 10–30 range of subset sizes increased to 0.988 (with the minimum and maximum values in this range of 0.984 and 0.992, respectively). The cost of this specificity increase was the decrease in the average sensitivity, in the same range of subset sizes, to 0.9765 (with the minimum and maximum values of 0.975 and 0.977). For easy visual comparison, the range of the vertical axis was kept the same as in Figure 17.2.

changes in the average values of OOB performance metrics between RF-RFE experiments with and without balancing the proportions of class observations is presented in Table 17.2. Observe that the average AUC was almost the same for both approaches.

17.3.5 Selecting Optimal Multivariate Biomarkers

The OOB estimates of performance resulting from RF-RFE experiments implementing class rebalancing were used to identify the optimal size of multivariate liver cancer biomarker for Substudy 2. These results are already visualized in Figure 17.3; however, Figure 17.4 zooms in on those results, as well as adds a smoothing trendline for specificity.

The average OOB estimates of AUC, sensitivity, and accuracy are relatively stable for subsets with ten to thirty variables (for example, AUC values in this range are between 0.9962 and 0.9977). Since the very local ups and downs in OOB specificity

Table 17.2 Summary of changes in the values of average OOB performance metrics between RF-RFE experiments with and without balancing, averaged subsequently over subsets including ten to thirty variables. Rebalancing the proportion of tumor-to-normal observations from 7.5:1 in the original training data to 2:1 in the random subsamples used for RF-RFE experiments with balancing increased average specificity by 6.9 percent while simultaneously decreasing average sensitivity by 1.1 percent. The rebalancing had a very small impact on average accuracy, and a negligible impact on average AUC.

Substudy	Average OOB Estimates of Performance for Cardinality between 10 and 30			
	AUC	**Sensitivity**	**Specificity**	**Accuracy**
1. RF-RFE with no balancing	0.9980	0.9871	0.9248	0.9797
2. RF-RFE with rebalancing	0.9973	0.9765	0.9884	0.9767
Change	–0.0007	–0.0106	0.0637	–0.0030
Percent Change	–0.07%	–1.07%	6.88%	–0.31%

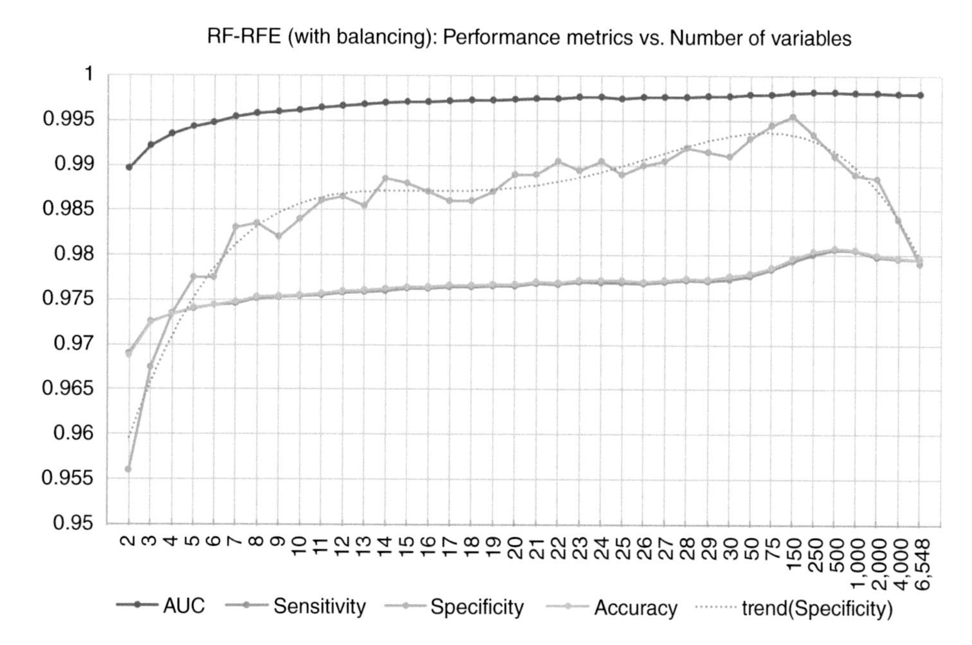

Figure 17.4 A closer look at the OOB performance metrics averaged over 500 RF-RFE feature selection experiments performed on random subsamples of the training data **with rebalancing** the ratio of tumor to normal observations (Substudy 2). In the cardinality range of 10–30, the OOB estimates of AUC, sensitivity, and accuracy are quite stable. Considering the polynomial specificity trendline, the same can be said about the specificity in the range of subsets with 10–20 variables.

estimates are irrelevant,[10] we were looking at the polynomial trendline (of the fourth order) smoothing specificity results – it is relatively flat in the cardinality range of ten to twenty (varies in this range between 0.9857 and 0.9874). Consequently, considering all four performance metrics, as well as applying the combined criterion of minimizing the subset size (parsimony promoting generalization) and maximizing estimated performance, the size of ten was identified as the optimal number of variables for this multivariate biomarker.

Using the `rfe` function from the `caret` package requires specifying a single performance metric; `rfe` returns a set of variables whose size and composition corresponds to the maximal value of this metric.[11] To retrieve from the RF-RFE results those that correspond to our optimal biomarker size, we needed to override this default `rfe` selection and appropriately update the `rfe` results.

When a size of ten is specified, the `update` function aggregates variable importance information from 500 random forests classifiers built on different sets of ten variables (identified during RF-RFE experiments) and provides an optimal set of ten variables. Then, a random forest classifier is built on the multivariate biomarker consisting of these ten variables.

```
# update rfe results
set.seed(369)
opt_rfe <- update(RF_RFE, x = Train,
                          y = Diagnosis,
                          size = 10)

opt_rfe$bestVar
## [1] "CLEC4G|339390"  "CLEC4M|10332"   "ECM1|1893"     "FCN2|2220"
## [5] "STAB2|55576"    "ADAMTS13|11093" "VIPR1|7433"    "SLC26A6|65010"
## [9] "CXCL14|9547"    "CXCL12|6387"
```

The same logic and criteria were applied to the results of RF-RFE experiments in Substudy 1, performed without balancing class proportions (see Figure 17.2). Although average specificity is significantly lower than for experiments with balancing, conclusions about the stability of performance metrics were the same. The average OOB estimates of AUC, sensitivity, and accuracy are also relatively stable for subsets with ten to thirty variables (for example, AUC values in this range are between 0.9972 and 0.9983), and average specificity is between 0.92 and 0.926 in the cardinality range of ten to twenty. Consequently, for Study 1, the size of ten was also identified as the

[10] They will not be generalized to independent or future data.

[11] For example, maximizing AUC (when metric is "ROC") returns a set of 250 variables, which is not useful if we are interested in a parsimonious and generalizable biomarker (especially when the results visualized in Figure 17.3 indicate that the differences between AUC for a set of 250 variables and those for much smaller set sizes are very small).

optimal number of variables for its multivariate biomarker. Table 17.3 later in this chapter includes the names of genes selected into the optimal multivariate biomarkers identified for RF-RFE experiments with and without balancing class proportions.

17.3.6 Notes on Misclassification Costs

Recall that rebalancing of class proportions was prompted by "technical" reasons: severe imbalance of classes in the training data and the sensitivity of the random forests learning algorithm (and later SVM) to such imbalance. Since it was not prompted by misclassification costs, rebalancing may or may not be aligned with such costs. As discussed in Chapter 4, misclassification costs should be provided by a domain expert, and they would depend on the purpose of the biomarker. For example, if we wanted to identify a screening biomarker for a target population with a very low prevalence of liver cancer, then rebalancing that boosts specificity would most likely be aligned with misclassification costs. However, if our goal was a diagnostic marker for a population with a high prevalence of liver cancer, then such a rebalancing would be counterproductive.

17.4 Feature Selection Experiments with Recursive Feature Elimination and Support Vector Machines (Substudy 3)

In addition to RF-RFE substudies described in the previous section, another multivariate biomarker discovery substudy was performed on the same training data. This time recursive feature elimination experiments were coupled with linear support vector machines. The design of this SVM-RFE substudy was the same as for the RF-RFE substudy with rebalancing of class proportions. The R script from the RF-RFE Substudy 2 was reused for all parts of Substudy 3, with the exception of feature selection part, where the default (for the `rfe` function in `caret`) random forests learning algorithm was replaced with support vector machines. This R script is also based on a script by Kuhn and Johnson (2013). However, since `caret` does not provide a function for calculating multivariate variable importance for recursive feature elimination with support vector machines, a custom function was created and implemented.[12,13] Furthermore, since building SVM classifiers requires tuning of their hyperparameters (for the linear SVM used in this substudy, there is only one hyper-parameter – regularization parameter C), running 500 SVM-RFE experiments would

[12] A generic `caret` function when used with SVM-RFE would calculate the univariate importance of each variable, which is not appropriate for multivariate feature selection experiments. The custom function `svmRank_custom` will calculate the multivariate importance of a variable as the absolute value of its weight coefficient in vector $\boldsymbol{\beta}$, the vector that defines the orientation of the optimal separating hyperplane identified for an SVM classifier built for a currently considered set of variables (see Chapter 11).

[13] Another custom function, `selectVar_custom`, also used for RF-RFE substudies, will select variables into the optimal biomarker by considering the frequency of their occurrence across 500 sets of the optimal size (Viscio 2017).

take much more time than running 500 RF-RFE experiments. To decrease the time necessary to complete this substudy, parallel (multicore) processing was utilized.[14]

```
# Multiple RFE runs on the randomly generated subsamples

library(kernlab)
# Parallel processing
cores <- parallel::detectCores(logical = FALSE)
if(cores > 1){
  library(doParallel)
  library(parallel)
  cl <- makeCluster(cores)
  registerDoParallel(cl)
}

# custom function for selecting optimal set of variables
selectVar_custom <- function(y, size){
  library(dplyr)
  finalImp <-
    dplyr::filter(y, Variables == size) %>%
    dplyr::select(c(Overall, var)) %>%
    group_by(var) %>%
    summarize(frequency = n()) %>%
    arrange(desc(frequency))
  as.character(finalImp$var[1:size])
}

# custom variable importance function for linear svm
svmRank_custom <- function (object, x, y){
  vimp <- varImp(object, scale = FALSE)$importance
  if (!is.data.frame(vimp))
    vimp <- as.data.frame(vimp, stringsAsFactors = TRUE)
  if (all(levels(y) %in% colnames(vimp)) & !("Overall" %in% colnames(vimp))){
    beta <- t(object$finalModel@coef[[1]] %*% object$finalModel@xmatrix[[1]])
    vimp$Overall <- abs(beta)
  }
  vimp <- dplyr::arrange(vimp, desc(Overall))
  vimp$var <- rownames(vimp)
  vimp
}
```

[14] Additional R packages used by this part of the script are `kernlab` (Karatzoglou et al. 2004), `dplyr` (Wickham et al. 2022), and `doParallel` (Microsoft Corporation and Weston 2022).

```
svmFuncs <- caretFuncs
svmFuncs$summary <- function(...)
                    c(twoClassSummary(...), defaultSummary(...))
svmFuncs$selectVar <- selectVar_custom
svmFuncs$rank <- svmRank_custom

# RFE with linear SVM

svmGrid <- expand.grid(.C = 2^(seq(-3, 3)))
cardinality <- c(2:30, 50, 75, 150, 250, 500, 1000, 2000, 4000)

# seeds for parallel processing
set.seed(369)
b <- length(index)
seeds_rfe <- vector(mode="list", length=b)
for(i in 1:b){
  seeds_rfe[[i]] <- sample.int(n=3333, length(cardinality)+1)
}
seeds_rfe[[b+1]] <- sample.int(n=3333, 1)

ctrl <- rfeControl(method = "repeatedcv",
                   number = 10,      # folds - ignored when index is used
                   repeats = 50,     # repeats - ignored when index is used
                                    # but both are reported in rfe results
                   verbose = TRUE,
                   functions = svmFuncs,
                   seeds = seeds_rfe,
                   returnResamp = "all",
                   rerank = TRUE, # recalculate var importance for each size
                   index = index
)

SVM_RFE <- rfe(x = Train,
               y = Diagnosis,
               sizes = cardinality,
               metric = "ROC",
               rfeControl = ctrl,
               method = "svmLinear",
               tuneGrid = svmGrid,
               trControl = trainControl(method = "cv",
                                        savePredictions = TRUE,
                                        verboseIter = FALSE,
```

```
                                       classProbs = TRUE,
                                       allowParallel = TRUE)
)
SVM_RFE
```

Figure 17.5 visualizes the main results of SVM-RFE experiments – the OOB estimates of four performance metrics (AUC, sensitivity, specificity, and accuracy) for the considered subset sizes. Each point on the graph represents the value of a performance metric, for a specific cardinality, calculated by averaging the metric values across the 500 SVM-RFE feature selection experiments.

The average OOB estimates of AUC were relatively stable for subsets with ten to thirty variables (with values between 0.9981 and 0.9986). The estimates of accuracy and sensitivity are almost identical, and they slightly increased in this range. Specificity is high, and when we look at the polynomial trendline smoothing its results, the trendline maximizes in the cardinality range of thirteen to twenty-six (with values between 0.9968 and 0.9976). Consequently, applying again the combined criterion of minimizing the subset size and maximizing its estimated performance, the size of

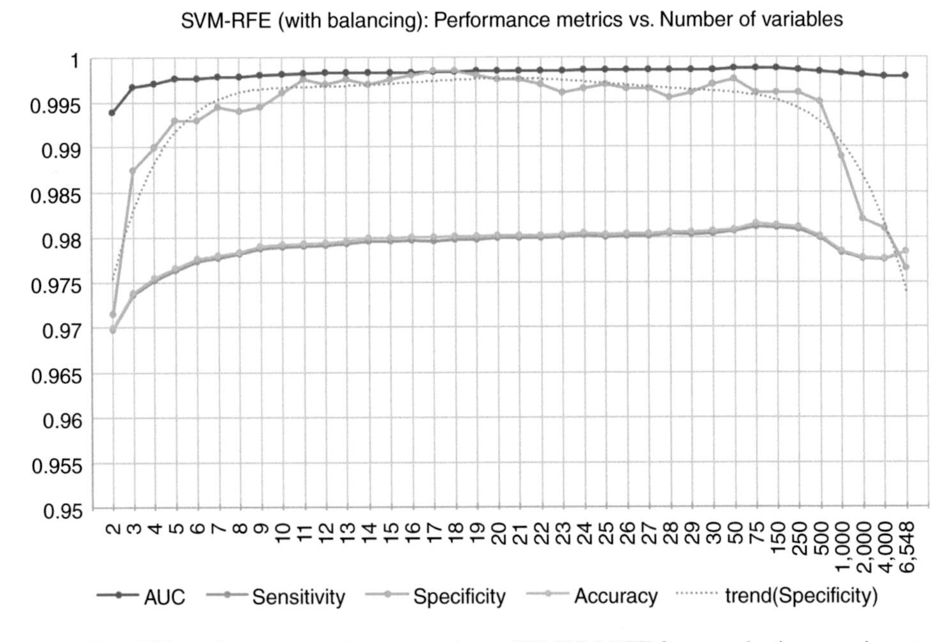

Figure 17.5 OOB performance metrics averaged over 500 SVM-RFE feature selection experiments performed on random subsamples of the training data **with rebalancing** the ratio of tumor to normal observations from 7.5:1 in the training data to 2:1 in each subsample generated for Substudy 3. In the cardinality range of 10–30, the OOB estimates of AUC are quite stable. The values of sensitivity and accuracy are very close to each other and increase slightly in this range. Based on its polynomial trendline, the specificity values maximize in the range of subsets with 13–26 variables.

thirteen was deemed the optimal number of variables for the multivariate liver cancer biomarker identified via SVM-RFE experiments. The `rfe` results of SVM-RFE experiments were updated accordingly.

```
# update rfe results
set.seed(369)
opt_rfe <- update(SVM_RFE, x = Train,
                  y = Diagnosis,
                  size = 13)

opt_rfe$bestVar
##   [1] "COL15A1|1306"   "CD34|947"       "DBH|1621"         "PLVAP|83483"
##   [5] "COL4A1|1282"    "CLEC4M|10332"   "ADAMTS13|11093"   "CPEB3|22849"
##   [9] "CLEC4G|339390"  "HBA2|3040"      "TGM3|7053"        "SLC26A6|65010"
##  [13] "CCL2|6347"
```

17.5 Summary of the Three Biomarker Discovery Substudies

With the aim of identifying multivariate gene expression biomarkers for liver cancer, this study comprised three substudies, each implementing 500 parallel feature selection experiments performed on random subsamples of the training data. To evaluate the efficiency of class rebalancing, two substudies were designed to perform recursive feature elimination with random forests (RF-RFE), one with and one without rebalancing class proportions. The third substudy coupled RFE with support vector machines. In each of the three substudies, an optimal multivariate biomarker was identified by applying a combined criterion of minimizing the subset size and maximizing its estimated performance. Each of the two optimal biomarkers identified in RF-RFE substudies included ten genes, and in the SVM-RFE substudy, a thirteen-gene optimal biomarker was identified. The genes selected into the three biomarkers are listed in Table 17.3.

Biomarkers 1 and 2 (the optimal multivariate biomarkers identified in the two RF-RFE substudies) share the same six genes (marked in Table 17.3 with the green font). Three of those genes are also included in Biomarker 3 resulting from the SVM-RFE substudy. Furthermore, Biomarker 3 also includes one gene from Biomarker 1 and one gene from Biomarker 2 genes (these two genes are marked with the red font). Recall that the experiments were performed on data with 6,548 variables, that each of the two RF-RFE substudies was using different sets of 500 random subsamples of the training data (one generated with and one without rebalancing the class proportions), and that the SVM-RFE substudy used a different – specific to SVM – method of calculating multivariate variable importance. Hence, the overlaps among the biomarkers can hardly be seen as coincidence; rather, they suggest that the biomarkers may tap into some common biological phenomena underlying the differences between the tumor and normal classes.

Table 17.3 Genes selected into the three multivariate biomarkers: the two ten-gene Biomarkers 1 and 2 identified via RF-RFE feature selection experiments (with and without balancing class proportions), and the thirteen-gene Biomarker 3 identified in SVM-RFE experiments. Biomarkers 1 and 2 share six genes (their names are in green font); three of these genes are also included in Biomarker 3. In addition, Biomarker 3 shares one more gene with Biomarker 1, and another gene with Biomarker 2 (whose names are in red font). The order of genes in each column corresponds to the frequency of their occurrence across the 500 classifiers built on different sets (of ten or thirteen genes, respectively) considered during feature selection experiments.

Multivariate biomarker 1 Substudy 1 (RF-RFE without balancing)	Multivariate biomarker 2 Substudy 2 (RF-RFE with balancing)	Multivariate biomarker 3 Substudy 3 (SVM-RFE with balancing)
CLEC4G	CLEC4G	COL15A1
ADAMTS13	CLEC4M	CD34
CLEC4M	ECM1	DBH
FCN2	FCN2	PLVAP
COL15A1	STAB2	COL4A1
APOF	ADAMTS13	CLEC4M
ECM1	VIPR1	ADAMTS13
STAB2	SLC26A6	CPEB3
VPS45	CXCL14	CLEC4G
CYP4A11	CXCL12	HBA2
		TGM3
		SLC26A6
		CCL2

Among the twenty-two genes included in the three biomarkers, six belong to tissue expression clusters, which are primarily associated with liver metabolism or hemostasis. Two other genes belong to clusters associated mainly with two tissue types, one being liver in either case. The remaining genes have been detected in many tissues, including liver. Table 17.4 provides information on all of the biomarker genes, based on The Human Protein Atlas (Uhlén et al. 2015).[15] In the Atlas, genes are clustered by their expression patterns, and each cluster is annotated according to the most common function of its members. Figure 17.6 shows heat map visualizations of the three multivariate biomarkers - the results of hierarchical clustering of the genes of each biomarker, as well as the training observations, using correlation distance and average linkage.

Apart from the sizes and compositions of the optimal biomarkers identified in the three substudies, the most important results are the OOB estimates of performance of

[15] Human Protein Atlas: www.proteinatlas.org.

Table 17.4 Information about the twenty-two genes selected into the three multivariate liver cancer biomarkers. The eight genes included in more than one of these biomarkers have their symbols printed in the same font color as in Table 17.3. The genes are ordered alphabetically.

Gene symbol	Gene description	Tissue expression cluster (RNA)
ADAMTS13	ADAM metallopeptidase with thrombospondin type 1 motif 13	Liver – metabolism (mainly)
APOF	Apolipoprotein F	Liver – hemostasis and lipid metabolism (mainly)
CCL2	C-C motif chemokine ligand 2	Adipose tissue – ECM organization (mainly)
CD34	CD34 molecule	Adipose tissue – ECM organization (mainly)
CLEC4G	C-type lectin domain family 4 member G	Liver – hemostasis and lipid metabolism (mainly)
CLEC4M	C-type lectin domain family 4 member M	Liver – hemostasis and lipid metabolism (mainly)
COL15A1	Collagen type XV alpha 1 chain	Placenta – pregnancy (mainly)
COL4A1	Collagen type IV alpha 1 chain	Placenta – pregnancy (mainly)
CPEB3	Cytoplasmic polyadenylation element binding protein 3	Nonspecific – mitochondria (mainly)
CXCL12	C-X-C motif chemokine ligand 12	Fibroblasts – ECM organization (mainly)
CXCL14	C-X-C motif chemokine ligand 14	Skin – epidermis development (mainly)
CYP4A11	Cytochrome P450 family 4 subfamily A member 11	Liver – metabolism (mainly)
DBH	Dopamine beta-hydroxylase	Pituitary gland – hormone signaling (mainly)
ECM1	Extracellular matrix protein 1	Esophagus – epithelial cell function (mainly)
FCN2	Ficolin 2	Liver – metabolism (mainly)
HBA2	Hemoglobin subunit alpha 2	Neutrophils – humoral immune response (mainly)
PLVAP	Plasmalemma vesicle associated protein	Thyroid gland – transcription (mainly)
SLC26A6	Solute carrier family 26 member 6	Intestine and liver – lipid metabolism (mainly)
STAB2	Stabilin 2	Macrophages – immune response (mainly)
TGM3	Transglutaminase 3	Esophagus – epithelial cell function (mainly)
VIPR1	Vasoactive intestinal peptide receptor 1	Liver and placenta – transport via ER (mainly)
VPS45	Vacuolar protein sorting 45 homolog	Heart – cardiac muscle contraction (mainly)

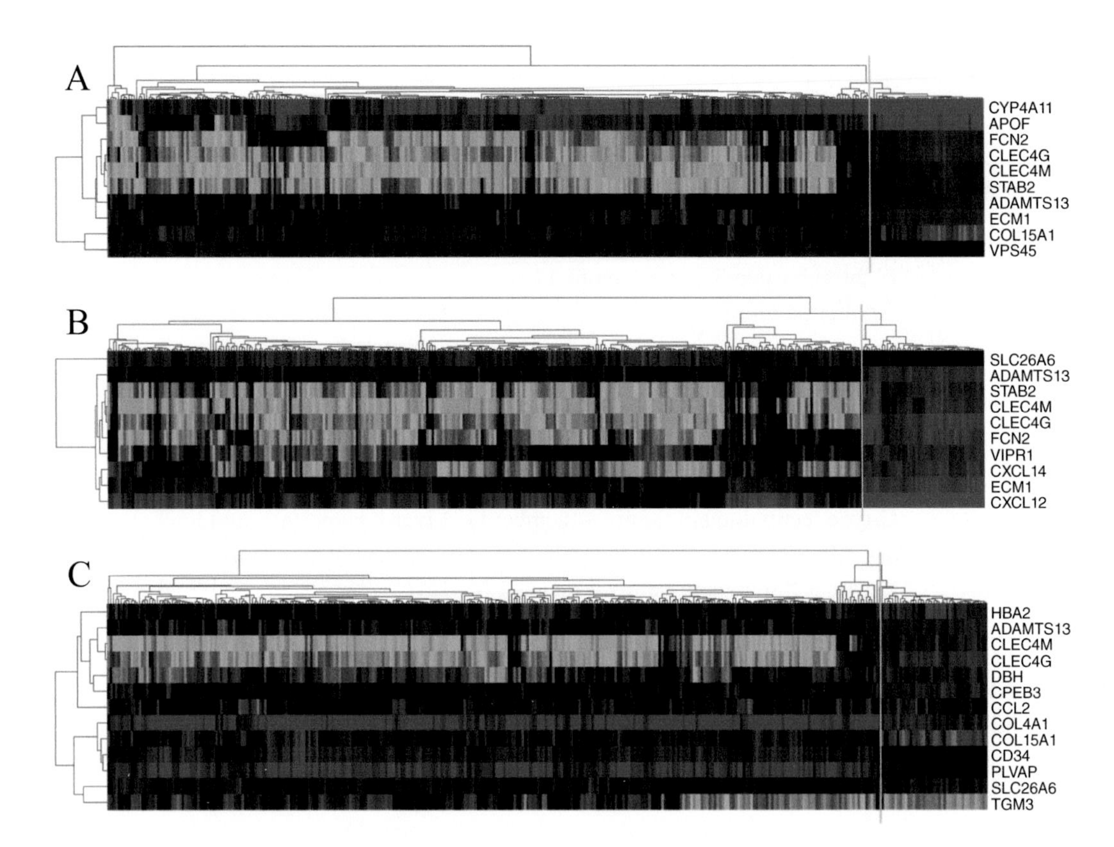

Figure 17.6 Heat maps visualizing the gene expression level of the three multivariate liver cancer biomarkers. Panel A represents ten-gene Biomarker 1 (RF-RFE without balancing), panel B represents ten-gene Biomarker 2 (RF-RFE with balancing the class proportions), and panel C represents thirteen-gene Biomarker 3 (SVM-RFE). The genes, as well as the training observations, have been clustered using the correlation distance and the average linkage. The dendrograms on the left side of the heat maps represent results of hierarchical clustering of genes, and the dendrograms across the top of the heat maps are results of clustering of the 339 training observations. An orange vertical line has been superimposed over each of the heat maps, in such a way that, in the block to the right of the line, all 40 normal training observations are clustered and only a few tumor ones (four in heat map A, seven in B, and only one in C). Color intensities represent gene expression levels – green corresponds to lower expression levels, and red to higher ones (with black representing expression values around the median).

the three final classifiers built on these biomarkers. Of course, we cannot evaluate their performance on the same training data, as it would amount to useless internal validation (see Chapter 4). However, in each substudy, the optimal biomarker was identified by aggregating OOB performance information across 500 classifiers of the same size, built during the RFE process. Averaging the performance metrics of those 500 classifiers provides good performance estimates of our optimal biomarker and classifier (for more information, see Chapter 4). These performance estimates are presented in Table 17.5.

Table 17.5 OOB estimates of performance for the three identified multivariate biomarkers. The estimates are not based on these specific biomarkers (as this would amount to internal validation), but – for each of the three substudies – based on averaging OOB estimates over 500 different biomarkers (of either ten or thirteen genes) identified during feature selection experiments. Such estimates are usually quite robust and considered as reliable as the results of testing on an independent test data set.

Multivariate biomarker	Number of genes	Average OOB estimates of performance			
		AUC	Sensitivity	Specificity	Accuracy
Biomarker 1	10	0.997	0.985	0.920	0.978
Biomarker 2	10	0.996	0.975	0.984	0.976
Biomarker 3	13	0.998	0.979	0.998	0.980

17.6 Testing the Three Final Multivariate Biomarkers and Classifiers

After identifying an optimal multivariate biomarker in each of the three substudies and building a classifier implementing the biomarker, the three classifiers were tested on the test data set. Recall that this test set was set aside before analysis and did not in any way influence the analysis.

```
# testing a random forest classifier
testResults <- predict(opt_rfe, newdata=Test)
confusionMatrix(testResults[,1], Test_Class, positive="tumor")

# testing the SVM classifier
testResults <- predict(opt_rfe$fit, newdata=Test[, opt_rfe$bestVar])
confusionMatrix(testResults, Test_Class, positive="tumor")
```

The confusion matrices produced by the three tests are presented in Table 17.6, and the test performance of the three classifiers is summarized in Table 17.7.

All the three classifiers misclassified only one of the seventy-four tumor test observations and thus they have the same sensitivity to tumor. The only difference in classification results is that the SVM classifier assigned all normal test observations to their true class, while both random forests classifiers misclassified one of them. Does this mean that the multivariate biomarker used by the SVM classifier (with 100 percent specificity on the test data) is better than the other two biomarkers (both with 90 percent specificity)? Not necessarily, if such a conclusion were based only on these test results. Though, this would be a valid conclusion if the number of normal test observations was much larger than ten. With only ten, the granularity in evaluating specificity is 10 percent, which means that misclassifying just one normal observation will decrease specificity by 10 percent, and that no specificity values between 90 percent and 100 percent are possible from this test (which also means that, in this situation, a result

Table 17.6 Confusion matrices resulting from testing the three final biomarkers (and the classifiers) on the holdout test set including eighty-four observations. Out of the seventy-four tumor test observations, seventy-three were properly classified, by each of the classifiers, into the tumor class (true positives). The SVM classifier properly classified all of the ten normal observations (true negatives), while both random forests classifiers misclassified one of the normal test observations.

A: Biomarker 1 (RF-RFE with no balancing)

True Class	Predicted Class: Tumor	Predicted Class: Normal
Tumor	73	1
Normal	1	9

B: Biomarker 2 (RF-RFE with rebalancing)

True Class	Predicted Class: Tumor	Predicted Class: Normal
Tumor	73	1
Normal	1	9

C: Biomarker 3 (SVM-RFE with rebalancing)

True Class	Predicted Class: Tumor	Predicted Class: Normal
Tumor	73	1
Normal	0	10

Table 17.7 Classification performance of the three final multivariate biomarkers as measured by testing them on the eighty-four observations of the holdout test data set, which was not used in any step of the biomarker discovery analyses.

Results of testing the three final multivariate biomarkers on the test data set			
Biomarker/classifier	**Sensitivity (%)**	**Specificity (%)**	**Accuracy (%)**
Biomarker 1 (RF)	98.7	90.0	97.6
Biomarker 2 (RF)	98.7	90.0	97.6
Biomarker 3 (SVM)	98.7	100.0	98.8

of neither 90 percent nor 100 percent should be fully trusted). Therefore, since these test results are based on such a small number of normal test observations, they should be interpreted in the context of OOB estimates, wherein averaging the OOB performance of 500 classifiers (built on different sets of variables of the same size as each of the final biomarkers) will have much smaller granularity.

Let's first consider Study 1 (see Chapter 16), in which its final biomarker was tested on the test data including 634 tumor and 66 normal observations, and thus the granularity in evaluating specificity was 1.5 percent. With two normal test observations misclassified, the test-based specificity was 97 percent, which was only 0.4 percent lower than the OOB estimate of the classifier's specificity. Observing also that the difference between test sensitivity (based on 634 tumor observations) and its OOB estimate was only 0.3 percent, we can conclude that the test results of Study 1 well aligned with the OOB estimates.

Looking at the current study, the differences in sensitivity evaluated on the test data and their corresponding OOB estimates are – for the three biomarkers – between 0.2 percent and 1.2 percent. Thus, with seventy-four tumor test observations (and thus granularity about 1.35 percent), the test-based sensitivity results are very much in agreement with the OOB estimates. We should expect the same for specificity if the number of normal test observations was similar. Consequently, since the tests on the larger number of tumor observations confirm the reliability of OOB estimates, we may conclude that when the classifiers are tested on a larger set of independent observations from the target population, their expected specificity would most likely be similar to the OOB estimates of their specificity.

Recall from Chapter 3 that results of testing biomarkers (and their predictive models) on test data should *not* be used to select one of the biomarkers. If we wanted to have only one final predictive model, its selection should take place *before* involving the test data (using, for example, the OOB estimates of models' performance, or testing them on a validation set, different from the test set). A possible advantage of having three final predictive models – especially when all of them together include only twenty-two genes – is the option of using them as a committee of classifiers (see Chapter 3) that would, for example, assign each new patient into a class receiving at least two of the three votes (however, due to similarity between Biomarker 1 and Biomarker 2, it would be preferable to use weighted voting, in which the SVM-based classifier would be assigned a higher weight).

References

Aggarwal, R., Sounderajah, V., Martin, G. et al. (2021). Diagnostic accuracy of deep learning in medical imaging: A systematic review and meta-analysis. *NPJ Digital Medicine*, **4**(65). https://doi.org/10.1038/s41746-021-00438-z.

Ahmad, F. B., Cisewski, J. A., and Anderson, R. N. (2022). Provisional mortality data: United States, 2021. *Morbidity and Mortality Weekly Report*, April 29, 2022.

Ambroise, C. and McLachlan, G. J. (2002). Selection bias in gene extraction on the basis of microarray gene-expression data. *Proceedings of the National Academy of Sciences*, **99**(10), 6562–6.

American Cancer Society. (2023). *Cancer Facts & Figures 2023*. Atlanta: American Cancer Society.

Azuaje, F. (2010). *Bioinformatics and Biomarker Discovery: "Omic" Data Analysis for Personalized Medicine*. Hoboken: Wiley-Blackwell.

Belkin, M., Hsu, D., Ma, S., and Mandal, S. (2019). Reconciling modern machine-learning practice and the classical bias-variance trade-off. *Proceedings of the National Academy of Sciences*, **116**(32), 15849–54.

Bellman, R. E. (1961). *Adaptive Control Processes: A Guided Tour*. Princeton: Princeton University Press.

Biomarkers Definitions Working Group. (2001). Biomarkers and surrogate endpoints: Preferred definitions and conceptual framework. *Clinical Pharmacology & Therapeutics*, **69**(3), 89–95.

Bishop, C. M. (1995). *Neural Networks for Pattern Recognition*. New York: Oxford University Press.

(2006). *Pattern Recognition and Machine Learning*. New York: Springer.

Boser, B. E., Guyon, I., and Vapnik, V. N. (1992). A training algorithm for optimal margin classifiers. In *Fifth Annual Workshop on Computational Learning Theory*. Pittsburgh: ACM, pp. 144–52.

Breiman, L. (1996a). Bagging predictors. *Machine Learning*, **24**, 123–40.

Breiman, L. (1996b). *Out-of-Bag Estimation: Technical Report*. Berkeley: Department of Statistics, University of California.

Breiman, L. (2001). Random forests. *Machine Learning*, **45**(5), 5–32.

Breiman, L., Friedman, J., Olshen, R., and Stone, C. (1984). *Classification and Regression Trees*. New York: Chapman & Hall.

Cichosz, P. (2015). *Data Mining Algorithms: Explained Using R*. Hoboken: Wiley-Blackwell.

Cohen, J. (1960). A coefficient of agreement for nominal scales. *Educational and Psychological Measurement*, **20**(1), 37–46.

Cortes, C. and Vapnik, V. (1995). Support-vector networks. *Machine Learning*, **20**(3), 273–97.

De Jong, K. (2005). Genetic algorithms: A 30-year perspective. In L. Booker, S. Forrest, M. Mitchell, and R. Riolo (eds.), *Perspectives on Adaptation in Natural and Artificial Systems*. New York: Oxford University Press, pp. 11–31.

De Jong, S. (1993). SIMPLS: An alternative approach to partial least squares regression. *Chemometrics and Intelligent Laboratory Systems*, **18**, 251–63.

Di Ruscio, D. (2000). A weighted view on the partial least-squares algorithm. *Automatica*, **36**, 831–50.

Domingos, P. (1999). MetaCost: A general method for making classifiers cost-sensitive. In *Proceedings of the Fifth ACM SIGKDD International Conference on Knowledge Discovery and Data Mining*. San Diego: Association for Computing Machinery, pp. 155–64.

Drori, I. (2023). *The Science of Deep Learning*. Cambridge: Cambridge University Press.

Duda, R. O., Hart, P. E., and Stork, D. G. (2001). *Pattern Classification*, 2nd ed. New York: Wiley.

Dziuda, D. (2010). *Data Mining for Genomics and Proteomics: Analysis of Gene and Protein Expression Data*. Hoboken: Wiley-Interscience.

Efron, B. (1979). Bootstrap methods: Another look at the jacknife. *Annals of Statistics*, **7**(1), 1–26.

Efron, B. and Hastie, T. (2016). *Computer Age Statistical Inference: Algorithms, Evidence, and Data Science*. Cambridge: Cambridge University Press.

Efron, B., Hastie, T., Johnstone, I., and Tibshirani, R. (2004). Least angle regression. *Annals of Statistics*, **32**(2), 407–51.

Efron, B. and Tibshirani, R. (1993). *An Introduction to the Bootstrap*. New York: Chapman & Hall.

Engelbrecht, A. P. (2007). *Computational Intelligence: An Introduction*, 2nd ed. Hoboken: John Wiley & Sons.

Etzioni, R., Gulati, R., and Weiss, N. S. (2022). Multicancer early detection: Learning from the past to meet the future. *Journal of the National Cancer Institute*, **114**(3), 349–52.

Fahrmeir, L., Kneib, T., Lang, S., and Marx, B. (2013). *Regression: Models, Methods and Applications*. New York: Springer.

Fan, J., Li, R., Zhang, C.-H., and Zou, H. (2020). *Statistical Foundations of Data Science*. London: CRC Press.

FDA-NIH Biomarker Working Group. (2021). *BEST (Biomarkers, EndpointS, and other Tools) Resource*. Silver Spring: Food and Drug Administration, Bethesda: National Institutes of Health.

Fernandez, A., Garcia, S., Galar, M., Prati, R. C., Krawczyk, B., and Herrera, F. (2018). *Learning from Imbalanced Data Sets*. Cham: Springer Nature.

Fisher, R. A. (1936). The use of multiple measurements in taxonomic problems. *Annals of Eugenics*, **7**, 179–88.

Fisher, R. A. (1938). The statistical utilization of multiple measurements. *Annals of Eugenics*, **8**, 376–86.

Frank, I. E. and Friedman, J. H. (1993). A statistical view of some chemometrics regression tools. *Technometrics*, **35**(2), 109–35.

Friedman, J. H. (1989). Regularized discriminant analysis. *Journal of the American Statistical Association*, **84**(405), 165–75.

Gómez-Verdejo, V., Parrado-Hernández, E., and Tohka, J. (2019). Sign-consistency based variable importance for machine learning in brain imaging. *Neuroinformatics*, **17**, 593–609.

Goodfellow, I., Bengio, Y., and Courville, A. (2016). *Deep Learning*. Cambridge, MA: MIT Press.

Guyon, I., Weston, J., Barnhill, S., and Vapnik, V. N. (2002). Gene selection for cancer classification using support vector machines. *Machine Learning*, **46**(1–3), 389–422.

Hair, J. F., Black, W. C., Babin, B. J., and Anderson, R. E. (2014). *Multivariate Data Analysis*. New York: Pearson.

Hanahan, D. and Weinberg, R. A. (2011). Hallmarks of cancer: The next generation. *Cell*, **144**(5), 646–74.

Hartigan, J. A. (1972). Direct clustering of data matrix. *Journal of the American Statistical Association*, **67**(337), 123–9.

Hastie, T., Montanari, A., Rosset, S., and Tibshirani, R. J. (2022). Surprises in high-dimensional ridgeless least squares interpolation. *Annals of Statistics*, **50**(2), 949–86.

Hastie, T., Tibshirani, R., and Friedman, J. H. (2009). *The Elements of Statistical Learning: Data Mining, Inference, and Prediction*, 2nd ed. New York: Springer.

Heaton, J. (2015). *Artificial Intelligence for Humans, Volume 3: Deep Learning and Neural Networks*. St. Louis: Heaton Research.

Hebb, D. O. (1949). *The Organization of Behavior: A Neuropsychological Theory*. New York: Wiley.

Helland, I. S. (1988). On the structure of partial least squares regression. *Communications in Statistics: Simulation and Computation*, **17**(2), 581–607.

Henze, N. and Zirkler, B. (1990). A class of invariant consistent tests for multivariate normality. *Communications in Statistics: Theory and Methods*, **19**(10), 3595–617.

Hinton, G. E., Srivastava, N., Krizhevsky, A., Sutskever, I., and Salakhutdinov, R. (2012). Improving neural networks by preventing co-adaptation of feature detectors. *ArXiv.1207.0580*.

Hoerl, A. E. and Kennard, R. W. (1970). Ridge regression: Biased estimation for nonorthogonal problems. *Technometrics*, **12**(1), 55–67.

Holland, J. H. (1992). *Adaptation in Natural and Artificial Systems: An Introductory Analysis with Applications to Biology, Control, and Artificial Intelligence*, 1st ed. Cambridge, MA: MIT Press.

Hotelling, H. (1951). A generalized T test and measure of multivariate dispersion. In J. Neyman (ed.), *Proceedings of the Second Berkeley Symposium on Mathematical Statistics and Probability (July 31–August 12, 1950)* (vol. 2). Berkeley: University of California Press, pp. 23–41.

Hoyert, D. L. and Xu, J. (2012). Deaths: Preliminary data for 2011. *National Vital Statistics Reports*, **61**(6), 1–51.

Huber, P. J. (1964). Robust estimation of a location parameter. *Annals of Mathematical Statistics*, **35**(1), 73–101.

Huberty, C. J. and Olejnik, S. (2006). *Applied MANOVA and Discriminant Analysis*. Hoboken: Wiley.

Hunger, S. P., Lu, X., Devidas, M. et al. (2012). Improved survival for children and adolescents with acute lymphoblastic leukemia between 1990 and 2005: A report from the Children's Oncology Group. *Journal of Clinical Oncology*, **30**(14), 1663–9.

Izenman, A. J. (2008). *Modern Multivariate Statistical Techniques: Regression, Classification, and Manifold Learning*. New York: Springer.

James, G., Witten, D., Hastie, T., and Tibshirani, R. (2014). *An Introduction to Statistical Learning with Applications in R*. New York: Springer.

Johnson, R. A. and Wichern, D. W. (2007). *Applied Multivariate Statistical Analysis*, 6th ed. Upper Saddle River: Prentice Hall.

Karatzoglou, A., Smola, A., Hornik, K., and Zeileis, A. (2004). kernlab: An S4 package for kernel methods in R. *Journal of Statistical Software*, **11**(9), 1–20.

Kennedy, J. and Eberhart, R. (1995). Particle swarm optimization. In *Proceedings of the IEEE International Conference on Neural Networks: Volume 4*. Piscataway: IEEE, pp. 1942–8.

Kennedy, J., Eberhart, R. C., and Shi, Y. (2001). *Swarm Intelligence*. San Diego: Morgan Kaufmann.

Kirkpatrick, S., Gelatt, C. D., and Vecchi, M. P. (1983). Optimization by simulated annealing. *Science,* **220**(4598), 671–80.

Kohonen, T. (1982). Self-organized formation of topologically correct feature maps. *Biological Cybernetics,* **43**(1), 59–69.

Kohonen, T. (1995). *Self-Organizing Maps*. Berlin: Springer.

Korot, E., Guan, Z., Ferraz, D. et al. (2021). Code-free deep learning for multi-modality medical image classification. *Nature Machine Intelligence,* **3**, 288–98.

Kosinski, M. (2022). RTCGA: The Cancer Genome Atlas data integration. R package version 1.29.0. https://bioconductor.org/packages/RTCGA.

Kuhn, M. (2022). caret: Classification and regression training. R package version 6.0-93. https://CRAN.R-project.org/package=caret.

Kuhn, M. and Johnson, K. (2013). *Applied Predictive Modeling*. New York: Springer.

Kuhn, M. and Johnson, K. (2020). *Feature Engineering and Selection: A Practical Approach for Predictive Models*. Boca Raton: CRC Press.

Kumar, C. and Van Gool, A. J. (2013). Biomarkers in translational and personalized medicine. In P. Horvatovich and R. Bischoff (eds.), *Comprehensive Biomarker Discovery and Validation for Clinical Application*. Cambridge: The Royal Society of Chemistry, pp. 3–39.

Lal, T. N., Chapelle, O., Weston, J., and Elisseeff, A. (2006). Embedded methods. In I. Guyon, S. Gunn, M. Nikravesh, and L. A. Zadeh (eds.), *Feature Extraction: Foundations and Applications*. Berlin: Springer-Verlag, pp. 137–65.

Lawley, D. N. (1938). A generalization of Fisher's z test. *Biometrika,* **30**(1–2), 180–7, correction 467–9.

Lazzeroni, L. and Oven, A. O. (2002). Plaid model for gene expression data. *Statistica Sinica,* **12**, 61–86.

Lennon, A. M., Buchanan A. H., Kinde I. et al. (2020). Feasibility of blood testing combined with PET-CT to screen for cancer and guide intervention. *Science,* **369**(6499), eabb9601.

Lindgren, F., Geladi, P., and Wold, S. (1993). The kernel algorithm for PLS. *Journal of Chemometrics,* **7**, 45–59.

Ling, C. X. and Sheng, V. S. (2017). Cost-Sensitive Learning. In C. Sammut and G. I. Webb (eds.), *Encyclopedia of Machine Learning and Data Mining*. New York: Springer, pp. 285–9.

Liu, X., Faes, L., Kale, A. U. et al. (2019). A comparison of deep learning performance against health-care professionals in detecting diseases from medical imaging: A systematic review and meta-analysis. *Lancet. Digital Health,* **1**(6), e271–97.

Lu, H., Plataniotis, K. N., and Venetsanopoulos, A. (2014). *Multilinear Subspace Learning: Dimensionality Reduction of Multidimensional Data*. New York: Chapman & Hall.

Mardia, K. V. (1970). Measures of multivariate skewness and kurtosis with applications. *Biometrika,* **57**, 519–30.

McCulloch, W. and Pitts, W. (1943). A logical calculus of the ideas immanent in nervous activity. *Bulletin of Mathematical Biophysics,* **5**, 115–33.

Microsoft Corporation and Weston, S. (2022). doParallel: Foreach parallel adaptor for the "parallel" package. R package version 1.0.17. https://CRAN.R-project.org/package=doParallel.

Mitchell, M. (1996). *An Introduction to Genetic Algorithms*. Cambridge, MA: MIT Press.

Nuffield Council on Bioethics. (2010). *Medical Profiling and Online Medicine: The Ethics of "Personalised Healthcare" in a Consumer Age*. London: Nuffield Council on Bioethics.

Orestes Cerdeira, J., Duarte Silva, P., Cadima, J., and Minhoto, M. (2023). subselect: Selecting variable subsets. R package version 0.15.4. https://CRAN.R-project.org/package=subselect.

Pepe, M. S., Etzioni, R., Feng, Z. et al. (2002). Elements of study design for biomarker development. In E. Diamondis, H. A. Fritsche, H. Lilja et al. (eds.), *Tumor Markers: Physiology, Pathobiology, Technology, and Clinical Applications*. Washington, DC: AACC Press, pp. 141–50.

Prainsack, B. (2017). *Personalized Medicine: Empowered Patients in the 21st Century?* New York: New York University Press.

R Core Team. (2022). *R: A Language and Environment for Statistical Computing*. Vienna: R Foundation for Statistical Computing. www.R-project.org.

Rakotomamonjy, A. (2003). Variable selection using SVM-based criteria. *Journal of Machine Learning Research,* **3**, 1357–70.

Rana, J. S., Khan, S. S., Lloyd-Jones, D. M., and Sidney, S. (2021). Changes in mortality in top 10 causes of death from 2011 to 2018. *Journal of General Internal Medicine,* **36**, 2517–18.

Rännar, S., Lindgren, F., Geladi, P., and Wold, S. (1994). A PLS kernel algorithm for data sets with many variables and fewer objects. Part 1: Theory and algorithm. *Journal of Chemometrics,* **8**, 111–25.

Rencher, A. C. (2002). *Methods of Multivariate Analysis*, 2nd ed. New York: Wiley.

Ripley, B. D. (1996). *Pattern Recognition and Neural Networks*. Cambridge: Cambridge University Press.

Rish, I. and Grabarnik, G. (2015). *Sparse Modeling: Theory, Algorithms, and Applications*. Boca Raton: CRC Press.

Rocks, J. W. and Mehta, P. (2022). Memorizing without overfitting: Bias, variance, and interpolation in overparameterized models. *Physical Review Research,* **4**(1), 013201.

Rosenblatt, F. (1958). The perceptron: A probabilistic model for information storage and organization in the brain. *Psychological Review,* **65**(6), 386–408.

Rosipal, R. and Krämer, N. (2006). Overview and recent advances in partial least squares. In C. Saunders, M. Grobelnik, S. Gunn, and J. Shawe-Taylor (eds.), *Subspace, Latent Structure and Feature Selection. SLSFS 2005. Lecture Notes in Computer Science* (vol. 3940). Berlin: Springer.

Royston, J. P. (1983). Some techniques for assessing multivariate normality based on the Shapiro–Wilk W. *Applied Statistics,* **32**, 121–33.

Rumelhart, D., Hinton, G., and Williams, R. (1986). Learning representations by back-propagating errors. *Nature,* **323**, 533–6.

Sanger, F., Nicklen, S., and Coulson, A. R. (1977). DNA sequencing with chain-terminating inhibitors. *Proceedings of the National Academy of Sciences,* **74**(12), 5463–7.

Schölkopf, B. and Smola, A. J. (2002). *Learning with Kernels: Support Vector Machines, Regularization, Optimization, and Beyond.* Cambridge, MA: MIT Press.

Schölkopf, B., Smola, A. J., Williamson, R. C., and Bartlett, P. L. (2000). New support vector algorithms. *Neural Computation,* **12**(5), 1207–45.

Siegel, R. L., Miller, K. D., Fuchs, H. E., and Jemal, A. (2022). Cancer statistics, 2022. *CA Cancer Journal for Clinicians,* **72**(1), 7–33.

Sievert, C. (2020). *Interactive Web-Based Data Visualization with R, plotly, and shiny*. New York: Chapman & Hall.

Smola, A. J. and Schölkopf, B. (2004). A tutorial on support vector regression. *Statistics and Computing,* **14**, 199–222.

Srivastava, N., Hinton, G. E., Krizhevsky, A., Sutskever, I., and Salakhutdinov, R. (2014). Dropout: A simple way to prevent neural networks from overfitting. *Journal of Machine Learning Research,* **15**, 1929–58.

Srivastava, S., Koay, E. J., Borowsky, A. D. et al. (2019). Cancer overdiagnosis: A biological challenge and clinical dilemma. *Nature Reviews: Cancer,* **19**(6), 349–58.

Tibshirani, R. (1996). Regression shrinkage and selection via the lasso. *Journal of the Royal Statistical Society: Series B (Methodological),* **58**(1), 267–88.

Tibshirani, R., Hastie, T., Eisen, M. et al. (1999). *Clustering Methods for the Analysis of DNA Microarray Data: Technical Report.* Stanford: Department of Statistics, Stanford University.

Uhlén, M., Fagerberg, L., Hallström, B. M. et al. (2015). Proteomics: Tissue-based map of the human proteome. *Science,* **347**(6220), 1260419.

Vapnik, V. N. (1998). *Statistical Learning Theory.* New York: Wiley.

Vapnik, V. N. (2000). *The Nature of Statistical Learning Theory*, 2nd ed. New York: Springer.

Viscio, J. A. (2017). *Diagnosis of Alzheimer's Disease Based on a Parsimonious Serum Autoantibody Biomarker Derived from Multivariate Feature Selection.* New Britain: Central Connecticut State University.

Walsh, J. E. (1962). *Handbook of Nonparametric Statistics.* Princeton: Van Nostrand.

Welch, B. L. (1939). Note on discriminant functions. *Biometrika,* **31**(1–2), 218–20.

Werbos, P. J. (1974). *Beyond Regression: New Tools for Prediction and Analysis in the Behavioural Sciences.* Ph.D. thesis, Harvard University.

Wickham, H., François, R., Henry, L., and Müller, K. (2022). dplyr: A grammar of data manipulation. R package version 1.0.10. https://CRAN.R-project.org/package=dplyr.

Wold, H. (1966). Estimation of principal components and related models by iterative least squares. In P. R. Krishnajah (ed.), *Multivariate Analysis.* New York: Academic Press, pp. 391–420.

Wold, H. (1975). Path models with latent variables: The NIPALS approach. In H. M. Blalock (ed.), *Quantitative Sociology.* New York: Academic Press, pp. 307–57.

Wolters, M. A. (2015). A genetic algorithm for selection of fixed-size subsets, with application to design problems. *Journal of Statistical Software,* **68**(1), 1–18.

Zou, H. and Hastie, T. (2005). Regularization and variable selection via the elastic net. *Journal of the Royal Statistical Society: Series B (Statistical Methodology),* **67**(2), 301–20.

Index

Printed in the United States
by Baker & Taylor Publisher Services